FE/EIT EXAM PREPARATION

MECHANICAL ENGINEERING

Third Edition

Jerry H. Hamelink, EdD, PE & Lloyd M. Polentz, MS, PE

KAPLAN AEC EDUCATION

President: Roy Lipner
Vice President & General Manager: David Dufresne
Vice President of Product Development and Publishing: Evan M. Butterfield
Editorial Project Manager: Laurie McGuire
Director of Production: Daniel Frey
Production Editor: Caitlin Ostrow
Creative Director: Lucy Jenkins

Published by Kaplan AEC Education
30 South Wacker Drive
Chicago, IL 60606-7481
(312) 836-4400
www.kaplanaecengineering.com

Printed in the United States of America.

11 10 9 8 7 6 5 4 3

CONTENTS

Introduction

OUTLINE

HOW TO USE THIS BOOK

Mechanical Engineering: FE/EIT Exam Preparation is designed to help you prepare for the Fundamentals of Engineering/Engineer-in-Training exam. The book covers the discipline-specific afternoon exam in mechanical engineering. For the morning exam, Kaplan AEC offers the comprehensive review book, *Fundamentals of Engineering: FE/EIT Exam Preparation.*

This book covers the major topics on the afternoon exam in mechanical engineering, reviewing important terms, equations, concepts, analysis methods, and typical problems. After reviewing the topic, you can work the end-of-chapter problems to test your understanding. Complete solutions are provided so that you can check your work and further refine your solution methodology.

After reviewing individual topics, you should take the Sample Exam at the end of the book. To alleviate anxiety about the actual test, you should simulate the exam experience as closely as possible. Answer the 60 questions in the Sample Exam in an uninterrupted four-hour period, without looking back at any content in the rest of the book. You may wish to consult the *Fundamentals of Engineering Supplied-Reference Handbook*, which is the only reference you are allowed to use in the actual exam.

When you've completed the Sample Exam, check the provided solutions to determine your correct and incorrect answers. This should give you a good sense of topics you may want to spend more time reviewing. Complete solution methods are shown, so you can see how to adjust your approach to problems as needed.

The following sections provide you with additional details on the process of becoming a licensed professional engineer and on what to expect at the exam.

BECOMING A PROFESSIONAL ENGINEER

To achieve registration as a Professional Engineer, there are four distinct steps: (1) education, (2) the Fundamentals of Engineering/Engineer-in-Training (FE/EIT) exam, (3) professional experience, and (4) the professional engineer (PE) exam. These steps are described in the following sections.

Education

Generally, no college degree is required to be eligible to take the FE/EIT exam. The exact rules vary, but all states allow engineering students to take the FE/EIT exam before they graduate, usually in their senior year. Some states, in fact, have no education requirement at all. One merely need apply and pay the application fee. Perhaps the best time to take the exam is immediately following completion of related coursework. For most engineering students, this will be the end of the senior year.

Fundamentals of Engineering/ Engineer-In-Training Examination

This eight-hour, multiple-choice examination is known by a variety of names—Fundamentals of Engineering, Engineer-in-Training (EIT), and Intern Engineer—but no matter what it is called, the exam is the same in all states. It is prepared and graded by the National Council of Examiners for Engineering and Surveying (NCEES).

Experience

States that allow engineering seniors to take the FE/EIT exam have no experience requirement. These same states, however, generally will allow other applicants to substitute acceptable experience for coursework. Still other states may allow a candidate to take the FE/EIT exam without any education or experience requirements.

Typically, four years of acceptable experience is required before one can take the Professional Engineer exam, but the requirement may vary from state to state.

Professional Engineer Examination

The second national exam is called Principles and Practice of Engineering by NCEES, but many refer to it as the Professional Engineer exam or PE exam. All states, plus Guam, the District of Columbia, and Puerto Rico, use the same NCEES exam. Review materials for this exam are found in other engineering license review books.

FUNDAMENTALS OF ENGINEERING/ ENGINEER-IN-TRAINING EXAMINATION

Laws have been passed that regulate the practice of engineering in order to protect the public from incompetent practitioners. Beginning in 1907 the individual states began passing *title* acts regulating who could call themselves engineers and offer services to the public. As the laws were strengthened, the practice of engineering was limited to those who were registered engineers, or to those working under the supervision of a registered engineer. Originally the laws were limited to civil engineering, but over time they have evolved so that the titles, and some-times the practice, of most branches of engineering are included.

There is no national licensure law; licensure is based on individual state laws and is administered by boards of registration in each state. You can find a list of contact information for and links to the various state boards of registration at the Kaplan AEC Web site: *www.kaplanaecengineering.com.* This list also shows the exam registration deadline for each state.

Examination Development

Initially, the states wrote their own examinations, but beginning in 1966 NCEES took over the task for some of the states. Now the NCEES exams are used by all states. Thus it is easy for engineers who move from one state to another to achieve licensure in the new state. About 50,000 engineers take the FE/EIT exam annually. This represents about 65% of the engineers graduated in the United States each year.

The development of the FE/EIT exam is the responsibility of the NCEES Committee on Examination for Professional Engineers. The committee is composed of people from industry, consulting, and education, all of whom are subject-matter experts. The test is intended to evaluate an individual's understanding of mathematics, basic sciences, and engineering sciences obtained in an accredited bachelor degree of engineering. Every five years or so, NCEES conducts an engineering task analysis survey. People in education are surveyed periodically to ensure the FE/EIT exam specifications reflect what is being taught.

The exam questions are prepared by the NCEES committee members, subject matter experts, and other volunteers. All people participating must hold professional licensure. When the questions have been written, they are circulated for review in workshop meetings and by mail. You will see mostly metric units (SI) on the exam. Some problems are posed in U.S. customary units (USCS) because the topics typically are taught that way. All problems are four-way multiple choice.

Examination Structure

The FE/EIT exam is divided into a morning four-hour section and an afternoon four-hour section. There are 120 questions in the morning section and 60 in the afternoon.

The morning exam covers the topics that make up roughly the first $2\frac{1}{2}$ years of a typical engineering undergraduate program.

Seven different exams are in the afternoon test booklet, one for each of the following six branches: civil, mechanical, electrical, chemical, industrial, and environmental. A general exam is included for those examinees not covered by the six engineering branches. Each of the six branch exams consists of 60 problems covering coursework in the specific branch of engineering. The general exam, also 60 problems, has topics that are similar to the morning topics. If you are taking the FE/EIT as a graduation requirement, your school may compel you to take the afternoon exam that matches the engineering discipline in which you are obtaining your degree. Otherwise, you can choose the afternoon exam you wish to take. There are two approaches to deciding.

One approach is to take the general afternoon exam regardless of your engineering discipline. Because the topics covered are similar to those in the morning exam, this approach may streamline your review time and effort. If you are still in college or recently graduated, these general topics may be very fresh in your mind.

The second approach is to take the afternoon exam that matches the discipline you majored in. Particularly if you have been out of college for several years, practicing this discipline in your daily work, you will be very familiar and comfortable with the topics. This may be to your advantage during your review time and in the pressure of the exam itself.

At the beginning of the afternoon test period, examinees will mark the answer sheet as to which branch exam they are taking. You could quickly scan the test, judge the degree of difficulty of the general versus the branch exam, then choose the test to answer. We do not recommend this practice, as you would waste time in determining which test to write. Further, you could lose confidence during this indecisive period.

Table 1 summarizes the major subjects for the mechanical engineering afternoon exam, including the percentage of problems you can expect to see on each one.

Table 1 Mechanical Engineering Afternoon Exam

NCEES Defined Topics	Number of Exam Questions	Review Chapter in Text
Fluid Mechanics and Fluid Machinery	15	Chapter 5, Chapter 6
Heat Transfer	10	Chapter 4
Kinematics, Dynamics, and Vibrations	15	Chapter 8
Materials and Processing	10	Chapter 7, Chapter 11
Measurement, Instrumentation, and Controls	10	Chapter 9, Chapter 10
Mechanical Design and Analysis	15	Chapter 7, Chapter 12
Refrigeration and HVAC	10	Chapter 3
Thermodynamics and Energy Conversion Processes	15	Chapter 1, Chapter 2

Taking the Examination

The National Council of Examiners for Engineering and Surveying (NCEES) prepares FE/EIT exams for use on a Saturday in April and October each year. Some state boards administer the exam twice a year; others offer the exam only once a year. The scheduled exam dates for the next ten years can be found on the NCEES Web site (*www.ncees.org/exams/schedules/*).

Those wishing to take the exam must apply to their state board several months before the exam date.

Examination Procedure

Before the morning four-hour session begins, the proctors pass out exam booklets and a scoring sheet to each examinee. Space is provided on each page of the examination booklet for scratchwork. The scratchwork will *not* be considered in the scoring. Proctors will also provide each examinee with a mechanical pencil for use in recording answers; this is the only writing instrument allowed. Do not bring your own lead or eraser. If you need an additional pencil during the exam, a proctor will supply one.

The examination is closed book. You may not bring any reference materials with you to the exam. To replace your own materials, NCEES has prepared a *Fundamentals of Engineering (FE) Supplied-Reference Handbook.* The handbook contains engineering, scientific, and mathematical formulas and tables for use in the examination. Examinees will receive the handbook from their state registration board prior to the examination. The *FE Supplied-Reference Handbook* is also included in the exam materials distributed at the beginning of each four-hour exam period.

There are three versions (A, B, and C) of the exam. These have the major subjects presented in a different order to reduce the possibility of examinees copying from one another. The first subject on your exam, for example, might be fluid mechanics, while the exam of the person next to you may have electrical circuits as the first subject.

The afternoon session begins following a one-hour lunch break. The afternoon exam booklets will be distributed along with a scoring sheet. There will be 60 multiple choice questions, each of which carries twice the grading weight of the morning exam questions.

If you answer all questions more than 15 minutes early, you may turn in the exam materials and leave. If you finish in the last 15 minutes, however, you must remain to the end of the exam period to ensure a quiet environment for all those still working, and to ensure an orderly collection of materials.

Preparing For and Taking the Exam

Give yourself time to prepare for the exam in a calm and unhurried way. Many candidates like to begin several months before the actual exam. Target a number of hours per day or week that you will study, and reserve blocks of time for doing so. Creating a review schedule on a topic-by-topic basis is a good idea. Remember to allow time for both reviewing concepts and solving practice problems.

In addition to review work that you do on your own, you may want to join a study group or take a review course. A group study environment might help you stay committed to a study plan and schedule. Group members can create additional practice problems for one another and share tips and tricks.

You may want to prioritize the time you spend reviewing specific topics according to their relative weight on the exam, as identified by NCEES, or by your areas of relative strength and weakness. Engineering departments at different schools place defferent emphasis on topics and courses, so there may be a topic identified for the exam that you have little or no exposure to. This would be a good area to focus on, time permitting, assuming you feel strong in other areas.

Those familiar with the psychology of examinations have several suggestions for examinees:

1. There are really two skills that examinees can develop and sharpen. One is the skill of illustrating one's knowledge. The other is the skill of familiarization with examination structure and procedure. The first can be enhanced by a systematic review of the subject matter. The second, exam-taking skills, can be improved by practice with sample problems—that is, problems that are presented in the exam format with similar content and level of difficulty.
2. Examinees should answer every problem, even if it is necessary to guess. There is no penalty for guessing.
3. Plan ahead with a strategy and a time allocation. A time plan gives you the confidence of being in control. Misallocation of time for the exam can be a serious mistake. You might allocate a little less time per problem for the areas in which you are most proficient, leaving a little more time in subjects that are more difficult for you. Your time plan should include a reserve block for especially difficult problems, for checking your scoring sheet, and finally for making last-minute guesses on problems you did not work. Your strategy might also include time allotments for two passes through the exam—the first to work all problems for which answers are obvious to you, the second to return to the more complex, time-consuming problems and the ones at which you might need to guess.
4. Read all four multiple-choice answers options before making a selection. All distractors (wrong answers) are designed to be plausible. Only one option will be the best answer.
5. Do not change an answer unless you are absolutely certain you have made a mistake. Your first reaction is likely to be correct.
6. If time permits, check your work.
7. Do not sit next to a friend, a window, or other potential distraction.

License Review Books

To prepare for the FE/EIT exam you need two or three review books.

1. A general review book for the morning exam, such as *Fundamentals of Engineering: FE/EIT Exam Preparation*, also from Kaplan AEC. That book will also prepare you for the general afternoon exam if you choose that option.
2. A review book, such as this one, for the afternoon exam, if you plan to take one of the discipline-specific exams.
3. *Fundamentals of Engineering (FE) Supplied-Reference Handbook*. At some point this NCEES-prepared book will be provided to applicants by their State Registration Board. You may want to obtain a copy sooner so you will have ample time to study it before the exam. You must, however, pay close attention to the *FE Supplied-Reference Handbook* and the notation used in it, because it is the only book you will have at the exam.

Textbooks

If you still have your university textbooks, they can be useful in preparing for the exam, unless they are out of date. To a great extent the books will be like old friends with familiar notation. You probably need both textbooks and license review books for efficient study and review.

Examination Day Preparations

The exam day will be a stressful and tiring one. You should take steps to eliminate the possibility of unpleasant surprises. If at all possible, visit the examination site ahead of time. Try too determine such items as

1. How much time should I allow for travel to the exam on that day? Plan to arrive about 15 minutes early. That way you will have ample time, but not too much time. Arriving too early, and mingling with others who are also anxious, can increase your anxiety and nervousness.
2. Where will I park?
3. How does the exam site look? Will I have ample workspace? Will it be overly bright (sunglasses), or cold (sweater), or noisy (earplugs)? Would a cushion make the chair more comfortable?
4. Where are the drinking fountains and lavatory facilities?
5. What about food? Most states do not allow food in the test room (exceptions for ADA). Should I take something along for energy in the exam? A light bag lunch during the break makes sense.

Items to Take to the Examination

Although you may not bring books to the exam, you should bring the following:

- *Calculator*—Beginning with the April 2004 exam, NCEES has implemented a more stringent policy regarding permitted calculators. For a list of permitted models, see the NCEES Web site (*www.ncees.org*). You also need to determine whether your state permits preprogrammed calculators. Bring extra batteries for your calculator just in case, and many people feel that bringing a second calculator is also a very good idea.
- *Clock*—You must have a time plan and a clock or wristwatch.
- *Exam Assignment Paperwork*—Take along the letter assigning you to the exam at the specified location to prove that you are the registered person. Also bring something with your name and picture (driver's license or identification card).
- *Items Suggested by Your Advance Visit*—If you visit the exam site, it will probably suggest an item or two that you need to add to your list.
- *Clothes*—Plan to wear comfortable clothes. You probably will do better if you are slightly cool, so it is wise to wear layered clothing.

Special Medical Condition

If you have a medical situation that may require special accommodation, you need to notify the licensing board well in advance of exam day.

Examination Scoring

The questions are machine-scored by scanning. The answer sheets are checked for errors by computer. Marking two answers to a question, for example, will be detected and no credit will be given.

Your state board will notify you whether you have passed or failed roughly three months after the exam. Candidates who do not pass the exam the first time may take it again. If you do not pass you will receive a report listing the percentages of questions you answered correctly for each topic area. This information can help focus the review efforts of candidates who need to retake the exam.

The FE/EIT exam is challenging, but analysis of previous pass rates shows that the majority of candidates do pass it the first time. By reviewing appropriate concepts and practicing with exam-style problems, you can be in that majority. Good luck!

CHAPTER 1

Thermodynamics

OUTLINE

Thermodynamics is truly one of the most basic of all engineering sciences because it deals with the relationship of heat, mechanical energy, work, and the conversion of one into the other. Thermodynamics consists of three basic laws:

1. **The First Law of Thermodynamics** is the basic law of conservation of energy. It states that energy cannot be created nor destroyed, but that it can be changed in form. The first law examines only the total energy units at the beginning and the ending of the process and does not look at the possibility of the energy form changes.
2. **The Second Law of Thermodynamics** states that heat cannot be transferred from a cold body to a hot body without an input of work. It similarly states that heat cannot be converted 100% into work. The bottom line is that an engine must operate between a hot and a cold reservoir. Also indicated is that energy has different levels of potential to do work and that energy cannot naturally move from a realm of lower potential to a realm of higher potential.
3. **The Third Law of Thermodynamics** states that the entropy of a substance is zero at absolute zero.

The first two laws are the basis of all studies in engineering thermodynamics. Even though these two laws are not actually laws (they have never been proven), they are accepted upon the basis that, despite all the attempts to disprove them, these statements have remained valid.

GENERAL ENERGY EQUATION

The general energy equation is an equation that relates all of the various forms of energy as a process moves from one set of state properties to another. The equation may be stated in a very simplistic form such as:

$$Q - W = \Delta E$$

where

Q = Heat transferred during the process
W = Work done on or by the system
ΔE = Change of total energy in the system

The equation can be expanded into different forms depending upon the energy types to be considered. The general energy equation is for a closed system, that is, a system that has a finite amount of mass that does not cross the boundary. It therefore remains a constant mass.

$$\text{Heat transfer} = \Delta(\text{Internal energy}) + \Delta(\text{Kinetic energy}) + \Delta(\text{Potential energy}) + \text{Work}$$

$$Q + u_1 + v_1^2/2 + Z_1 = u_2 + v_2^2/2 + Z_2 + W$$

where

u = Internal energy, or energy due to temperature
$v^2/2$ = Kinetic energy, or energy due to the velocity of motion
Z = Elevation

There may be various other forms of energy included in special problem types. These various other forms may include potential energy of springs, potential energy due to chemical composition, nuclear energy, and potential energy due to stretched membranes such as balloons and drum heads. The ones labeled above, however, are the most commonly used forms of energy.

Example **1.1**

A rigid container contains a liquid-vapor mixture of water in which the quality at the initial state is 25%. The original temperature of the 10 kg of water mixture in the tank is 100°C.

Determine

a) Volume of the tank

b) Heat transfer required to vaporize all of the liquid in the tank

c) Final temperature in the tank

Solution

a) The volume of the tank is determined by first calculating the specific volume of the liquid-vapor mixture of water in the tank. Since the initial conditions are known, the specific volume of the mixture at state one shall be calculated.

Using the quality equation given in the *Fundamentals of Engineering Supplied-Reference Handbook:*

$$v_1 = xv_{g1} + (1 - x)V_{f1}$$

Substituting in the values obtained from the "saturated water-temperature table" at T = 100°C yields

$$v_1 = (0.25)(1.673) + (1 - 0.25)(0.0010435) = 0.4190 \text{ m}^3/\text{kg}$$

Therefore total volume equals:

$$V_1 = mv_1 = (10 \text{ kg})(0.4190) = 4.19 \text{ m}^3$$

b) To determine the heat transfer required to vaporize the liquid in the tank requires the use of the general energy equation.

$$Q/m + u_1 + v_1^2/2 + Z_1 = u_2 + v_2^2/2 + Z_2 + W/m$$

Note now that in the statement of the problem there was no indication that there might be any kinetic energy exchange or potential energy change. In addition, there was no indication that work was crossing the boundary or that the boundary was expanding or contracting. The equation therefore reduces to:

$$Q = m(u_2 - u_1)$$

where u_1 is obtained using the quality equation and u_2 is obtained from the saturated water tables by interpolation, knowing that the substance in the tank is saturated water vapor with a specific volume of .4190 m^3/kg.

$$u_1 = xu_g + (1 - x)u_f = (0.25)(2506.5) + (0.75)(418.94)$$
$$u_1 = 940.83 \text{ kJ/kg},\ u_2 = 2557.36,\ T_2 = 147.7°\text{C}$$

Therefore,

$$Q = 10 \text{ kg } (2557.36 - 940.83) \text{ kJ/kg} = 16{,}165 \text{ kJ}$$

c) $T_2 = 147.7°\text{C}$

The second form of the general energy equation is for the case of mass flowing into and out of the control volume. When mass is flowing the energy equation becomes:

$$Q/m + h_1 + v_1^2/2 + Z_1 = h_2 + v_2^2/2 + Z_2 + W/m$$

The only difference between the first general energy equation and this form of the general energy equation is that h's are used instead of u's. The h, called **enthalpy**, is equal to u plus a PV term. This PV term (pressure multiplied by volume) has the same units as work and is actually the work of a fluid flowing from one location to another.

$$h = \text{enthalpy}$$
$$= u + PV$$

Example 1.2

A fluid at a pressure of 700 kPa absolute enters a heat engine with a velocity of 190 m/s. At the entrance condition the fluid has a specific volume of 300 l/kg. The fluid leaves the engine at a velocity of 335 m/s and a specific volume of 1.200 m^3/kg at a pressure of 35 kPa absolute. The work output of the engine is 2600, and 85 kW are lost due to radiation and convection. The mass flow rate through the engine is 4.5 kg/s.

Determine

a) Change of kinetic energy

b) Change in flow work

c) Change in internal energy

Solution

The best approach to a problem of this nature is to write a general energy balance equation for a control volume. Then, after the equation is written, one may solve for the various parts of the problem.

$$Q/m + h_1 + v_1^2/2 + Z_1 = h_2 + v_2^2/2 + Z_2 + W/m$$

There was no indication regarding any potential energy change due to elevations, and, as a consequence, the *Z* terms may be eliminated. Since an objective of this problem is to solve for the flow work, the equation must be written in a slightly different format.

$$Q/m + u_1 + P_1V_1 + v_1^2/2 = u_2 + P_2V_2 + v_2^2/2 + W/m$$

a) Solving for the change in kinetic energy:

$$\Delta\text{KE} = (v_2^2 - v_1^2)/2 = \{[(335)^2 - (190)^2]/2\}\text{m}^2/\text{s}^2(\text{N})/(\text{kg-m/s}^2)(\text{kJ})/(10^3\,\text{N-m})$$
$$\Delta\text{KE} = 38.06\ \text{kJ/kg}$$

b) Solving for the change in *PV*:

$$\Delta PV = P_2V_2 - P_1V_1 = (35\ \text{kPa})(1.2\ \text{m}^3/\text{kg}) - (700)(0.3)$$
$$\Delta PV = -168\ \text{kPa-m}^3/\text{kg} = -168\ \text{kJ/kg}$$

c) To solve for internal energy we merely have to substitute the calculated values into the general energy equation.

$$Q/m = \Delta u + \Delta\text{KE} + \Delta PV + W/m$$
$$(-85\ \text{kJ/s})/(4.5\ \text{kg/s}) = \Delta u + 38.06\ \text{kJ/kg} + -168\ \text{kJ/kg} + (2{,}600\ \text{kJ/s})/(4.5\ \text{kg/s})$$
$$\Delta u = -466.73\ \text{kJ/kg}$$

The internal energy has been reduced by 466.7 kJ/kg.

Example 1.3

A piston-cylinder system originally has 10 m^3 of gas at 10 bar pressure absolute and a temperature of 20 °C. The gas is compressed polytropically according to $n = 1.15$ until the pressure is 30 bar.

Determine

a) Mass of the gas if the molecular weight of the gas is 30

b) V_2

c) Work required to compress the gas

Solution

a) Select the perfect gas law and solve for the mass.

$$PV = mRT$$

$$m = PV/RT = \frac{[(10\ \text{bar})(10^5\ \text{N/m}^2)/(\text{bar})](10\ \text{m}^3)}{\{[(8.314\ \text{kJ/(kol-K)}]/30\ \text{kg/kmol}\}(293\ \text{K})(1000\ \text{Nm/kJ})}$$
$$m = 123.1\ \text{kg}$$

b) The volume at state two is obtained by using the polytropic relationship listed in the *FE Handbook*.

$$P_1V_1^n = P_2V_2^n$$
$$(V_2/V_1)^{1.15} = P_1/P_2 = 10/30 = 0.333$$
$$V_2/V_1 = 0.384$$
$$V_2 = (10)(.384) = 3.84 \text{ m}^3$$

c) The work required to compress the gas in a polytropic, closed-system process uses the equation from the *FE Handbook:*

$$W = (P_2V_2 - P_1V_1)/(1 - n)$$
$$W = \{[(30 \text{ bar})(3.84 \text{ m}^3) - (10)(10)]/(1 - 1.15)\} \times[(10^5 \text{ N/m}^2)/(\text{bar})][\text{kJ}/(10^3 \text{ N-m})]$$
$$W = -10{,}133 \text{ kJ}$$

Example 1.4

Air originally filling 0.5 m^3 is compressed isentropically from an initial pressure of 7.0 kPa gage pressure to a final temperature of 200°C. Assume that the gas acts as an ideal gas; that is, it has a constant specific heat and obeys the ideal gas laws. The work done in compressing the gas is equal to 100 kJ.

Determine

a) Mass of the air

b) Original temperature

c) Final pressure

d) Final volume

e) Change of internal energy of the air

Solution

a) The mass of the gas has to be obtained by using the ideal gas law, $PV = mRT$. The temperature at state one is not known; therefore, it has to be determined using the isentropic work equation found in the *FE Handbook*, with the substitution of the perfect gas law.

The isentropic work equation is shown below. $P_2V_2 = mRT_2$ is then substituted into the equation to eliminate the two unknowns P_2 and V_2.

$$W = \frac{P_2V_2 - P_1V_1}{1-k} = \frac{mRT_2 - P_1V_1}{1-k}$$

Rearrange the equation in order to solve for mass. Then, check to see if the units are all comparable: work is in kJ and P_1V_1 is in (kPa)(m^3). Since 1 Pa = 1 N/m^2, the PV term is in units of kN-m, which is equal to kJ. The numerator is therefore in consistent units. Checking the denominator: (8.314)/29 is (kJ/kmol-K)/(kg/kmol) = kJ/kg-K. This means that all of the units are consistent, and the mass may be calculated.

Solving for the mass yields

$$m = \frac{W(1-k)+P_1V_1}{RT_2} = \frac{100\,\text{kJ}(1-1.4)+(7.0+101)\,\text{kPa}(0.5\text{ m}^3)}{\left(\frac{8.314}{29}\right)\text{kJ/kg-K}\,(200+273)\,\text{K}}$$

$m = 0.693$ kg

b) To solve for original temperature one merely has to take the ideal gas law at original conditions and solve for the temperature, or

$$T_1 = (P_1V_1)/(mR)$$
$$T_1 = [(7.0 + 101)\text{ kPa }(0.5\text{ m}^3)]/[(0.693\text{ kg})(8.314/29)\text{ kJ/kg-K}]$$
$$= 272\text{ K}$$

c) The final pressure is found by using the isentropic pressure relationship:

$$P_2/P_1 = (T_2/T_1)^{k/(k-1)}$$
$$P_2 = (108\text{ kPa})(473/272)^{1.4/(1.4-1)} = 749\text{ kPa absolute}$$

d) The final volume can be determined from the isentropic relationship:

$$T_2/T_1 = (V_2/V_1)^{(k-1)}$$
$$V_2 = V_1(T_2/T_1)^{1/(k-1)} = 0.5\text{ m}^3(473/272)^{1/(1.4-1)} = 1.99\text{ m}^3$$

e) The change in internal energy can be obtained merely by using $\Delta u = c_p\Delta T$, where the c_p value can be obtained from the *FE Handbook*.

$$\Delta u = 0.718(473 - 272) = 144.3\text{ kJ/kg}$$

If the total change in internal energy is wanted then we must multiply by the mass.

$$\Delta U = (144.3)(0.693) = 100\text{ kJ}$$

ENTROPY

Entropy is a property that is defined as the amount of disorder in a process or a system. It is a direct application of the second and third laws of thermodynamics. Entropy indicates that real processes and real cycles cannot run backwards and that complete systems are *running down.*

The change in entropy is defined as $\int dQ/dT$ of a thermodynamic process and indicates that if isothermal heat transfer takes place at absolute zero the entropy production is zero. If a process moves from a state one to a different state two with no production of entropy the process is considered to be isentropic. This process may also be defined as a reversible adiabatic process.

CARNOT CYCLE

The discussion of entropy logically leads to the discussion of ideal cycles and the term **efficiency**. Efficiency is the measure of the ability of the cycle or process to do what it is designed to do. In the area of cycles, the Carnot cycle is the most efficient cycle known to man. It consists of two reversible isothermal processes and two reversible adiabatic or isentropic processes. The temperature-entropy, or *T-S*, diagram may be represented by a rectangle, as can be seen in the *FE Handbook*.

The thermal efficiency of a cycle may be stated as ***the energy produced divided by the energy input***. For a general cycle it may be represented by

$$\eta_{th} = (Q_{in} - Q_{out})/Q_{in}$$

where η_{th} is the thermal efficiency. For a Carnot cycle the equation may be modified further to

$$\eta_{th} = (T_H - T_L)/T_H$$

Other efficiency equations are used to compare the isentropic process with actual processes. These efficiency equations are often called ***mechanical efficiency*** equations.

Example 1.5

Steam enters a nozzle at a low velocity 10 bar and 500°C and exits at 1 bar and at a velocity of 1000 m/s. Heat is transferred from the nozzle at the rate of 5 kJ/kg.

Determine

a) The temperature of the steam leaving the nozzle.

b) The exit velocity if the nozzle were isentropic

Solution

a) The solution to this problem is aided by the writing of the general energy equation. The equation may then be simplified by noting that the elevation does not change, i.e., $\Delta Z = 0$, and that the entrance velocity may be considered to be equal to zero. In addition, nozzles do no work. Obtain the enthalpy values from the steam tables in the *FE Handbook*.

$$Q/m + h_1 + v_1^2/2 + Z_1 = h_2 + v_2^2/2 + Z_2 + W/m$$
$$-5 \text{ kJ/kg} + 3478.5 \text{ kJ/kg} = h_2 + [(1000)^2]/2 \text{ (m/s)}^2[\text{N/(kg-m)/s}^2][\text{kJ/}(10^3 \text{ N-m})]$$
$$h_2 = 2973.5 \text{ kJ/kg}$$

Go to the superheated water tables in the *FE Handbook* and look under 1 bar pressure (0.10 MPa) and find that the temperature is $T_2 = 250$°C.

b) If the nozzle is ***isentropic*** it means that no heat transfer is occurring and the change in velocity from input to output is maximum. This will occur when and only when $s_1 = s_2 = 7.7622$. This means that interpolation using the superheated water tables in the *FE Handbook* at 0.10 MPa and entropy of 7.7622 kJ/kg-K is required to obtain the leaving enthalpy value.

$$h_2 = 2843.0 \text{ kJ/kg}$$

Now consider the general energy equation in the absence of heat transfer, work, ΔZ, and incoming velocity:

$$h_1 = h_2 + v_2^2/2$$

Rearrange,substitute the values of enthalpy, and solve for the velocity:

$$v_2 = [(h_1 - h_2) \text{ kJ/kg } (2)(10^3 \text{ N-m})/(\text{N/kg-m})/\text{s}^2]^{1/2}$$
$$v_2 = [(3478.5 - 2843)(2)(10^3)]^{1/2} = 1127 \text{ m/s}$$

Example 1.6

A heat engine operates on the Carnot cycle between temperatures of 500 °C and 30 °C, and produces 500 kJ of work.

Determine

a) Thermal efficiency of the engine

b) Quantity of heat supplied to the engine

c) Amount of entropy produced during the heat rejection portion of the process

Solution

a) Since this heat engine operates on a Carnot cycle, we may use the temperature form of the efficiency equation.

$$\eta_{th} = (T_H - T_L)/T_H = (500 - 30)/(500 + 273) = 60.8\%$$

b) To solve for the heat supplied to the engine we need to use the form of the efficiency equation that uses heat transfer.

$$\eta_{th} = (Q_{in} - Q_{out})/Q_{in}$$
$$0.608 = (500 \text{ kJ})/Q_{in}$$

or

$$Q_{in} = 822 \text{ kJ}$$

c) To determine the entropy production during the heat rejection process we must calculate the heat rejection, which is:

$$Q_{out} = 822 - 500 = 322 \text{ kJ}$$
$$\Delta s = (322 \text{ kJ})/(273 + 30) = 1.0627 \text{ kJ/K}$$

Example 1.7

A heat pump operating on a Carnot cycle transfers heat from a source at −10°C to the interior of a house to be maintained at 21 °C. The house requires an energy input of 40,000 kJ per hour.

Determine

a) Coefficient of performance

b) Work input in hp

Solution

The efficiency of a heat pump or refrigeration cycle may still be written as ***energy output divided by energy input***. In this case it would be written as

$$\text{Efficiency} = \text{Heat to the house/Work to the compressor}$$

or as

$$\text{Efficiency} = Q_H/(Q_H - Q_L)$$

Since the heat pump is based upon the Carnot cycle the efficiency may be written in terms of absolute temperatures. One also may notice that the efficiency term shall be larger than unity, which makes no sense, as it indicates that a violation of the first law of thermodynamics has occurred. However, upon closer inspection, one notices that this type of efficiency term actually is not an efficiency term at all but is another indication of performance. This efficiency term is therefore called the coefficient of performance (C.O.P.).

a) C.O.P. $= T_H/(T_H - T_L) = (21 + 273)/[21 - (-10)] = 9.48$

b) To determine the work in hp of the motor powering the compressor one has to use the energy form of the Carnot equation.

$$\text{C.O.P.} = Q_H/W$$
$$9.48 = (40{,}000 \text{ kJ/hr})/W$$
$$W = 40{,}000/9.48 = 4{,}219 \text{ kJ/hr} = 4{,}219/3600 = 1.17 \text{ kW}$$
$$W = (1.17 \text{ kW})(1.1341 \text{ hp/kW}) = 1.33 \text{ hp}$$

COMBUSTION

Power-producing engines use mainly hydrocarbon fuels that are burned in air to produce the energy used in prime movers. Since hydrocarbon fuels are used the products of combustion generally consist of carbon dioxide and water. If the hydrocarbon fuel also consists of sulfur then, ideally, the products will include sulfur dioxide. The amount of air included is usually equal to or more than the minimum for stoichiometric combustion.

Example 1.8

Propane, C_3H_8, is burned completely in oxygen.

Determine

a) Ideal stoichiometric balanced equation

b) A/F ratio

c) The actual equation if burned in 150% theoretical air

d) A/F ratio

Solution

a) The ideal stoichiometric equation is one in which the products consist only of CO_2 and H_2O.

$$C_3H_8 + a(O_2) = bCO_2 + dH_2O$$

Setting up the balance equations:

$$\text{C: } 3 = b$$
$$\text{H: } 8 = 2d \qquad d = 4$$
$$\text{O: } 2a = 2b + d \qquad a = 5$$

The equation then becomes:

$$C_3H_8 + 5O_2 = 3CO_2 + 4H_2O$$

b) The A/F ratio is equal to the number of moles of air (oxygen in this case) divided by the number of moles of the fuel.

$$A/F = 5/1 = 5$$

c) When the fuel is burned in air one must include the other gases that are in the air mixture. When one considers the combustion to be in dry air the relative humidity, the water in the air, does not have to be considered. In addition, it is usually assumed that air consists only of oxygen (21%) and nitrogen (79%). Therefore there are 3.76 molecules of nitrogen per molecule of oxygen. The 150% theoretical air indicates how much extra air is included in the reaction—which is 50% more than the theoretical required amount.

Let us first look at the stoichiometric equation of propane burning in air.

$$C_3H_8 + 5[O_2 + 3.76N_2] = 3CO_2 + 4H_2O + 18.8N_2$$

Now consider 150% theoretical air:

$$C_3H_8 + 1.5(5)[O_2 + 3.76N_2] = 3CO_2 + 4H_2O + 2.5O_2 + 28.2N_2$$

d) The molar air-to-fuel ratio is

$$A/F = [1.5(5)(4.76)]/1 = 35.7 \text{ moles air/mole of fuel}$$

The mass air/fuel ratio is:

$$A/F = 35.7(29)/[3(12) + 8] = 23.5 \text{ kg air/kg fuel}$$

Example 1.9

A dry analysis of the exhaust products of combustion of a hydrocarbon fuel C_aH_b burned in dry air is 8.5% CO_2, 2.1% CO, 4.8% O_2 and 84.6% N_2.

Determine

a) Fuel composition

b) A/F actual by mass

c) % theoretical air

Solution

a) Write an equation for the combustion of a hydrocarbon fuel. Put in letters for the unknown values.

$$C_aH_b + d(O_2 + 3.76N_2) = 8.5CO_2 + 2.1CO + 4.8O_2 + 84.6N_2 + eH_2O$$

Set up the balance equations for the various elements:

$$\begin{aligned} &C: a = 8.5 + 2.1 && a = 10.6 \\ &H: b = 2e && b = 32.6 \\ &N: 3.76(2)d = 2(84.6) && d = 22.5 \\ &O: 2(22.5) = 2(8.5) + 2.1 + 2(4.8) + e && e = 16.3 \end{aligned}$$

The fuel composition may be written as: $C_{10.6}H_{32.6}$.

b) A/F = (moles of air)(molecular weight of air)/(moles of fuel)(molecular weight of fuel)

$$A/F = [22.5(1 + 3.76)][29]/[(1)\{(10.6)(12) + (32.6)(1)\}] = 19.4$$

c) To find the percent theoretical air one has to obtain the stoichiometric equation of the fuel.

$$C_{10.6}H_{32.6} + a(O_2 + 3.76N_2) = bCO_2 + dH_2O + eN_2$$

$$\begin{array}{ll} \text{C: } 10.6 = b & \\ \text{H: } 32.6 = 2d & d = 16.3 \\ \text{N: } 3.76a = e & e = 70.5 \\ \text{O: } 2a = 2(10.6) + (16.3) & a = 18.75 \end{array}$$

Therefore, the stoichiometric equation may be written as

$$C_{10.6}H_{32.6} + 18.75(O_2 + 3.76N_2) = 10.6CO_2 + 16.3H_2O + 70.5N_2$$

The A/F ideal may now be obtained on a mass basis:

$$A/F = [18.75(1 + 3.76)(29)]/[(10.6)(12) + (32.6)(1)] = 16.2$$

The percent theoretical air may now be found by dividing the actual A/F by the theoretical A/F.

$$\% = [A/F \text{ actual}]/[A/F \text{ ideal}] \times 100 = [19.4/16.2] \times 100 = 120\%$$

PROBLEMS

The following information is to be used for questions 1.1 through 1.3.

A piston acts upon 1 kg of air at 300 K and 5 bars. Heat is transferred in a constant-pressure process until the volume has doubled. The final temperature is 400 K.

1.1 The initial volume is most nearly:

a. 1.7 m^3 c. 0.17 m^3
b. 2.5 m^3 d. 0.017 m^3

1.2 The work done done by the gas in kJ is most nearly:

a. 86 c. 128
b. 104 d. 35

1.3 The heat transfer to the gas is most nearly:

a. 86 kJ c. 135 kJ
b. 115 kJ d. 158 kJ

1.4 Steam at 500 °C and 0.80 MPa is flowing through a line that is connected to an evacuated tank. The valve to the tank is carefully opened and the steam is allowed to flow in. Once the tank is filled the valve is closed. Consider that the tank is insulated and that kinetic energy and potential energy changes are minimal. The final temperature of the steam in the tank is most nearly:

a. 400 °C c. 600 °C
b. 500 °C d. 700 °C

The following information is to be used for questions 5 and 6.

An insulated, open-feedwater heater is used to heat 40 °C water at a rate of 1000 kg/min, at 1 MPa coming from the pump and supplying the boiler. Steam at 500 °C is mixed with the 40 °C water, and the water mixture leaves as saturated water at 1 MPa.

1.5 Mass flow rate of the steam to be mixed with the water in kg/min is most nearly:

a. 220 c. 120
b. 160 d. 250

1.6 If the control volume contains the feedwater heater, the production of entropy (kJ/kg-min-K) of the entering water is most nearly:

a. 1.335 c. 0.765
b. 1.005 d. 0.336

1.7 A 75 kW internal combustion engine is being tested by loading it with a water-cooled Prony brake. When the engine delivers the full-rated 75 kW to the shaft, the Prony brake being cooled with tap water absorbs and transfers to the cooling water 95 percent of the 75 kW. Which of the following most nearly equals the rate at which tap water passes through the Prony brake, if the water enters at 18 °C and leaves at 55 °C?

a. 18 l/min
b. 28 l/min
c. 35 l/min
d. 42 l/min

1.8 Assume that 2 kg of a gas mixture is 30 percent CO and 70 percent O_2, by mass fractions. It is compressed from (1) a pressure of 100 kPa absolute to (2) a pressure of 1.05 MPa absolute by a process characterized by the equation PV^n = constant, where the value of n is 1.0. The temperature of the mixture is 15 °C at the beginning of the process. Which of the following most nearly equals the work needed to compress the 2 kg of mixture?

a. 217 kJ
b. 252 kJ
c. 287 kJ
d. 323 kJ

1.9 Saturated steam is used in a mixing tank to heat and mix 750 liters of solution. The solution is to be heated from 15 to 65 °C by introducing steam directly into the solution. The specific heat of the solution is about 3.56 kJ/(kg-K) at 15 °C and remains essentially constant over the temperature range stated. The density of the solution is 1.00 kg/l. Which of the following most nearly equals the amount of saturated steam at a gage pressure of 138 kPa, which will be required for heating if all the steam is condensed in the solution and the heat loss to the surroundings is assumed to be negligible?

A steam table gives the following data for saturated steam at 138 kPa gage:

Temperature, 126.7 °C
Enthalpy (sat. liquid), 530 kJ/kg
Enthalpy (sat. vapor), 2.715 MJ/kg

a. 55 kg
b. 60 kg
c. 65 kg
d. 68 kg

1.10 Given 280 l of a gas at 63.5 cm Hg. The gas has a specific heat at constant pressure with a c_p of 0.847 kJ/kg·K and a specific heat at constant volume of 0.659 kJ/kg-K. Which of the following most nearly equals the volume the gas would occupy at a final pressure of 5 atm if the process is adiabatic?

a. 62 liters
b. 70 liters
c. 77 liters
d. 82 liters

1.11 The volumetric analysis of a gas mixture shows that it consists of 70 percent nitrogen, 20 percent carbon dioxide, and 10 percent carbon monoxide. If the pressure and temperature of the mixture are 140 kPa absolute and 40 °C, respectively, which of the following most nearly equals the mass percent of the carbon monoxide?

a. 13%
b. 11%
c. 9%
d. 7%

1.12 The relationship between mass and energy as derived by Einstein is $e = mc^2$, where e = energy, m = mass, and c = speed of light. If 1 lb of uranium is caused to fission and there is a loss of one-tenth of 1 percent of the mass in the process, which of the following most nearly equals the amount of energy that would be liberated?

a. 6.5 GW-hours
b. 7.8 GW-hours
c. 9.1 GW-hours
d. 11.3 GW-hours

1.13 A steam turbine receives 1600 kg of steam per hour at 35 m/s velocity and 3550 kJ/kg. The steam leaves at 250 m/s and 3020 kJ/kg. Which of the following most nearly equals the power output?

a. 198 kW
b. 210 kW
c. 223 kW
d. 247 kW

1.14 Air in the cylinder of a diesel engine is at 30 °C and 138 kPa abs. in compression. If it is further compressed to one-eighteenth of its original volume, which of the following most nearly equals the work done in compression if the displacement volume of the cylinder is 14.2 l?

a. 8 kJ
b. 11 kJ
c. 14 kJ
d. 17 kJ

1.15 An indicator card of a four-cycle automobile engine taken with a 114 kPa/mm spring is 76 mm long and has an area of 806.5 mm^2. If the engine has 8 cylinders, each 9.00 cm in diameter with a 76-mm stroke, and turns at 2800 revolutions/min, which of the following most nearly equals the developed power?

a. 81 kW
b. 89 kW
c. 97 kw
d. 109 kW

1.16 A steam turbine receives 1630 kg of steam per hour at 340 km/s velocity and 3550 kJ/kg enthalpy. The steam leaves at 259 m/s and 3020 kJ/kg. Which of the following most nearly equals the power output?

a. 242 kW
b. 251 kW
c. 269 kW
d. 269 kW

1.17 If 0.5 kg of nitrogen with an initial volume of 170 l and a pressure of 2070 kPa abs. expands in accordance with the law PV^n = constant to a final volume of 850 l and a pressure of 275 kPa abs., which of the following most nearly equals the work done by the gas?

a. 465 kJ
b. 472 kJ
c. 481 kJ
d. 489 kJ

1.18 An automobile cooling system of 25 l of water and antifreeze has a freezing point of −24 °C. Which of the following most nearly equals the amount of the above mixture that must be drained out in order to add sufficient anti-freeze to protect the engine at −26 °C? The table printed on the antifreeze container indicates that 2.4 l will protect a 12-l cooling system to −24 °C and 4.3 l to −36 °C

a. 0.617 l
b. 0.714 l
c. 0.825 l
d. 0.896 l

1.19 A balloon has a mass of 193 kg and has a volume of 993 m^3 when filled with hydrogen gas with a density of 89.8 g/m^3. Which of the following most nearly equals the load that the balloon will support in air with a density of 1.30 kg/m^3?

a. 1009 kg
b. 987 kg
c. 897 kg
d. 1205 kg

1.20 What temperature has been assumed for the air in which the balloon will float in the preceding problem?

a. 8.2°C
b. 4.7°C
c. 2.0°C
d. −1.4°C

1.21 A gas company buys gas at 620 kPa gage and 24°C and sells it at 9.65 cm of water pressure and −2.0°C. Disregarding the losses in distribution, which of the following most nearly equals the number of cubic meters sold for each cubic meter purchased?

a. 2.3
b. 4.1
c. 6.4
d. 7.1

1.22 An indicator card of a four-cycle automobile engine taken with a 115.37-MPa/m spring is 71 mm long and has an area of 774.2 mm^2. If the engine has six cylinders, each 89 mm in diameter with a stroke of 76 mm, and is operating at 3000 rpm, which of the following most nearly equals the developed power of the engine?

a. 92 kW
b. 98 kW
c. 88 kW
d. 79 kW

1.23 Steam expands behind a piston doing 67,800 joules of work. If 12.7 kJ of heat is radiated to the surrounding atmosphere during the expansion, which of the following most nearly equals the change of internal energy?

a. 80.5 kJ
b. 76.2 kJ
c. 91.4 kJ
d. 84.5 kJ

1.24 Suppose that 1.0 kg of a dry and saturated steam is confined in a tank. A gauge on the tank indicates a pressure of 1.0 MPa. Barometric pressure is 736.6 mm Hg. Which of the following most nearly equals the enthalpy of the steam?

a. 2445.9 kJ/kg
b. 2579.2 kJ/kg
c. 2678.3 kJ/kg
d. 2781.5 kJ/kg

1.25 A barometer tube 88 cm long contains some air above the mercury. It gives a reading of 68 cm when upright. When the tube is tilted to an angle of 45°, the length of the mercury column becomes 80 cm. Which of the following most nearly equals the reading of an accurate barometer?

a. 76.0 mm Hg
b. 75.8 mm Hg
c. 75.6 mm Hg
d. 75.2 mm Hg

1.26 Which of the following most nearly equals the resulting quality of the steam if 5.0 kg of water is injected into an evacuated vessel that is maintained at 175°C, if the volume of the vessel is 283 l?

a. 26%
b. 32%
c. 21%
d. 35%

1.27 If a 75-kW heat engine operates on the ideal Carnot cycle between the temperatures of 280 °C and 60 °C, which of the following most nearly equals the amount of heat rejected?

a. 41 kJ/s
b. 49 kJ/s
c. 54 kJ/s
d. 57 kJ/s

1.28 An air-speed indicator of a plane is operated on the same principle as the pitot tube. It is calibrated to read knots. If the indicator reads 200 knots at an altitude where the density of the air is one-half the density at which it was calibrated, which of the following most nearly equals the true speed of the plane?

a. 200 knots
b. 187 knots
c. 250 knots
d. 283 knots

SOLUTIONS

1.1 **c.** The initial volume is solved by using the equation of state, $PV = mRT$. Putting this into the proper format yields

$$V_1 = mRT_1/P_1$$
$$= [1\ \text{kg}(8.314/29)\ \text{kJ/kg-K}(10^3\ \text{N-m/kJ}) \times(300\ \text{K})]/(5\ \text{bar})(10^5\ \text{N/m}^2\text{-bar})$$
$$V_1 = 0.172\ \text{m}^3$$

1.2 **a.** The work done by the gas may be obtained using the $\int PdV$ equation.

$$W = \int PdV = P(V_2 - V_1)$$
$$= [5\ \text{bar}(10^5\ \text{N/m}^2\text{-bar})(\text{kJ}/10^3\ \text{N-m})](0.172\ \text{m}^3)$$
$$W = 86\ \text{kJ}$$

1.3 **d.** The heat transfer is obtained by using the general energy equation.

$$Q/m + u_1 + \text{KE}_1 + \text{PE}_1 = u_2 + \text{KE}_1 + \text{PE}_1 + W/m$$

In the absence of ΔKE and ΔPE the equation may be written as

$$Q/m = c_v(T_2 - T_1) + 86\ \text{kJ} = 0.718(400 - 300) + 86 = 157.8\ \text{kJ}$$

1.4 **d.** The solution may be best obtained by writing the general generic energy equation that includes all of the terms for non-steady-state, non-steady-flow case.

$$Q/m + h_{\text{in}} + \text{KE}_{\text{in}} + \text{PE}_{\text{in}} = \Delta u + \Delta\text{KE} + \Delta\text{PE} + h_{\text{out}} + \text{KE}_{\text{out}} + \text{PE}_{\text{out}} + W/m$$

In the absence of kinetic energy, potential energy, work, and heat transfer, the equation simplifies to

$$h_{\text{in}} = (u_2 - u_1) + h_{\text{out}}$$

Knowing that the chamber was initially evacuated means that $u_1 = 0$, and knowing that there is no mass flow out of the chamber reduces the equation to

$$h_{\text{in}} = u_2 = 3480.6\ \text{kJ/kg}$$

Interpolation from the superheated water table of the *FE Handbook* yields

$$T_2 = 702°\text{C}$$

1.5 **a.** A mass balance and an energy balance are to be written for the feedwater heater.

$$m_3 = m_1 + m_2$$
$$m_1h_1 + m_2h_2 = m_3h_3$$

Substituting the mass equation into the energy equation and the values from the saturated and superheated water tables in the *FE Handbook*, one gets the following equation:

$$\begin{aligned} h_1 &= h_T + \int V dP \\ &= 167.57 \text{ kJ/kg} + (0.001008 \text{ m}^3\text{/kg}) \\ &\quad \times (1000 - 7.384)\text{kPa (kN/m}^2\text{-kPa)(kJ/kN-m)} \\ h_1 &= 167.57 + 1.0 = 168.57 \text{ kJ/kg} \end{aligned}$$

$$\begin{aligned} &1000 \text{ kg/min } (168.57 \text{ kJ/kg}) + m_2(3478.5 \text{ kJ/kg}) \\ &\quad = (1000 + m_2) \text{ kg/min } (762.81) \text{ kJ/kg} \\ &m_2 = 218.8 \text{ kg/min} \end{aligned}$$

1.6 **d.** An entropy balance equation has to be written for control volume, including the feedwater heater. The basic equation may be obtained from the *FE Handbook*.

$$0 = \Sigma Q/T + m_1 s_1 + m_2 s_2 + m_3 s_3 + \sigma$$

Since the feedwater heater is insulated ($Q = 0$), the equation may be simplified and we may solve for σ. Since the problem requires that the entropy production to be found as a function of entering water, the equation may be written as:

$$\begin{aligned} \sigma/m_1 &= (m_3/m_1)s_3 - (m_1/m_1)s_3 - (m_2/m_1)s_2 \\ \sigma/m_1 &= [(1000 + 218.8)/1000](2.1387) - 0.5725 \\ &\quad - [218.8/1000](7.7622) \\ \sigma/m_1 &= 0.3358 \text{ kJ/kg-min-K} \end{aligned}$$

1.7 **b.** The heat to be absorbed by the Prony brake equals $75 \times 0.95 = 71.25$ kW or 71.25 kJ/s.

The specific heat of water equals 4.18 kJ/kg-K (see the *FE Handbook*).

The temperature change equals $55 - 18 = 37$ K.

So each kg of water will absorb $37 \times 4.18 = 154.66$ kJ of heat.

The water flow must then equal $71.25/154.66 = 0.4607$ kg/s or 27.64 kg/min.

Since 1 liter of water has a mass of 1 kg, the flow of cooling water must equal 27.64 l/min.

1.8 **d.** This is a polytropic process and work equals $(P_2V_2 - P_1V_1)/(1 - n)$ as given in the *FE Handbook*.

But the problem states that $n = 1$, which would give an infinite amount of work using this relationship. However, if PV = constant, it is a constant-temperature process, and work is given by $W = RT \times \ln(P_1/P_2)$, as given in the *FE Handbook*. Volume is given by $V = MRT/P$. For 100 kg of mixture, $30/44 = 0.682$ kg mol of CO_2, and $70/32 = 2.188$ kg-mol of O_2.

$$\begin{aligned} &2.188 + 0.682 = 2.870 \text{ mol of gas mixture per 100 kg} \\ &\quad 100/2.870 = 34.843 \text{ kg per kg} \cdot \text{mol} \end{aligned}$$

Then R for the mixture equals 8,314/34.843 = 238.67 (see *Handbook*).

$$\text{work} = 238.67 \times (273 + 15) \times \ln(100/1050) = 68{,}737 \times (-2.351)$$
$$= -161.601 \text{ joules work per kg done on the gas}$$

The total amount of work done on the gas equals 2 × 161,601 = 323,202 joules = 323 kJ

1.9 **a.** The pressure of the steam is given as 138 kPa. To this must be added atmospheric pressure of 101.3 kPa, giving an absolute pressure of 239.3 kPa. Each kg of steam will give up 2,715 − 530 = 2185 kPa as it condenses. The temperature of the condensate will equal 126.7 °C, which will cool to 65 °C, giving an additional 61.7 calories of heat, or 61.7 × 4.185 = 258 kJ/kg of heat (one calorie is equivalent to 4.18 J) as it cools to the desired temperature of 65 °C. So each kg of steam will give up 2185 + 258 = 2416 kJ of heat as it cools to the desired temperature of 65 °C. There are 750 liters of solution with a density of 1.00 kg/l, giving a total mass of 750 kg. Each kg of solution will absorb (65 − 15) × 3.56 = 178 kJ as it warms up to the desired 5 °C. The total amount of heat required will thus equal 178 × 750 = 133,500 kJ. The amount of steam required will equal 133,500/2,416 = 55.26 kg.

1.10 **b.** The process is adiabatic so PV^k = constant. The constant is given by $c_p/c_v = k = 0.847/0.659 = 1.285$.

$$V_2 = V_1 \times (P_1/P_2)^{1/k}$$
$$P_2 = 5 \times 76 = 380 \text{ cm Hg}$$

So

$$V_2 = (63.5/380)^{0.778} \times V_1 = 0.249 \times 280 = 69.7 \text{ l}$$

1.11 **c.** Assume one mole in volume V and calculate the mass of each gas in that volume. Tabulate the data in a chart. As an example, take nitrogen.

$$PV = mRT \text{ or } m = PV/RT$$

The molecular weight of nitrogen equals 28 (see *Handbook*), so R for nitrogen = 8,314/28 = 296.9. The pressure is given as 140 kPa abs. and the temperature is 40 °C or 313 K, so the mass of one mole of nitrogen under the stated conditions equals m = 140,000 × V/(296.9 × 313) = 1.507 × V. The mass of 70% of 1 mol equals 1.055 kg. The masses of the other gases are calculated in a similar fashion.

Gas	Vol %	MW	R	Mass	Mass %
N_2	70	28	297	1.055 × V	63
CO_2	20	44	189	0.473 × V	28
CO	10	28	297	0.151 × V	9
				Total mass = 1.679 × V	

1.12 **d.**

$$\Delta\text{mass} = 0.001 \times 454 = 0.454 \text{ g}$$
$$E = 0.454/1{,}000 \times 299{,}000^2 = 40{,}588{,}0654\ 10^6 \text{ kg-m}^2/\text{s}^2$$

or

$$E = 40.6 \times 10^6 \text{ MN-m or } 40.6 \times 10^6 \text{ MJ}$$

A joule equals one watt-second or 1/3600 watt-hour

$$40.6 \times 10^6 \text{ MJ} = 40{,}600{,}000 \times 10^6/3600$$
$$= 11{,}278 \times 10^6 \text{ watt-hours or } 11.3\ 10^6 \text{ kWh}$$

1.13 **c.** The rate of steam flow equals 1600/3600 = 0.444 kg/s. The change in enthalpy equals 3500 – 3020 = 530 kJ/kg or 0.444 × 530 = 235.3 kJ/s. The change in kinetic energy of the steam equals $\frac{1}{2} \times 0.444 \times (35^2 - 250^2) = -13{,}603$ kg-m^2/s^2 or –13.6 kJ/s. The energy extracted from the steam thus equals 235.3 – 13.6 = 221.7 kJ/s or 223 kW.

1.14 **b.** The compression would be rapid and can be considered as adiabatic—no heat absorbed or given up. However it should be noted that it would not be isentropic.

$$P_2 = P_1 \times (V_1/V_2)^k = 138 \times 18^{1.4} = 138 \times 57.20 = 7894 \text{ kPa}$$
$$V_1 = 14.2 \text{ l} = 0.0142 \text{ m}^3$$
$$V_2 = 0.000789 \text{ m}^3$$

Adiabatic work (see *Handbook*) equals $(P_2V_2 - P_1V_1)/(1 - k)$. Work = (7894 × 0.000789 – 138 × 0.0142)/(1.0 – 1.4) = –10.672 kN-m = –10.672 kJ where the negative sign indicates that work was done on the gas.

1.15 **d.** The indicator diagram is 76 mm long and has an area of 80.6 mm^2, so the average height of the diagram equals 806.5/76 = 10.61 mm. This corresponds to an average pressure of 10.61 × 114 = 1.210 Mpa gauge pressure. It is an eight-cylinder, four-cycle engine, so there are four power strokes per revolution or (2800/60) × 4 = 186.67 power strokes per second. The volume per power stroke (per cylinder) equals

$$V = (\pi/4) \times 9.00^2 \times 7.60 = 483.5 \text{ cm}^3 \text{ or } 483.5 \times 10^{-6} \text{ m}^3$$
$$(483.5 \times 10^{-6} \text{ m}^3 \times 186.67) \text{ m}^3/\text{s} \times 1{,}210{,}000 \text{ N/m}^2 =$$
$$109{,}208 \text{ watts or } 109 \text{ kW}$$

1.16 **b.** The entering steam contains heat energy (enthalpy) and kinetic energy. The amount of the change in the energy contained in the steam equals the energy given to the turbine.

The change in enthalpy equals 3550 – 3020 = 530 kJ/kg

The change in kinetic energy equals $\frac{1}{2}(340^2 - 259^2) = 24{,}260$ kg·m^2/s^2 per kg = 24.3 kJ/kg

The total energy given up by the steam equals 530 + 24.3 = 554.3 kJ/kg or

$$(1630/3600) \times 554.3 = 251 \text{ kJ/s or } 251 \text{ kW}$$

1.17 **a.** This is an example of polytropic non-flow work since it takes place in a closed container, i.e., no gas is added and no gas is removed. From the *Handbook* polytropic work equals $(P_2V_2 - P_1V_1)/(1 - n)$. The polytropic exponent can be calculated from the data given.
Since $P_1V_1^n = P_2V_2^n$

$$n = [\ln(P_1/P_2)]/[\ln(V_2/V_1)] = (\ln 7.527)/(\ln 5.00) = 1.254$$

$$\text{work} = (275 \times 0.850 - 2070 \times 0.170)/(1.000 - 1.254)$$
$$= (233.8 - 351.9)/(-0.254) = 465 \text{ kJ}$$

1.18 **c.** The cooling system initially contains 25 l total mixture of water and antifreeze in the ratio 2.4/12.

This is (2.4 l coolant and 9.6 l of water), so the cooling system initially contains (2.4/12) × 25 = 5.00 l of coolant or 20 percent of the total mixture. Assuming a straight-line relationship, 20% coolant protects to −24°C and 35.83% coolant protects to −36°C, since −26°C is 2/12 the way from −24°C to −36°C the concentration of coolant would have to equal 20% plus (2/12) × (35.83 − 20.00) = 22.64% Thus the 25 l of coolant would have to contain

$$0.2264 \times 25 = 5.660 \text{ l of antifreeze. } 5 - 0.20 \times X + X = 5.660 \text{ l}$$

which gives $X = 0.660/0.800 = 0.825$ l to be drawn out of the cooling system and replaced with antifreeze to lower the freezing point to −26°C from the original freezing temperature of −24°C.

1.19 **a.** The buoyancy per cubic meter of volume equals 1.30 − 0.0898 = 1.2102 kg, so the total buoyancy of the balloon equals 993 × 1.2102 = 1202. The balloon has a mass of 193 kg, so the load that the balloon will support equals 1,202 − 193 = 1009 kg.

1.20 **d.** The density of the air is given as 1.30 kg/m^3. From the gas law we have that $T = PV/Rm$. For air $R = 8314/29 = 286.699$ (see *Handbook*).

For a volume of one cubic meter, pressure equals 101.3 kPa (see *Handbook*), and the mass equals 1.30 kg. So

$$T = 101{,}300 \times 1.00/(286.89 \times 1.30) = 271.8 \text{ K or } -1.4°\text{C}$$

1.21 **c.** The initial pressure of the gas equals 620 + 101.3 = 721.3 kPa abs. temperature = 273 + 24 = 297 K.

Final pressure equals 0.0965 meters of water gage, which equals 0.0965 × 1,000 × 9.807 = 946 N/m^2 or 946Pa. Adding one atmosphere gives a pressure of 102.26 kPa abs.

Temperature = 271 K

$$P_1V_1/T_1 = P_2V_2/T_2$$

So,

$$V_2 = V_1 \times (T_2/T_1) \times (P_1/P_2) = V_1 \times 0.913 \times 7.054 = 6.44 \times V_1$$

1.22 **a.** The average height of the diagram equals 774.2/71 = 10.904 mm, which indicates an average, or mean, effective pressure (MEP) of 0.010904 × 115,370,000 = 1,258,000 N/m^2 = 1.258 MPa

Power strokes per second = (6/4) × 2 × (3,000/60) = 150

The volume of one cylinder, cylinder displacement, is $\pi/4 \times 0.089^2 \times 0.076 = 0.000473$ m^3

Power = 1,285,000 × 150 × 0.000473 = 91,171 N·m/s = 91.2 kJ/s or 91.2 kW

1.23 **a.** Work equals 67.8 kJ, and 12.7 kJ of heat is lost in the process, so a total of 67.8 + 12.7 = 80.5 kJ have been extracted in the process, or the change in internal energy of the gas equals 80.5 kJ.

1.24 **d.** One standard atmosphere equals 760 mm of Hg (see *Handbook*), which equals 101.3 kPa. Then, 736.6 mm Hg would equal (736.6/760) × 101.3 = 98.18 kPa. This would be added to the gage pressure to obtain the absolute pressure of the steam, which would equal 1.0982 MPa. The problem then becomes one of interpolating between the two values since the problem states that the steam is dry and saturated. The listed values of enthalpy for saturated vapor for the nearest greater and lesser pressures given in the *Handbook* are

P = 1.0021 MPa Enthalpy (h_g) = 2778.2 kJ/kg T = 180°C

1.1227 MPa Enthalpy (h_g) = 2782.4 kJ/kg T = 185°C

The difference in the pressures equals 120.6 kPa, and the steam pressure of 1.0982 MPa is 96.1 kPa greater than the lower pressure. The difference in the higher and lower enthalpies equals 4.2 kJ/kg.

The enthalpy of the 1.0982 MPa steam would then be

$$h = (96.1/120.6) \times 4.2 + 2778.2 = 2781.5 \text{ kJ/kg}$$

1.25 **c.** When the barometer tube is upright the air volume is 20 × tube area, when it is tilted at 45° the volume equals 8 × tube area. The system will be at a constant temperature so the pressure of the trapped air will differ at the two conditions in the ratio of

$$20/8 = 2.5 \text{ or } P_2 = 2.5P_1$$

One atmosphere = P_1 + 68 cm Hg = P_2 + 80 × 0.707 cm Hg = 2.5 × P_1 + 56.56 cm Hg.

Combining the first and last equations gives 1.5 × P_1 = 11.44 cm Hg and P_1 = 7.627 cm Hg.

The correct barometric pressure would thus equal 68 + 6.627 = 75.627 cm Hg.

1.26 **a.** The volume of the tank, and of the resulting mixture, is 283 l, or 0.283 m^3. The resulting specific volume of the mixture is 0.283/5.0 = 0.0566 m^3/kg. From the steam tables in the *Handbook*, the specific volume of saturated steam at 175 °C = 0.2168 m^3/kg. The specific volume of saturated liquid at 175 °C = 0.001121 m^3/kg. If the percent of steam is taken as x, then

$0.2168x + (1 - x) \times 0.001121 = 0.0566$ m^3/kg, which gives $x = 0.257$

1.27 **b.** The engine would put out 75,000 joules per second of energy. A Carnot cycle consists of two constant-temperature lines and two constant-entropy lines that form a rectangle as shown in the *Handbook*. The power output is given by

$$T_1 \times (S_2 - S_1) - T_2 \times (S_2 - S_1)$$

The heat input equals $T_1 \times (S_2 - S_1)$ and the heat lost equals $T_2 \times (S_2 - S_1)$, where T_1 is the higher temperature and T_2 is the lower temperature, and S_2 is the higher entropy and S_1 is the lower entropy. The efficiency equals the work done (power output) divided by the amount of heat input, or

$$\text{efficiency} = [T_1 \times (S_2 - S_1) - T_2 \times (S_2 - S_1)]/[T_1 \times (S_2 - S_1)]$$
$$= (T_1 - T_2)/T_1, \text{ where the temperatures are absolute, or expressed in Kelvin.}$$

The two temperatures are 280 + 273 = 553 K and 60 + 273 = 333 K.

$$\text{efficiency} = (553 - 333)/553 = 39.78\%$$

The heat rejected would thus equal $(1 - 0.3978) \times 75{,}000 = 49{,}098$ J/sec or 49.1 kJ/s.

1.28 **d.** For a Pitot tube, the velocity of the fluid, or relative velocity between the Pitot tube and the fluid (in this case, the plane), equals $C/(2gh)$, which can be expressed as

$$h = v^2/(C^2 \times 2g)$$

But, since we are concerned with a comparison of the readings of the same Pitot tube at different conditions, the coefficient will be the same for both conditions (same Pitot Tube). We can disregard the C factor, which gives

$$h = v^2/2g$$

If the density of the air at the new altitude is one half that at the calibration altitude, then h at the new altitude will be one half that at the calibration altitude for the same speed. For the same speed,

$$h_{elev} = \tfrac{1}{2} h_{calib}$$

So, at the new altitude, the same reading on the indicator would mean that

$$v_{elev} = v_{calib} \times \sqrt{(2h/h)} = 1.414 \times v_{calib}$$

The true speed of the plane would thus equal $200 \times 1.414 = 282.8$ knots.

CHAPTER 2

Energy Conversion and Power Plants

OUTLINE

The conversion of energy to useful power constitutes a major portion of the endeavors of the mechanical engineering profession. Energy is available in many forms of organic fuels, including coal, oil, natural gas, wood and peat moss. In addition, energy is available from other sources, such as ocean water in the form of tides and temperature differentials due to ocean currents, water flowing in rivers, the sun, nuclear reactors, and underground thermal sources, to mention a few. The major types of conversion systems include internal combustion engines, gas and vapor turbines, hydroelectric turbine systems, and reciprocating steam engines.

The largest amount of energy conversion is done in the many electric-generating facilities throughout the United States and the world. The energy is produced by electric generators that are powered by prime movers. Other things being equal, these systems are selected generally on the basis of their thermal efficiency. Probably the best definition of thermal efficiency is:

$$\eta_{th} = \text{(Energy wanted)/(Energy that costs \$)}$$

Normally this equation is then written in terms of output power in kW divided by the input energy in kJ/s.

Example 2.1

A power plant for a small company produces 54 kW of useful electric power. The generator has an efficiency of 96%, and the pump that is used to circulate the fluid takes 2.5 kW. The heat source supplies 10.2 MJ/min of energy to the system.

Determine

a) Power output of the heat engine

b) Thermal efficiency of the power plant

c) Amount of rejected heat from the plant

Solution

a) Since the power that is produced by the plant is equal to the useful (or net) power plus any other power need for electric generation, we must add the power that is used by the circulating pump and then divide this total output of the generator by the generator efficiency to obtain the power produced by the prime mover.

$$\text{Power produced} = (54 \text{ kW} + 2.5 \text{ kW})/(0.96) = 58.9 \text{ kW}$$

b) Thermal efficiency is 54×10^3 J/s divided by $10.2 \times 10^6/60$ J/s

$$\eta = 31.8\%$$

c) Whatever energy that is not used is rejected; that is, 68.2% of the energy that was supplied to the system is lost.

$$\text{Heat lost} = 0.682 \times 10.2 \text{ MJ/min} = 6.96 \text{ MJ/min.}$$

Example 2.2

A windmill is directly coupled to a generator that is coupled to a battery. If the generator is able to produce 10 kW of power and the losses, in the form of heat losses due to line resistance and battery heat transfer, are 1.0 kW,

a) Determine the total amount of energy, in kJ, stored during a 10-hour period.

Solution

$$\begin{aligned}\text{Total energy produced} &= (\text{net production}) \times (\text{hours}) \\ &= (10 - 1)\text{kW}(10 \text{ hrs})(3600 \text{ s/hr})(1 \text{ (kJ/s)/kW}) \\ &= 3.24 \times 10^5 \text{ kJ}\end{aligned}$$

SOLAR ENERGY

Example 2.3

One of the least exploited forms of available energy is the energy from the sun. This is a renewable form of energy, with a fresh supply each day. Consider a flat plate solar collector used to supply energy to the working fluid (HFC 134a) in a 10-MW developmental power plant operating on a Rankine cycle. Saturated vapor enters the turbine at 20 bar and exits at 0.4 bar. Consider the turbine to be 85% efficient, the pump to be 70% efficient, and the rate of energy from the sun to the working fluid to be 0.3 kW/(m^2 of collector area). Cooling water at 20°C is available for the condenser.

Determine

a) η_{th} of the cycle

b) Mass flow rate through the turbine

c) Mass flow rate of cooling water through condenser if $\Delta T = 10°C$

d) Size of collector in m^2

Solution

Property values are found on the HFC 134a chart in the *FE Handbook*

	1	2s	2	3	4s	4
P	20 bar	0.4bar	⇒	⇒	20 bar	⇒
T	Sat. Vap		Sat. Liq			
h	430	351	362.9	143	144.37	144.96

In order to find the enthalpy of the HFC 134a leaving the turbine, one must use the turbine efficiency equation:

$$\eta_t = \text{(Work turbine actual)/(Work turbine ideal)}$$
$$= (h_1 - h_2)/(h_1 - h_{2s}).$$
$$0.85 = (430 - h_2)/(430 - 351)$$
$$h_2 = 362.9 \text{ kj/kg}$$

To calculate the enthalpy of the HFC 134a at state 4, one must use the work of a pump equation.

$$h_{4s} = h_3 + W_p$$
$$= 143 + \int V dP$$
$$= 143 \text{ kJ/kg} + (0.0007)\text{m}^3\text{/kg}(20 - 0.4)\text{bar}[(10^5\text{N/m}^2)/(\text{bar})][1 \text{ kJ}/(10^3\text{Nm})]$$
$$= 143 + 1.37$$
$$= 144.37 \text{ kJ/kg}$$

To calculate the enthalpy at the actual state one must use the pump efficiency equation, which is:

$$\eta_p = \text{(Isentropic work of pump)/(Actual work of the pump)}$$
$$= 0.70 = (1.37)/W_p$$

Therefore,

$$W_p = 1.96 \text{ kJ/kg}$$
$$h_4 = h_3 + W_p = 143 + 1.96 = 144.96$$

a)

$$\eta_{th} = \text{[Energy wanted]/[Energy that costs]}$$
$$= [W_t - W_p]/[Q_{in}]$$
$$= [(h_1 - h_2) - 1.96]/[h_1 - h_4]$$
$$= [430 - 362.9 - 1.96]/[430 - 144.96]$$
$$= 22.8\%$$

b) Mass flow rate $= [\text{Power out}]/[W_{net}]$

$$= (10\text{MW})(1000\text{kJ/kW})(3600 \text{ s/hr})/(65.14)\text{kJ/kg}$$
$$= 5.527 \times 10^5 \text{ kg/hr}$$

c) Mass flow rate of cooling water through the condenser requires an energy balance through the condenser.

$$m_t(h_2 - h_3) = m_{cond}(\Delta h_{cooling\ water})$$
$$[5.527 \times 10^5 \text{ kg/hr}][362.9 - 143]\text{kJ/kg} = [m_{cond}\text{kg/hr}][°\text{C kJ/kg}\cdot\text{K}][\Delta T]\text{K}$$
$$m_{cond} = [1.221 \times 10^8]/[4.179 \times 10] = 2.92 \times 10^6 \text{ kg/hr}$$

d) Collector size in square meters is obtained by dividing the heat into the working fluid by the rate into the collector per unit area.

$$\text{Area} = 10{,}000 \text{ kW}/0.3 \text{ kW/m}^2 = 33{,}333 \text{ m}^2$$

Example 2.4

Water is the working fluid in a power cycle that operates according to an ideal Rankine cycle. Superheated vapor enters the turbine at 10 MPa and 520 °C and is exhausted into the condenser at 8 kPa. The net output of the Rankine cycle is 120 MW.

Determine

a) Mass flow rate in kg/hr

b) Heat flow into the boiler in kJ/hr

c) Heat flow out of the condenser kJ/hr

Property values:

	1	2	3	4	
P	100 bar	8 kPa	8 kPa	100 bar	
T	520 C				
h	3425.1	2071.38		173.88	183.96
s	6.6622	6.6622			
v				1.0084×10^{-3}	

Obtaining quality at state 2:

$$s_2 = (x_2)(s_g) + (1 - x_2)(s_f) = 6.6622 = (x_2)(8.2287) + (1 - x_2)(0.5926)$$
$$x_2 = 0.7896$$

Therefore

$$h_2 = (x_2)(h_g) + (1 - x_2)(h_f)$$
$$= (0.7896)(2577.0) + (1 - 0.7896)(173.88)$$
$$= 2071.38 \text{ kJ/kg}$$

and

$$h_4 = h_3 + \int v\,dP$$

$$= 173.88 + (0.0010084)\text{m}^3/\text{kg } (100 - 0.08) \text{ bar}(10^5 \text{ N/m}^2)/\text{bar}(\text{kJ}/10^3 \text{ Nm})$$
$$= 173.88 + 10.08$$
$$= 183.96 \text{ kJ/kg}$$

a) Mass flow rate = $W_{net}/(W_t - W_p)$
$= 120{,}000\ \text{kW}/[(h_1 - h_2) - 10.08]$
$= (120{,}000\ \text{kJ/s})(3600\ \text{s/hr})/(3425.1 - 2071.38 - 10.08)\ \text{kJ/kg}$
$= 4.837 \times 10^6\ \text{kg/hr}$

b) Heat flow into the boiler = (mass flow rate) $(h_1 - h_4)$
$= (4.837 \times 10^6)\text{kg/hr}(3425.1 - 183.96)\ \text{kJ/kg}$
$= 1.568 \times 10^{10}\ \text{kJ/hr}$

c) Heat flow out of the condenser = (mass flow rate)$(h_2 - h_3)$
$= (4.837 \times 10^6\ \text{kg/hr})(2071.38 - 173.88\ \text{kJ/kg})$
$= 9.178 \times 10^9\ \text{kJ/hr}$

HEAT ENGINES

In thermodynamics there are two basic, fundamental engine classifications, which are based on the working fluid. The first classification is an engine that operates on a fluid that is alternately vaporized and then condensed. This type of engine may be of the heat engine type in which the Rankine cycle is utilized, as discussed previously in this chapter, or may be of the heat pump/refrigeration type (see Chapter 12). The second classification of engine uses a working fluid that remains in the gaseous state throughout the operation of the engine. This second engine classification may be further divided into internal combustion engines and external combustion engines. Let us first consider the internal combustion engine.

INTERNAL COMBUSTION ENGINES

Internal combustion engines are engines in which the fuel air mixture that powers the engine ignites and burns in a closed, limited, and very dynamic region called the cylinder. These engines may be spark ignition engines (gasoline powered engines) or they may be compression ignition engines (diesel powered engines).

MEASUREMENT OF POWER

Power produced may be measured in various ways. Generally, the measurement methods may be divided into two classifications, external measurement and internal measurement. The external measurement variety uses a device on the outside of the engine, and it tends to brake the engine until it begins to lose power. This method therefore measures *brake horsepower*. The internal method uses a device mounted internally to the cylinder to measure the pressure and an external device to measure the corresponding volume. This method measures *indicated horsepower*.

Example 2.5

A prony brake is one of the common devices used to measure the actual output power of engines. It does this by means of a braking system mounted on the output shaft. A moment arm is connected to the brake shoe and the force required to hold the arm in place is measured. The prony brake equation from the *PE Handbook*:

$$W_b = 2\pi TN$$

here

W_b = Brake power
T = Torque, N·m
N = Rotational speed, rev/s

A prony brake system, as shown in Exhibit 1, is used to measure the output of an engine. For this setup $L = 700$ mm, and the brake is tightened until the rotational speed starts to drop. The engine operates at 1,150 rpm, and the force is measured to be $F = 2{,}200$ N.

Determine the output power of the engine.

Solution

Using the prony brake equation:

$$\begin{aligned} W &= 2\pi TN \\ &= 2\pi(2{,}200 \times 0.700)\text{N}\cdot\text{m}(1{,}150/60)\ \text{rev/s} \\ &= 185{,}460\ \text{J/s} \\ &= 185.5\ \text{kW} \end{aligned}$$

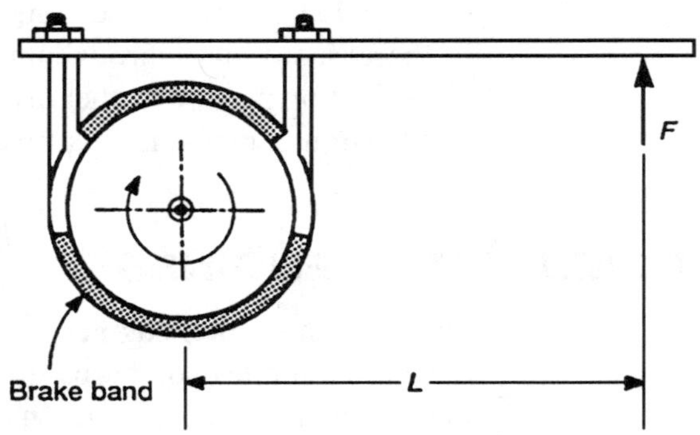

Exhibit 1

Example 2.6

A four-cylinder, four-cycle engine with 12.0-cm-diameter pistons and an 18-cm stroke operates at a speed of 500 rpm and yields an indicator diagram as shown in Exhibit 2. The area under the curve (PV diagram) is equal to 10.42 cm^2. The length of the diagram is 8.23 cm, and the spring constant of the indicator spring is 550 kPa/cm.

Determine

a) Mean Effective Pressure

b) Indicated power

Solution

The net area under the curve on the indicator card is proportional to the net work developed by the engine. The *Y*-axis represents a function of the pressure and the *X*-axis represents the stroke/volume of the cylinder.

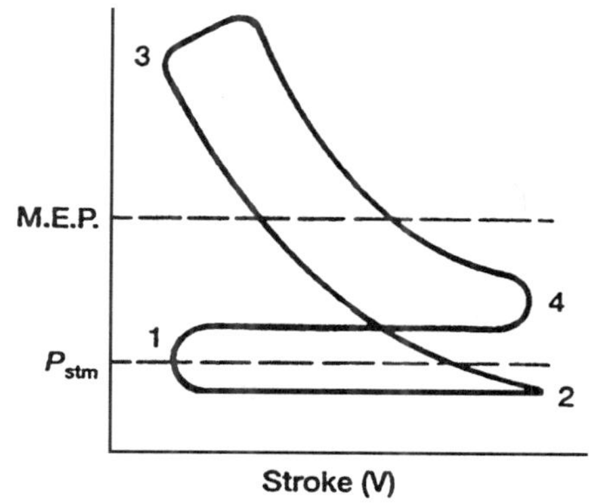

Exhibit 2

a) The *Mean Effective Pressure (M.E.P.)* is the average pressure felt by the piston over the range of the stroke.

$$\begin{aligned} \text{MEP} &= [(\text{Area under curve})/(\text{stroke})][\text{spring constant}] \\ &= [10.42\ \text{cm}^2/8.23\ \text{cm}][550\ \text{kPa/cm}] \\ &= 696.4\ \text{kPa} \end{aligned}$$

b) Work of a single stroke of a piston, $W/\text{cyl} = \text{M.E.P.} \times \text{Area} \times \text{Stroke}$

$$\begin{aligned} &= [696{,}400\ \text{Pa}][\pi D^2/4\ \text{cm}^2][\text{stroke cm}] \\ &= [696{,}400\ \text{Pa}][\pi(12.0)^2\ \text{cm}^2]/4\ [18\ \text{cm}]/10^6\ \text{cm}^3/\text{m}^3 \\ &= [1418\ \text{Pa}\cdot\text{m}^3][(\text{N/m}^2)/\text{Pa}][\text{J/N}\cdot\text{m}] \\ &= 1{,}418\ \text{J} \end{aligned}$$

Since this is a four-cycle engine it makes a power stroke once every two revolutions for each piston. Therefore, the total power output is:

$$\begin{aligned} W &= [1{,}416\ \text{J}][500\ \text{revolutions/min}]/[2\ \text{revolutions/(power stroke)}(4\ \text{cylinders})] \\ &= 1416\ \text{kJ/min} = 23.6\ \text{kJ/s} = 23.6\ \text{kW} \end{aligned}$$

OTTO CYCLE

The gasoline engines that are so familiar to us operate on the Otto cycle. This is a spark ignition engine in which the fuel-air mixture is compressed until the piston is near top dead-center and the spark plug fires, igniting the fuel in the cylinder. The fuel combines with the oxygen in the air producing CO_2 and H_2O, along with a great amount of heat that expands the gasses and forces the piston down rapidly, thus producing power that may be used to our advantage.

Example 2.7

An air-standard Otto cycle with a compression ratio of 8.5 has air entering at 105 kPa and 20 °C. There is an input of heat of 1500 kJ/kg during the heat addition process.

Determine

a) Thermal efficiency

b) Net work in kJ/kg

c) Maximum temperature

Solution

a) Select the thermal efficiency equation from the *FE Handbook*:

$$\eta_{th} = 1 - r^{1-k} = 1 - 8.5^{1-1.4} = 1 - 0.42 = 58\%$$

b) Set up the equation for thermal efficiency in terms of the heat addition and heat rejection.

$$\eta_{th} = [Q_{in} - Q_{out}]/Q_{in} = W_{net}/Q_{in}$$
$$0.58 = W_{net}/1500 \quad \text{or} \quad W_{net} = 870 \text{ kJ/kg}$$

c) To find the maximum temperature we need to calculate T_2 using the isentropic relationships,then equate the heat addition to the internal energy rise, and, finally, obtain the temperature using $c_v\Delta T$.

$$T_2 = T_1[v_1/v_2]^{k-1} = 293[8.5]^{1.4-1} = 689.7 \text{ K}$$
$$Q_{added} = 0.718\ \Delta T$$

or

$$T_3 - T_2 = 1500/0.718 = 2089 \text{ K}$$
$$T_3 = T_2 + 2089 = 689.7 + 2089 = 2778.7 \text{ K}$$

DIESEL/DUAL CYCLE

The diesel cycle also represents an internal combustion engine; however, it is a compression ignition engine. Air, alone, is compressed up to top dead-center when the fuel is sprayed into the cylinder. Since the temperature is quite high, the fuel ignites spontaneously and propels the piston downward in its power stroke. The *PV* and *TS* diagrams are shown in the *FE Handbook*. The dual cycle is a closer representation of what really happens in a diesel engine. After the compression stroke, the fuel is ignited and the pressure increases rapidly before the downward motion of the piston can level off the pressure rise.

Example **2.8**

An air-standard dual cycle receives air at 27°C and 1 atmosphere pressure. The compression ratio is 18, and the cutoff ratio is 3. Pressure doubles during the constant volume heat addition process and the maximum temperature is 2200 K.

Determine

a) Heat added in the constant volume process

b) Heat added in the constant pressure process

c) Thermal efficiency

Solution

a) The most effective way to attack this problem is to make a table of values. Notice the *PV* and *TS* diagrams (Exhibit 3 and 4).

	1	2	3	4	5
P	1.014 bar	58	116	116	9.4
T	300 K	953	1906	2200	1074

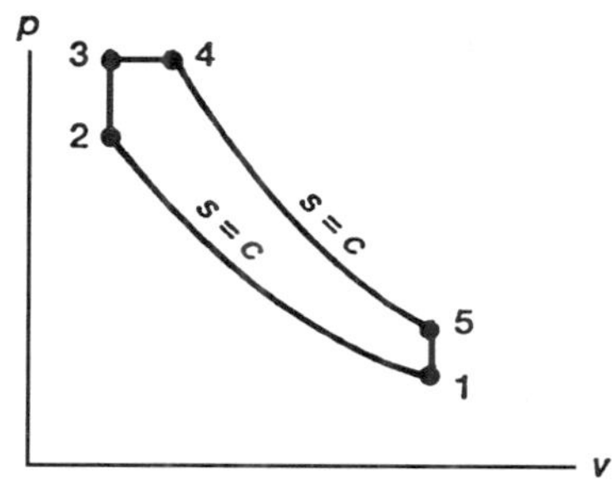

Exhibit 3 *PV* diagram

Since process 1 to 2 is isentropic, we use the isentropic relations listed in the *FE Handbook* to find T_2 and P_2.

$$P_2 = P_1[v_1/v_2]^k = 1.014[18]^{1.4} = 58 \text{ bar}$$
$$T_2 = T_1[v_1/v_2]^{k-1} = 300[18]^{1.4-1} = 953 \text{ K}$$

Process 2 to 3 is a constant volume ($V_2 = V_3$) process with $P_3 = 2P_2$, so Charles and Boyles law reduces to:

$$P_3/P_2 = T_3/T_2 = 2$$
$$T_3 = 2T_2 = 2(953) = 1906 \text{ K}$$

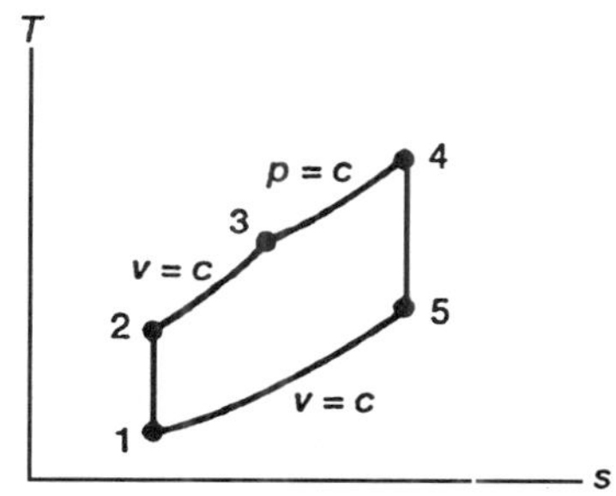

Exhibit 4 *T-s* diagram

Process 3 to 4 is a constant pressure process with $T_4 = 2200$ K. Process 4 to 5 is an isentropic expansion process. Note that $V_5 = V_1$; therefore:

$$V_4/V_5 = [V_4/V_3][V_2/V_1] = (\text{cut off ratio})/(\text{compression ratio}) = 3/18 = 1/6$$

Using the isentropic relations:

$$T_5/T_4 = [V_4/V_5]^{k-1} = (1/6)^{1.4-1} = 0.488$$
$$T_5 = 0.488 \times 2200 = 1074 \text{ K}$$
$$P_5/P_4 = [V_4/V_5]^k$$

Therefore

$$P_5 = [116][1/6]^{1.4} = 9.4 \text{ bar}$$

Now that all of the critical values have been obtained, it is a straightforward process to get the remaining answers.

a) The heat transfer during the constant volume process is:

$$Q_{2\text{–}3} = c_v[T_3 - T_2] = 0.718[1906 - 953] = 684 \text{ kJ/kg}$$

b) The heat transfer in the constant pressure process is:

$$Q_{3\text{–}4} = c_p[T_4 - T_3] = 1.00[2200 - 1906] = 294 \text{ kJ/kg}$$

c) The thermal efficiency is:

$$\begin{aligned}\eta_{th} &= [Q_{in} - Q_{out}]/Q_{in} = [Q_{2\text{–}3} + Q_{3\text{–}4} - Q_{5\text{–}1}]/[Q_{2\text{–}3} + Q_{3\text{–}4}]\\ &= [684 + 294 - c_v\Delta T]/[684 + 294] = [978 - 0.718(1074 - 300)]/978\\ &= 43.2\%\end{aligned}$$

BRAYTON CYCLE

The Brayton cycle is the basis for the operation of the jet engine. It is considered to be an external combustion engine because the fuel and air are mixed and burned in a combustion chamber and then the exhaust products are fed to the turbine. The turbine is directly coupled to the compressor and the excess power generated is used to power the aircraft. This excess power may be in the form of high velocity gasses that pass through a nozzle and thus propel the aircraft, or may be through extra rows of turbine blades that are used to power a propellor or a fan. The cycle is shown in the *FE Handbook* and consists of two isentropic processes and two constant pressure processes.

Example **2.9**

Air at 3 bar and 500 K is extracted from a jet engine compressor and is to be used in cabin cooling. The extracted air is cooled at a constant pressure heat exchanger down to 400 K. It then expands isentropically through a turbine down to 1 bar. The power developed by the turbine is used for cabin lighting. The mass flow rate is 5 kg/min.

Determine

a) Temperature of air to cabin

b) Power developed by the turbine in kW

c) Rate of heat transfer in the heat exchanger in kW

Solution

Exhibit 5 illustrates the processes discussed above.

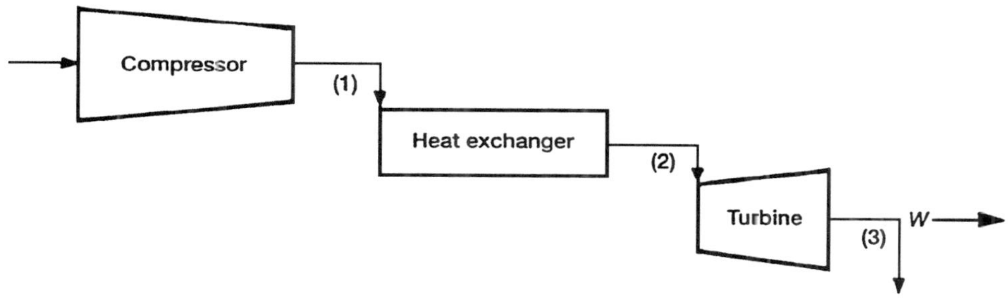

Exhibit 5

a) T_3 has to be found using the isentropic equation:

$$T_3 = T_2[P_2/P_3]^{(1-k)/k} = 400[3/1]^{(1-1.4)/1.4} = 292 \text{ K}$$

b)

$$\begin{aligned} W &= m[c_p(T_2 - T_3)]; \text{ therefore:} \\ &= 5\text{kg/min}[1.00(400 - 292)] \\ &= [540 \text{ kJ/min}]/60 \\ &= 9 \text{ kW} \end{aligned}$$

c)

$$\begin{aligned} \text{The heat transfer } Q &= m[h_1 - h_2] \\ &= 5 \text{ kg/min}[c_p \text{ kJ/kg}\cdot\text{K}(T_1 - T_2) \text{ K}] \\ &= 5[1.00(500 - 400)]/60 \\ &= 8.33 \text{ kW} \end{aligned}$$

PROBLEMS

The following information may be used for questions 2.1 and 2.2.

A large municipal power plant produces 1,000 MW of power. The overall plant efficiency is 41%. Coal with a heating value of 24 MJ/kg is used for heating the water in the boilers.

2.1 How much coal in kg is burned each day?
a. 5,000 c. 5,000,000
b. 8,778,000 d. 85,000

2.2 If cooling water is allowed to raise 12 °C as it goes through the condenser, how much cooling water is necessary in kg/min?
a. 450,000 c. 1,100,000
b. 850,000 d. 1,380,000

The following information may be used for questions 2.3 to 2.7.

Water is used as the working fluid in a power plant that produces a net output of power of 100 MW. The constant pressure condenser operates at 8 kPa and the constant pressure boiler operates at 1 MPa. The steam enters the turbine, which operates at 85% efficiency at 600 °C, and the water leaving the condenser at 8 kPa enters the 75% efficient pump.

2.3 The mass flow rate kg/hr of the working fluid is most nearly:
a. 310,000 c. 185,000
b. 256,000 d. 375,000

2.4 The heat supplied to the boiler in kJ/hr is most nearly:
a. 2.1×10^9 c. 8.5×10^6
b. 1.5×10^8 d. 1.1×10^9

2.5 The heat rejected to the condenser in kJ/hr is most nearly:
a. 7.4×10^8 c. 8.4×10^6
b. 6.0×10^7 d. 8.6×10^9

2.6 The work of the pump in kW is most nearly:
a. 250 c. 115
b. 400 d. 525

2.7 The thermal efficiency is most nearly:
a. 51% c. 38%
b. 45% d. 33%

The following information is to be used for questions 2.8 and 2.9.

Air enters the diffuser of a ramjet engine at 25 kPa and 250 K with a velocity of 4000 km/hr and is decelerated to a low velocity at the exit of the diffuser of the engine. The heat addition in the combustion chamber is 1000 kJ/kg. The air exits the nozzle at 25 kPa.

2.8 The pressure at the diffuser exit, in kPa, is most nearly:

a. 850
b. 1200
c. 1450
d. 1717

2.9 The velocity at the nozzle exit is most nearly:

a. 1000 m/s
b. 1200 m/s
c. 1400 m/s
d. 1600 m/s

SOLUTIONS

2.1 **b.** Thermal efficiency equation is *energy wanted/energy that costs$*. It may then be written as:

$$[1{,}000 \text{ MW}]/\eta_{th} = \text{energy in the coal} = 1{,}000/0.41 = 2439 \text{ MJ/s}$$
$$\text{kg coal needed} = \text{energy needed/energy/kg coal} = [2439 \text{ MJ/s}/24 \text{ MJ/kg}]$$
$$\text{kg coal needed} = 101.6 \text{ kg/s}[3600 \text{ s/hr} \times 24 \text{ hr/day}] = 8.778 \times 10^6 \text{ kg/day}$$

2.2 **d.** To find the amount of water needed for cooling we go back to the energy in and subtract the work obtained. The difference between the two is equal to the heat rejected. This amount of heat is to be divided by the enthalpy difference due to the increase in temperature of the cooling water.

$$Q_{in} - W_{out} = Q_{out}$$
$$2439 \text{ MJ/s} - 1000\text{MW} = 1439 \text{ MJ/s}$$
$$\text{Water needed} = Q_{out}/\Delta h$$
$$= [1439 \text{ MJ/s}]/[(4.18 \text{ kJ/kg-K})(15 \text{ K})][1000\text{M/k}]$$
$$= [22.95][1000] \text{ kg/s}$$
$$= 22{,}950 \text{ kg/s} \times 60 \text{ s/min}$$
$$= 1{,}377{,}000 \text{ kg/min}$$

Problems 2.3–2.7

The most efficient method of solution is to set up a table in which to put the calculated values. Once the table is completed then the problem is very straightforward to solve. The values for position (1 which is entering the turbine, can be obtained directly from the steam tables listed in the *FE Handbook*. Position 2s is the isentropic drop to 8 kPa pressure.

Knowing:

$$s_1 = s_{2s} = 8.0290 = xs_g + (1 - x)s_f = x(8.2287) + (1 - x)(0.5926)$$
$$x_2 = 0.97$$

	1	2s	2	3	4s	4
P	1 MPa	8 kPa	—	—	1 Mpa	—
T	600 °C					
h	3697.9	2504.9	2683.9	173.88	174.88	175.21
s	8.0290	—				
x		0.97				

Solving for h_{2s} using the quality equation:

$$h_{2s} = xh_g + [1 - x]h_f = 0.97[2577] + 0.03[173.88] = 2504.9 \text{ kJ/kg}$$

Use the turbine efficiency equation to determine h_2:

$$\eta_T = [h_1 - h_2]/[h_1 - h_{2s}]$$
$$h_2 = h_1 - \eta_T[h_1 - h_{2s}] = 3697.9 - 0.85[3697.9 - 2504.9] = 2683.9 \text{ kJ/kg}$$
$$h_3 = \text{the saturated water value} = 173.88 \text{ kJ/kg}$$

Use the pump equation and add the result to the enthalpy at Point 3 to give h_{4s}.

$$h_{4s} = h_3 + \int v\mathrm{d}P = 173.88 \text{ kJ/kg} + [1.0084/1000]\text{m}^3/\text{kg}$$
$$\times[1 \times 10^6 - 8{,}000]\text{Pa}/10^3 \text{ J/kJ}$$
$$h_{4s} = 173.88 + 1.0 = 174.88$$

To find h_4 we must use the efficiency equation for the pump:

$$\eta_p = [h_{4s} - h_3]/[h_4 - h_3]$$
$$h_4 = h_3 + [h_{4s} - h_3]/\eta_p = 173.88 + [174.88 - 173.88]/[0.75] = 175.21$$

Now that we have filled out the table, it is very straightforward to solve each of the questions.

2.3 **a.**

$$\text{Mass flow rate} = \text{Work output of system/Work output/kg}$$
$$\text{Mass} = 100{,}000 \text{ kW}/[(h_1 - h_2) - (h_4 - h_3)]\text{kJ/kg}$$
$$= 100{,}000/[(3697.9 - 2540.7) - (175.21 - 173.88)]$$
$$= 86.5 \text{ kg/s}$$
$$= 311{,}400 \text{ kg/hr}$$

2.4 **d.**

$$Q_{in} = m[h_1 - h_4] = 311{,}400 \text{ kg/hr } [3697.9 - 175.21] \text{ kJ/kg}$$
$$= 1.096 \times 10^9 \text{ kJ/hr}$$

2.5 **a.**

$$Q_{out} = m[h_2 - h_3] = 311{,}400[2540.7 - 173.88] = 7.37 \times 10^8 \text{ kJ/hr}$$

2.6 **c.**

$$W_p = m[h_4 - h_3]$$
$$= [311{,}400\text{kg/hr}]/3600 \text{ s/hr } [175.21 - 173.88] \text{ kJ/kg}$$
$$= 115 \text{ kW}$$

2.7 **d.**

$$\eta_{th} = W_{net}/Q_{in} = [(3697.9 - 2540.7) - 1.33]/[3697.9 - 175.21] = 32.8\%$$

Problems 2.8 and 2.9

2.8 **d.** Again, the best way to solve this problem is to set up a table of values, find the values, and then make the final calculations. Point 1 is entering the diffuser, Point 2 is exiting the diffuser, Point 3 is after combustion and just entering the nozzle and Point 4 is leaving the nozzle. It shall be assumed that the ramjet acts on an ideal Brayton cycle.

	1	2	3	4
P	25 kPa	1716.7	—	25
T	250 K	837	1837	548.7

Write a first law equation for a diffuser to get the condition at state 2:

$$h_1 + v_1^2/2 = h_2 + v_2^2/2$$
$$c_p[T_2 - T_1] = v_1^2/2$$

Noting that

$$[4000 \text{ km/hr}/3600 \text{ s/hr}][1000 \text{ m/km}] = 1111 \text{ m/s}$$

or

$$1.00 \text{ kJ/kg}\cdot\text{K}[T_2 - 220]\text{K} = [1111 \text{ m/s}]^2/2[\text{N/kg}\cdot\text{m/s}^2][\text{kJ}/10^3\text{N}\cdot\text{m}]$$
$$T_2 = 837 \text{ K}$$

Then, using the perfect gas relationship:

$$P_2 = P_1\ [T_1/T_2]^{k/(1-k)}$$
$$= 25 \text{ kPa}[250/837]^{1.4/(1-1.4)}$$
$$= 1716.7 \text{ kPa}$$

2.9 **d.** The heat addition into the engine occurs between states 2 and 3.

$$Q = c_p[T_3 - T_2];\ \text{therefore:}$$
$$1000\text{kJ/kg} = 1.00 \text{ kJ/kg}\cdot\text{K}[T_3 - 837]$$
$$T_3 = 1837 \text{ K}$$

To find T_4 one must sagain use the isentropic relationship, relating pressure and temperature.

$$T_4 = T_3[P_3/P_4]^{(1-k)/k} = 1837[1716.7/25]^{(1-1.4)/1.4} = 548.7 \text{ K}$$

Using the general energy equation for a nozzle we may find the exit velocity:

$$h_3 = v_3^2/2 = h_4 + v_4^2/2$$

The velocity entering the nozzle is taken as zero, thus:

$$v_4 = [2(h_3 - h_4)]^{1/2} = [2c_p(T_3 - T_4)]^{1/2} = [2(1)(1000)\ (1837 - 549)]^{1/2}$$
$$= 1604 \text{ m/s.}$$

CHAPTER 3

Refrigeration and Heating, Ventilating and Air Conditioning

OUTLINE

Refrigeration is a process whereby heat is extracted from a system. In the past it was done by means of ice stored in an ice box. The heat of fusion of the ice as it melts, came from the substance that was being cooled. This method has essentially been replaced by the mechanical refrigeration process, which operates on the Clausius statement of the *Second Law of Thermodynamics*. Mechanical refrigeration extracts heat from a cold temperature reservoir and, through the means of a work input, deposits this heat into the high temperature reservoir. It is interesting to note that the old concept of melting ice is still with us in terms of the rating of refrigeration units. A ton of refrigeration is the amount of heat that is absorbed by a ton of ice as it melts in a 24-hour period. This term was defined in the English system of units by:

$$\text{Mass of ice} \times \text{Heat of fusion} = \text{Heat required in a 24-hr day}$$
$$(200\ \text{lb ice})(144\ \text{Btu/lb}) = 288{,}000\ \text{Btu/day which is equal to } 12{,}000\ \text{Btu/hr}$$

The most common term is in Btu/min, which is 200 Btu/min. In the SI system this is 211 kJ/min or, dividing by 60 s/min, 3.517 kJ/sec.

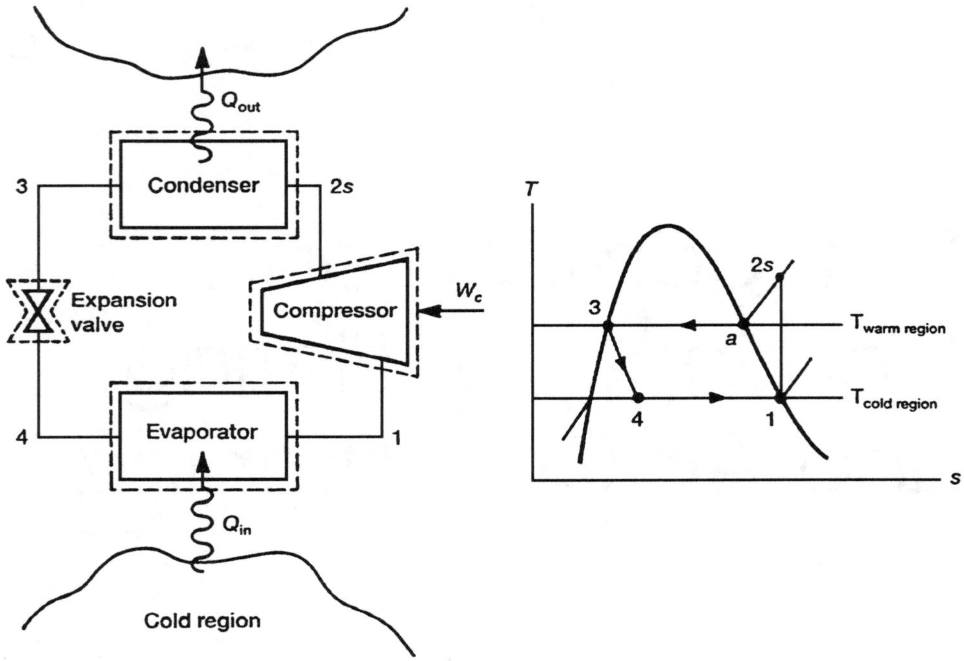

Figure 3.1

MECHANICAL REFRIGERATION

Mechanical refrigeration is based on the principle that, as a liquid vaporizes at a specific pressure, it will have a controlled absorption of the heat of vaporization. The principle of operation is patterned after the Carnot cycle with some changes due to physical constraints. The ideal refrigeration cycle is depicted in figure 3.1 and also in the Thermodynamics section of the *Handbook*. The mechanical system consists of a compressor that compresses the cold vapor coming out of the evaporator to a high temperature and high pressure vapor. This vapor then enters the condenser where the vapor is cooled to a saturated or possibly slightly subcooled liquid. The liquid then expands through a throttling valve to a lower pressure and, consequently, a lower temperature liquid-vapor mixture and enters the evaporator where it continues to vaporize to a saturated or slightly superheated vapor for entrance into the compressor.

ABSORPTION REFRIEGRATION

Another type of refrigeration system that is commonly used in facilities that have an abundant amount of heat energy available is the absorption system. This type of system uses a refrigerant such as lithium bromide or ammonia which, when evaporated, will absorb energy and, when cooled or condensed, will give off much energy. The advantage of this type of system is that the machinery operating costs are very low. All that is necessary are a few low-power pumps and a cheap source of energy and cooling water. This system is currently being researched for use in residential and commercial applications. The Coefficients of Performance (C.O.P.) of the absorption systems compare very favorably with the compression refrigeration systems.

EFFICIENCY OF REFRIGERATION SYSTEMS

Thermal efficiency has been defined as:

$$\eta_t = \text{(Energy wanted)/(Energy that costs \$)}$$

The same general philosophy can be used whether the system is an engine, pump, refrigeration system or almost any other system one can imagine. Considering an Air Conditioning system, the equation would be:

Efficiency = (Energy extracted from cold reservoir)/(Energy used in compression)

The difficulty with using the efficiency term is that efficiency can never be greater than one, or we would have a perpetual motion machine. Therefore, the term *Coefficient of Performance* has been coined to indicate relative efficiency terms. As can be determined from the equation, a higher C.O.P. is desired. These values typically range from about 2.0 to 5.0 or 6.0.

Example 3.1

A mechanical refrigeration system operates on a Carnot cycle. Saturated vapor leaves the compressor and enters the constant temperature condenser at 100 °C and exits from the constant temperature evaporator at –10 °C. If the unit provides 3 tons of air conditioning, determine

a) the C.O.P.

b) the amount of work needed to transfer this heat in kJ/min.

Solution

A Carnot cycle is the ideal cycle for heat engines and refrigeration systems. The schematic of the reverse cycle is in the Thermodynamics section of the *FE Handbook*. Using the refrigeration model of the efficiency equation:

a) C.O.P = $(T_L)/(T_H - T_L)$ where temperatures are in Kelvin.

$$\text{C.O.P.} = (-10 + 273)/(100 - [-10]) = 263/110 = 2.39$$

b) Now, using the efficiency equation in terms of heat transfer:

$$\text{C.O.P.} = (Q_L/(Q_H - Q_L)$$

where

Q_L = Heat transfer at low temperature
Q_H = Heat transfer at high temperature
$(Q_H - Q_L)$ = Work

Substituting in the known values:

$$2.39 = (211\text{kJ/min-ton})(3 \text{ tons})/W$$

Solving for W:

$$W = 264.8 \text{ kJ/min}$$

Example 3.2

An ideal vapor-compression refrigeration cycle operates with HFC 134a as the refrigerant. Saturated vapor enters the compressor at –10°C. Saturated liquid leaves the condenser at 2 MPa. The mass flow rate through the unit is 10 kg/min. Determine the work of the compressor in kW, the refrigeration capacity in tons and the Coefficient of Performance.

Solution

An ideal vapor compression refrigeration system is patterned after the *TS* diagram that is shown in the Thermodynamics section of the *FE Handbook*. Note that the problem above is for just a one-stage cycle where, as in the Mechanical Engineering section, a diagram is included for a two-stage cycle. The schematic diagram is also listed on the same page. One should then turn to the *P-h* diagram for HFC 134a to obtain the enthalpy values. Note that Points 1 and 3 are saturated vapor and saturated liquid, respectively. Point 2 represents the gas after it has been isentropically compressed. The table of values shown below is taken from the HFC 134a diagram.

	1	2	3	4
P		2 Mpa	2 Mpa	
T	–10			–10
h	390	445	300	300
s	1.74	1.74		

a)
$$\begin{aligned} W_c &= m(h_2 - h_1) \\ &= (10 \text{ kg/min})(445 - 390) \text{ kJ/kg}(1/60)(\text{min/sec}) \\ &= 4.3 \text{ kW} \end{aligned}$$

b)
$$\begin{aligned} Q_{\text{tons}} &= m(h_1 = h_2) \\ &= (10 \text{ kg/min})(390 - 300)\text{kJ/kg}(1/211)(\text{tons/kJ}) \\ &= 4.3 \text{ tons} \end{aligned}$$

c)
$$\begin{aligned} \text{C.O.P.} &= (Q_{in}/W) \\ &= (h_1 - h_4)/(h_2 - h_1) \\ &= (390 - 300)/(445 - 390) \\ &= 90/55 \\ &= 164 \end{aligned}$$

HVAC

HVAC stands for the words Heating, Ventilating and Air Conditioning, which is the science of creating the ideal environment for man and machine in a selected space. The creation of this ideal environment indicates that not only is temperature control required, but also humidity control. These controls are for comfort, for the dimensional stability of equipment and instrumentation, and for the stability of other characteristics.

BODY HEAT

The human body releases both sensible heat, since the body is usually higher in temperature than the surroundings, and latent heat due to perspiration. Perspiration tends to add water vapor to the air in the form of superheated steam. This water vapor, and the water vapor already existing in the air, can only be removed by means of condensation. The amount of heat that must be removed to maintain the room at a constant humidity is called the "latent heat gain" of the air-conditioned space. The rates of heat given off by the human body have been measured for various situations and conditions in order to determine the necessary size of air conditioners.

Example **3.3**

A theater holds 600 people for an evening performance. Measurements have indicated that an average person gives off 205.7 kJ of sensible heat per hour and 163.5 kJ of latent heat per hour, when sitting in a theater for an evening. What would the air-conditioning load be to maintain a comfortable temperature during this period of time? Give the answer in tons of air-conditioning.

Solution

Six hundred people will give off heat at the rate of:

$$\begin{aligned}\text{Total heat} &= (\text{Number of people}) \times (\text{Sensible heat} + \text{Latent heat})\\ &= (600)(205\ \text{kJ/hr} + 163.5\text{kJ/hr})\\ &= \frac{(221.100)\text{kJ/hr}}{(211\ \text{kJ/min ton})(60\ \text{min/hr})}\\ &= 17.46\ \text{tons, which is the amount of refrigeration required to counteract the heat supplied by the people.}\end{aligned}$$

GAS REFRIGERATION

The use of mechanical, vapor-compression air conditioning systems in aircraft is not a feasible alternative due to the added weight. Gas systems in aircraft utilize high temperature and high pressure air from the compressor. This air then passes through a heat exchanger, which exhausts the heat to the ambient air. The cooled, yet high-pressure, air then passes into an air turbine that acts as an auxiliary power source for the cabin of the aircraft. Upon exhausting from the turbine, the air is cooled considerably below the ambient temperature and, thus, may be exhausted into the cabin as cooling air. The Mechanical Engineering section of the *FE Handbook* gives a schematic diagram for the use of a Brayton cycle in the production of refrigeration.

Example **3.4**

Air at 3 bar and 500 K is extracted from a jet engine compressor to be used for cabin cooling and for the generation of auxiliary power for the cabin. The extracted air is cooled in a constant pressure heat exchanger down to 400 K. It then enters an isentropic turbine and expands to 1 bar before being rejected into the cabin. If the mass flow is 5 kg/min, determine:

a) The temperature of the gas as it leaves the turbine

b) The power developed by the turbine in kW

c) The rate of heat transfer out of the constant-pressure heat exchanger

Solution

Assuming that the air acts as an ideal gas, and using the ideal-gas relationships for each of the processes, a table of values can be constructed. Note that the schematic diagram for this process can be illustrated by reducing the air refrigeration cycle shown in the Mechanical Engineering section of the *FE Handbook* to include only the air from the compressor running through the heat exchanger and exhausting through the turbine into the air-conditioned space.

	1	**2**	**3**
P	3	3	1
T	500	400	257.8

a) Process 2–3 is isentropic, and T_3 must be obtained using the relationship:

$$T_3 = T_2(P_3/P_2)^{(k-1)/k} = 400(1/3)^{(1.4-1)/1} = 257.8 \text{ K}$$

b) Power developed by the turbine $= m(h_2 - h_3)$
$= mc_p(T_2 - T_3)$
$= (5 \text{ kg/min})(1.00 \text{ kJ/(Kg}\cdot\text{K)}(400 - 257.8)\text{K } (1 \text{ min/60 s})$
$= 11.85 \text{ kW}$

c) Heat transfer from the heat exchanger $= m(h_1 - h_3)$
$= (5 \text{ kg/min})(1.00 \text{ kJ/kg}\cdot\text{K})(500 - 400)\text{K } (1/60) \text{ min/s}$
$= 8.33 \text{ kW}$

INFILTRATION

Living and working spaces that need to be conditioned, whether by heating in the winter or by cooling in the summer, need to be analyzed for infiltration. This infiltration can be intentional, such as in maintaining a positive pressure in *clean rooms*, etc., or unintentional, such as from cracks under doors, around windows, etc. The *FE Handbook* lists formulas in the Mechanical Engineering section that can be used in each of the above cases.

Example 3.5

A computer chip manufacturing plant of 5000 square meters has 4-meter-high ceilings and changes its inside air three times per hour by maintaining a slightly positive air pressure (1.01 bar). The room temperature is maintained at 21°C. To assist in the sizing of a heating system it is assumed in the design conditions that the outside temperature is –25 °C. Basing the heating requirements on the indoor conditions, determine the heating required to compensate for the infiltration.

Solution

Select the equation for the air change method listed under *Infiltration* in the *FE Handbook*. Determine the density of air using the perfect gas law:

$$PV = RT$$

$$\rho = \left[\frac{(1.01\,\text{B})(10^5\ \text{N/m}^2\text{B})}{\frac{(8.314\ \text{kJ/kmol}\cdot\text{K})(292\ \text{K})}{(28.97\ \text{kg/kmol})}}\right] \times \left[1\frac{\text{kJ}}{10^3\ \text{N}\cdot\text{m}}\right]$$

$$= 1.205\ \text{kg/m}^3$$

Heat required, $Q = (1.205\ \text{kg/m}^3)(c_p\ \text{kJ/kg K})(\text{V m}^3)(n_{AC}\ \text{changes/hr})$
$(T_i - T_o)\text{K}/(3600\ \text{sec/hr})$
$= (1.205)(1.00)(5000 \times 4)(3)[21 - (-25)]/3600$
$= 307.9\ \text{kW}$

PSYCHROMETRICS

Psychrometrics is the study of systems consisting of dry air and water. This combination of air and water is called moist air. The particular composition of the water and air mixture may be indicated by means of the *humidity ratio*, which is the ratio of the mass of water vapor to the mass of dry air. The make-up of moist air can also be described in terms of the relative humidity, which is the ratio of the partial pressure of the water vapor in the moist air to the saturation pressure of water vapor at the temperature of the air. These values may be readily obtained from the psychrometric chart in the *FE Handbook* merely by knowing the wet-bulb and dry-bulb temperatures of the moist air. Note the schematic diagrams in the Mechanical Engineering section of the *FE Handbook* that illustrate various HVAC systems and the representation of these processes on the schematics of the psychrometric chart.

Example 3.6

Air at 30 °C, 1 atm, and 70% relative humidity enters a dehumidifier operating at steady state. Saturated air and water leave both leave at 10 °C in two different streams.

Determine

a) The heat transfer from the incoming air stream to the coils of the dehumidifier in kJ/kg dry air

b) The amount of water condensed in kg per kg of dry air.

Solution

Turn to the Thermodynamic Section of the *FE Handbook* where the psychrometric chart is shown. The 30 °C is the dry bulb temperature of the incoming air stream. Locate the incoming condition by sketching a vertical line from the 30 °C position up to the 70% *Relative Humidity* line. Record the humidity ratio from the right-hand side of the chart (19 grams moisture/kg dry air) and the enthalpy from the left side of the chart (78.2 kJ/kg dry air). Now run the line directly to the left-hand side, which lists the wet-bulb temperature, or the saturation temperature, of the moist air. Run down the curve until you reach 10 °C. Again record the humidity ratio (7.7 grams H_20/kg dry air, and pick off the value of the enthalpy of 29.5 kJ/kg dry air). Now you have all of the values necessary for the solution of the problem.

a) Heat transfer = $(h_1 - h_2) = (78.2 - 29.5) = 48.7$ kJ/kg dry air

b) Water out = (19 – 7.7) gm water/kg dry air = 11.3 gm/kg = 0.0113kg/kg.

Example 3.7

Air enters a steam humidifier at a temperature of 21°C and a wet-bulb temperature of 10°C. The mass flow rate of the dry air is 100 kg/min and saturated steam is injected at 100°C at 1 kg/min. Consider that there is no heat transfer to or from the surroundings. Consider that the pressure is constant at 1 atm throughout.

Determine

a) The exit humidity ratio, in (kg moisture)/(kg air)

b) The exit temperature, in °C

	State 1	**State 2**
$T_{wb} = 10°C$	$T_1 = 21°C$	
	Moist air	Moist air
	→	→

↗

Injected steam

$T = 100°C$, 1 kg/min

Solution

An energy balance equation should be written for the schematic diagram of the process shown above.

$$m_a h_{a1} + m_{w1} h_{g1} + m_{st} h_{st} = m_a h_{a2} + m_{w2} h_{g2}$$

Knowing that $m_{w1}/m_a = \omega_1$ and $m_{w2}/m_a = \omega_2$ the equation may be rewritten as:

$$(h_a + \omega_1 h_g)_1 + (\omega_2 - \omega_1) h_{g100} = (h_a + \omega_2 h_g)_2$$

The value for the enthalpy of the incoming water vapor—air mixture can be obtained from the psychrometric diagram in the *FE Handbook.*

$$h_1 = 30 \text{ kJ/kg dry air}$$
$$\omega_1 = 3.5 \text{ g water/kg dry air}$$

The steam injected = 1 kg/min steam

Therefore, $(\omega_2 - \omega_1)$ = (1000 g/min steam)/(100 kg/min air) = 10 g moisture/kg air

Therefore, $\omega_2 = 3.5 + 10 = 13.5$ g water/kg dry air, $h_{g100} = 2676.3$ kJ/kg

a) $\omega_2 = 13.5/1000 = 0.0135$ kg moisture/kg air

b) Substituting values into (i):

(30 kJ/kg) + 0.0135 kg moisture/kg dry air (2676.3)kJ/kg = h_2

$h_2 = 66.13$ kJ/kg; therefore, temperature = 30°C

Example 3.8

A research laboratory building is to be heated and ventilated. Outside air is to be supplied to a room 12 m by 18 m at the rate of 1 m^3/min of replacement air per square meter of floor space. The design conditions are 20°C inside air temperature with a –20°C outside air temperature. The air is to be pressurized in the room to a positive air pressure of 7.0 kPa to ensure that the room does not get contaminated from the outside. How much heat would be required to condition the air for this room?

Solution

The floor area is 12 m × 18 m = 216 m^2, so the air-flow rate required would be 216 m^3/min at a positive pressure of (101,300 + 7000) Pa absolute pressure. The mass of air equals:

$$
\begin{aligned}
m &= PV/RT \\
&= (108{,}300 \text{ Pa})(216 \text{ m}^3/\text{min}) \\
&\quad (1 \text{ N/m}^2 \cdot \text{Pa})/(8314/28.97)\text{J/kg}\cdot\text{K } (293 \text{ K}) \text{ } 1 \text{ N}\cdot\text{m/J} \\
&= 278 \text{ kg/min}
\end{aligned}
$$

Therefore, calculating the heat transfer:

$$
\begin{aligned}
Q &= mc_p(\Delta T) \\
&= 278 \text{ kg/min } (1.00 \text{ kJ/kg}\cdot\text{K})(T_i - T_o) \text{ K} \\
&= (278)(1.00)(40) \\
&= 11{,}120 \text{ kJ/min} \\
&= (11{,}120 \text{ kJ/min})(1 \text{ min/60 s}) \\
&= 185.3 \text{ kJ/s or } 185.3 \text{ kW heating}
\end{aligned}
$$

Example 3.9

An industrial air conditioning system mixes 70% conditioned air with 30% recirculated air before returning it to the plant. The system handles some 285 m^3/min of mixed air. The conditioned air leaves the cooling coils at 9 °C and 90% relative humidity and then is mixed with recirculated air at 25 °C and 50% relative humidity.

Determine

a) Dry bulb temperature of the air returning to the plant

b) Wet bulb temperatures of the air returning to the plant

Solution

Since the air is being mixed on a 70%:30% basis, the final temperature may be considered as a weighted average of these temperatures:

a) $T_{\text{final}} = (0.70)(9 + 273) + (0.30)(25 + 273) = 286.8 \text{ K or } 13.8\,°\text{C}$

b) The wet bulb temperature may be found from the psychrometric chart by first locating the points indicated by the two streams of air. Then draw a line from point 9 °C dry bulb and 90% relative humidity to point 25 °C dry bulb and 50% relative humidity. Where this line intersects, the dry-bulb temperature of 13.8 °C, marks the condition of the exit air. One should merely follow the constant wet bulb line to the saturation line and pick off the wet-bulb temperature of 12.1 °C.

Example 3.10

A line carrying dry, saturated steam at 790 kPa gage pressure passes through an instrument room measuring 6m × 6m × 4m. The line develops a leak and the steam seeps into the room at a rate of 1 kg/hr and diffuses throughout the room. If the room is initially at 20 °C and has a relative humidity of 50%, and the pressure of the room remains at one standard atmosphere, estimate what the relative humidity will be after one hour. Assume there is no exchange of heat between the room and the surroundings and that the steam leak displaces some of the air in the room.

Solution

Assume that the steam and the air both act as perfect gasses. Calculation of the volume of the room is 6m × 6m × 4m = 144 m^3.

Determine the mass of the dry air in the room and the mass of the superheated steam at initial condition. According to Dalton's law of partial pressures we know that the total pressure in the room is equal to the partial pressure of the steam plus the partial pressure of the air. The partial pressure of the steam may be estimated as 50% of the saturation pressure of steam at 20 °C, which is obtained from the steam tables in the *FE Handbook* as 2.339 kPa.

Therefore

$$P_{\text{steam}} = (0.50)(2.339)\text{kPa} = 1{,}170 \text{ Pa}$$

Thus, the partial pressure of the air is:

$$P_{\text{air}} = 101{,}300 \text{ Pa} - 1170 \text{ Pa} = 100{,}130 \text{ Pa.}$$

The original masses of air and water vapor are calculated as:

$$\begin{aligned} m_{\text{steam}} &= \text{PV/RT} \\ &= (1{,}170 \text{ Pa})(\text{N/m}^2\cdot\text{Pa})(144 \text{ m}^3)/ \\ &\quad [8.314 \text{ kJ/mol}\cdot\text{K})/(18 \text{ kg/kmol})][293]\text{K}(10^3\text{Nm/kJ}) \\ &= 1.2449 \text{ kg} \\ m_{\text{air}} &= (101{,}300 - 1{,}170)(144)/[8.314/28.97][293][10^3] = 171.5 \text{ kg} \end{aligned}$$

Assuming that the steam displaces the air in the room, the amount of heat that is dissipated into the room by the saturated steam at 891 kPa (175 °C from the steam tables) is approximated by:

$$Q = mc_p\Delta T = (1 \text{ kg})(1.87 \text{ kJ/kg}\cdot\text{K})(175 - 20)\text{K} = 290 \text{ kJ}$$

The addition of the 290 kJ into the room will raise the temperature in the room by:

$$\begin{aligned} \Delta T &= Q/mc_p \\ &= (290 \text{ kJ})/(171.5 \text{ kg})(1.00 \text{ kJ/kg}) \\ &= 1.69°\text{C} \end{aligned}$$

This change in temperature is not significant when compared with the original temperature. It may be determined that the temperature rise does not materially affect the humidity of the room for a first approximation.

It may now be estimated that the relative humidity after 1 kg of steam has been injected into the room is:

$$\text{Relative Humidity} = [(1.2449 + 1.0)/1.2449][50\%] = 90\%$$

PROBLEMS

3.1 To maintain a comfortable temperature, what would be the cooling requirement to offset the emission of body heat in a dance hall that had an attendance of 1500 people if it has been determined that 75% of the people will be dancing and the others will be seated. Tests have shown that an adult will give off 71.8 W of sensible heat while engaging in moderate dancing, and 57.1 W of sensible heat while sitting. The latent heat emission for similar situations has been found to be equal to 177.3 W for dancers and 45.4 for sitters.

a. 200kW
b. 267 kW
c. 301 kW
d. 319 kW

3.2 An evaporative cooling device is used in Mesa Arizona for the *conditioning* of the air. Outside air is brought in at 40 °C and a relative humidity of 10%. Water is then sprayed in at a temperature of 15 °C before it is introduced back into the room. The exit air is 20 °C.

Determine the exit relative humidity and the amount of water sprayed into the air (kg water/kg dry air).

a. RH = 75%, water added 0.1kg/kg dry air
b. RH = 85%, water added 0.0081 kg/kg dry air
c. RH = 65%, water added 0.0099 kg/kg dry air
d. RH = 55%, water added 0.011 kg/kg dry air

3.3 A residential heat pump provides some 3.2×10^6 kJ/day in order to maintain the house at 21 °C. The design condition is for the outside temperature to be –20 °C. If the electricity cost \$0.09/kW·h, determine (1) the minimum theoretical operating cost per day and (2) the ideal C.O.P. of the system.

a. Operating cost = \$5.48, C.O.P. = 12.2
b. Operating cost = \$10.50, C.O.P. = 8.5
c. Operating cost = \$7.56, C.O.P. = 6.8
d. Operating cost = \$9.23, C.O.P. = 9.48

SOLUTIONS

3.1 **d.** Total body heat = 0.75(1500)(71.8 + 177.3) + 0.25(1500)(57.1 + 45.4)
= 318,675 W = 319 kW

3.2 **b.** Write an energy balance of the system:

$$(m_a h_a + h_g)_{in} + m_f h_f = (m_a h_a + h_g)_{out}$$

Rearrange the terms and divide by the mass of the air:

$$c_p(T_1 - T_2) + \omega_1 h_{g1} + (\omega_2 - \omega_1)h_{f15} = \omega_{g2}$$

Obtain values from the steam tables and the psychrometric chart in the *FE Handbook*:

$$(1.00)(40 - 20) + (0.0045)(2574.3) + (\omega_2 - 0.0045)(62.99) = \omega_2(2538.1)$$

$$\omega_2 = 0.0126$$

Therefore, water added = (0.0126 – 0.0045) = 0.0081 kg water/kg dry air

Relative Humidity may be selected from the psychrometric chart as 85%

3.3 **d.** The ideal Coefficient of Performance is that of the Carnot cycle:

b) C.O.P. = $T_H/(T_H - T_L)$ = (21 + 273)/[(21) – (–10)] = 9.48

a) C.O.P. = $Q_H/(Q_H - Q_L)$ = 9.48 = (3.5 × 10^6)/work

Work = 3.69 × 10^5 kJ/day

Cost = (3.69 × 10^5)(0.09)(1/3600) = \$9.23/day

CHAPTER 4

Heat Transfer

OUTLINE

Heat transfer is that branch of thermodynamics that is concerned with the transfer of energy from one point to another by virtue of temperature difference. There are three modes of heat transfer: conduction, convection and radiation. Conduction and radiation are true forms of heat transfer since these modes depend only on temperature differential and the characteristics of the materials involved in the temperature exchange. Convection, however, not only depends upon the materials and temperature differentials, but is also dependent upon mass transport.

CONDUCTION

Conduction is the molecule-to-molecule transfer of energy. The one-directional equation of energy flow for a steady-state condition is given by Fourier's Law of Conduction:

$$Q = [kA/L][T_1 - T_2]$$

where

Q = rate of heat flow in watts or J/s
k = thermal conductivity W/(m-K)
A = area perpendicular to heat flow
L = thickness in direction of heat flow
$\Delta T = (T_1 - T_2)$ difference in temperature within the material in the direction of heat flow, may be given in terms of C or K

Example 4.1

What is the rate of heat flow through a wall constructed of brick and mortar, which is 25 cm thick and 3.0 m × 2.0 m in area. The temperature on one side is 165 °C and 55 °C on the other side. The average coefficient of thermal conductivity is equal to 0.750 W/m · K.

Solution

$Q = [kA/L][T_1 - T_2] = [(0.750 \text{ W/m} \cdot \text{K})(6 \text{ m}^2)/(25/100)\text{m}][165 - 55]$ K or °C
$Q = 1980$ watts $= 1.98$ kW

SERIES COMPOSITE WALL

A series composite wall is a wall that is made with layers of various materials sandwiched together. The concept is that the heat will transfer through the wall in a straight line perpendicular to the wall. The same amount of energy that goes into one side of the wall comes out the other side. Each of the various layers of material adds to the thermal resistance of the wall and thus reduces the amount of energy transferred through the wall.

Example 4.2

A series composite wall of a furnace consists of 22 cm of fire brick, 11 cm of a high temperature insulating material, 10 cm of ordinary brick, and, lastly, 6.5 mm of asbestos cement board. The inside of the furnace wall is 980 °C and the temperature of the outside wall is 90 °C. Determine the heat transfer through the wall per m^2.

Solution

Since a series composite wall is a wall that is layered with the heat transfer passing through each of the layers in series, the general equation for the solution of a series composite wall is:

$$Q = [\Delta T_{\text{overall}}]/\Sigma R_{\text{th}}$$

Since the furnace size was not given, it will be taken on a m^2 basis. It is also assumed that, since this is a wall, the area of conduction of every portion of the wall is the same. We may therefore write the equation:

$$Q = [980 - 90]/[L_1/k_1 + L_2/k_2 + L_3/k_3 + L_4/k_4]$$

where

Fire Brick	$k_1 = 1.4$ W/m-K
High Temperature Insulation	$k_2 = 0.22$ W/m-K
Ordinary Brick	$k_3 = 0.90$ W/m-K
Asbestos Cement board	$k_4 = 0.39$ W/m-K

$Q = [890 \text{ C}]/[(0.22/1.4) + (0.11/.22) + (0.10/0.90) + (0.065/0.39)]\text{W/m}^2 \cdot \text{K}$
$Q = 832 \text{ watts/m}^2$

CIRCULAR PIPES

An insulated circular pipe behaves much as a series composite wall. In this case, however, the heat is transferred in a straight line radially out from the center of the pipe. Calculation of the heat transfer, however, is modified from the simple equation used for a plane wall because, as the heat travels out from the center, it goes through a larger and larger area. As a consequence, the equation becomes a function of a logarithmic mean radius. The generic equation is shown both in the *FE Handbook* and below.

$$Q = \frac{2\pi kL[T_1 - T_2]}{\ln(r_2/r_1)}$$

Example 4.3

A 10-cm *OD* and 9.5-cm *ID* cast iron pipe with conductivity of 80 W/m · K has 3.5 cm of polystyrene insulation with a conductivity of 0.027 W/m · K. Saturated steam at 400 °C is flowing through the pipe and the outside surface temperature is 40 °C.

Determine

a) Heat transfer per 100 meters of pipe

b) The temperature of the outside surface of the pipe

Solution

a) The specific equation for the determination of the heat transfer is:

$$Q = \frac{T_1 - T_3}{\frac{\ln(r_2/r_1)}{2\pi k_1 L_1} + \frac{\ln(r_3/r_2)}{2\pi k_2 L_2}}$$

where

$$[\ln(r_2/r_1)]/2\pi k_1 L_1 = \ln(10/9.5)/[2\pi\, 80 \text{ W/m}\cdot\text{K}][100 \text{ m}] = 1.02 \times 10^{-6} \text{ K/W}$$

and $[\ln(r_3/r_2)]/2\pi k_2 L_2 = \ln(17/10)/[2\pi(0.027)][100] = 0.03128$ K/W;

therefore

$$Q = [400 - 40]/[1.02 \times 10^{-6} + 0.03128] = 11{,}509 \text{ W} = 10.5 \text{ kW}$$

b) The temperature of the surface of the pipe can be obtained by using the generic equation, knowing that the heat transfer does not change and by writing the equation across temperature difference from the surface of the pipe to the outside surface of the insulation. The thermal resistance includes only what is between the two temperatures.

$$Q = [T_2 - T_3]/[\ln(r_3/r_2)/2\pi k_2 L_2] = [T_2 - 40]/0.03128$$
$$T_2 = (0.03128)(11{,}509) + 40 = 399.99 °\text{C}$$

Notice that the pipe does not give any significant thermal resistance.

RADIATION

Radiation is the transfer of energy from one body to another without the necessity of a intermediate medium. In fact, the transfer is done much more efficiently without a medium. The transfer takes place due to the difference in the fourth power of the absolute temperatures of the bodies. Examining first the energy emitted by a body:

$$Q = \varepsilon\sigma A T^4$$

where

Q = Energy transferred in watts
ε = Emissivity of the radiating body (varies from 0 to 1)
σ = Stefan Boltzmann Constant (5.67×10^{-8} W/m$^2 \cdot$ K^4)
A = Surface area of the body in m^2
T = Absolute temperature K

If we now consider the case where the body radiates to outer space and that none of the radiation comes back, the equation above changes to a slightly different form.

$$Q = \varepsilon\sigma F_{1-2} A\left[T_1^4 - T_2^4\right]$$

where

F_{1-2} = Shape factor between body 1 and body 2 (varies from 0 to 1)
T_1 = Absolute temperature of body sending the radiation
T_2 = Absolute temperature of body receiving the radiation

Example **4.4**

A body at 20°C is set out on a roof top during the night. The body "sees" nothing but the sky which has an effective temperature of 120 K. Determine the heat transfer rate from the body to the sky if the body temperature is maintained at 20°C, the surface emissivity of the body is equal to 0.90, and none of the radiation going out of the body comes back.

Solution

Since the body sees nothing else than the sky, the shape factor is equal to one. Putting the temperatures into K, the heat transfer may be calculated using the equation above:

$$Q = (0.90)(5.67 \times 10^{-8})\text{W/m}^2 \cdot \text{k}^4(1.0)(1\ \text{m}^2)(293^4 - 120^4)\ \text{K}^4$$
$$Q = 365\ \text{W}$$

Example **4.5**

Consider the same body at the same conditions setting in a 20 m^2 room at 120 K and an emissivity of 0.4. Determine the heat transfer from body one to the room.

Solution

The situation may be best represented by an electric analog of a series of resistances as shown in Exhibit 1.

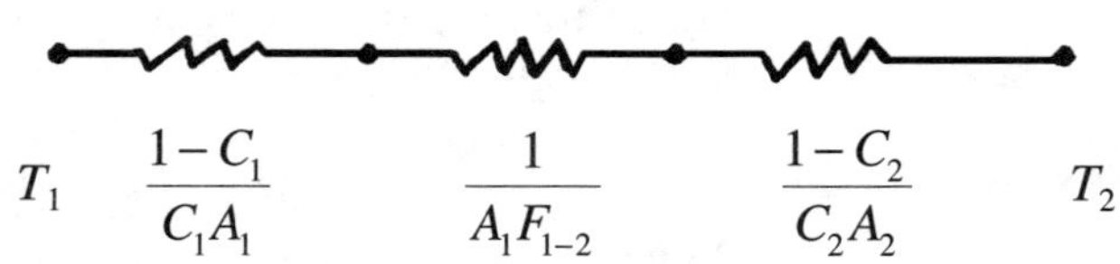

Exhibit 1

Therefore the equation for heat transfer becomes:

$$Q = \frac{\sigma(T_1^4 - T_2^4)}{\frac{1-\varepsilon_1}{\varepsilon_1 A_1} + \frac{1}{A_1 F_{1-2}} + \frac{1-\varepsilon_2}{\varepsilon_2 A_2}}$$

Substituting in the values:

$$Q = \frac{(5.67\times10^{-8})(293^4 - 120^4)}{\frac{1-.90}{(0.90)(1)} + \frac{1}{(1)(1)} + \frac{1-.40}{(0.40)(20)}}$$

$$Q = 342 \text{ W}$$

CONVECTION

Convection is a mode of heat transfer that combines molecule-to-molecule conduction with mass transport. There are two types of convection: natural and forced convection. Natural convection is circulation caused by differences of densities of fluids. These differences of densities are caused by expansion of the fluid due to elevated temperatures. Forced convection, on the other hand, is caused by mechanical circulation of the fluids. As the fluid moves faster over the heat transfer surface the film layer gets thinner and, as a consequence, the amount of heat transfer increases.

The generic heat transfer equation for convection may be written as:

$$Q = hA(T_s - T_\alpha)$$

where

Q = Heat transfer—watts
h = Convection coefficient, $W/m^2 \cdot K$
A = Surface area, m^2
T_s = Temperature of the heated surface
T_α = Temperature of the flowing fluid

Example 4.6

Strawberry blossom time is critical in the upper Midwest. If the blossoms get a touch of frost on them, the harvest can be reduced significantly. Consider a very clear night with a slight breeze of about 10 km/hr making the convection coefficient equal to 20 $W/m^2 \cdot K$. The outer space temperature may be assumed to be 120 K and the emissivity of the strawberry is 0.6. At what surrounding temperature will the strawberries freeze?

Solution

The approach to this problem is to write an energy balance equation.

$$\text{Heat in by convection} = \text{Heat out by radiation}$$

$$hA(T_{\text{air}} - T_{\text{berry}}) = \varepsilon\sigma F_{1-2}A(T_b^4 - T_{\text{o sp}}^4)$$

Notice that the area of convection to the strawberry is the same as the area of radiation away from the strawberry. Therefore, the areas may be eliminated.

$$20 \text{ W/m}^2\cdot\text{K}\ (T_{\text{air}} - 0°\text{C}) = (0.6)(5.67\times10^{-8})\text{W/m}^2\cdot\text{K}^4(1)(273^4 - 120^4)\text{K}^4$$

$$T_{\text{air}} = 9.1°\text{C}$$

Example 4.7

5 cm spheres at 300°C are dropped on a conveyer belt traveling 5 m/min through an environment that is 10°C. The convection coefficient of the spheres is 25 $W/m^2 \cdot K$ and the other properties are: $\rho = 8933$ kg/m^3, $c_p = 385$ $J/kg \cdot K$ and $k =$ 400 $W/m \cdot K$. How long must the belt be to have the spheres drop off the end at 35°C?

Solution

This is a conduction/convection system. The object is relatively small and has a large thermal conductivity compared to the convection coefficient. Since it is possible that the system is a *lumped capacity* type of problem it must be checked. The *FE Handbook* shows that if $Bi << 1$ the lumped capacity method may be used.

$$Bi = hV/kA_s << 1$$

where

h = Convection coefficient = 25 $W/m^2 \cdot K$
V = Volume = $4/3\pi r^3$ m^3
k = Thermal conductivity = 400 $W/m \cdot K$
A_s = Surface area = $4\pi r^2$ m^2

$$\begin{aligned} Bi &= [(25\ W/m^2 \cdot K)(4/3\ \pi r^3)]/[(400\ W/m \cdot K)(4\pi r^2)] \\ &= [(1/16)(0.025)]/3 = 0.00052 \end{aligned}$$

The problem fits the restrictions for the lumped-capacity type of problem.

Using the equation given in the *FE Handbook*:

$$Q = hA_s(T - T_\infty) = -\rho c_p V(dT/dt)$$

Integrating the equation and taking the natural log of both sides of the equation:

$$\ln\left(\frac{T - T_\infty}{T_o - T_\infty}\right) = -\left(\frac{hA}{\rho c_p V}\right)t = -\left(\frac{(h)(3)}{\rho c_p r}\right)t$$

$$\begin{aligned} \ln[(35 - 10)/(300 - 10)] &= -[(25\ W/m^2 \cdot K)(3)]/[(8933\ kg/m^3) \\ &\quad (385\ J/kg \cdot K)(0.025)m]t \\ t &= 46.8\ \text{min} \end{aligned}$$

Therefore, the belt must be 46.8 min × 5 m/min = 234 meters long.

HEAT EXCHANGERS

A heat exchanger is a device that is used to transfer heat. There are various types of heat exchangers, such as the counterflow heat exchanger, in which the cold fluid comes in one end and the hot fluid comes in the other end. There is also the parallel-flow heat exchanger where the cold fluid and the hot fluid come in on the same end and go out on the same end. In general, it has been found that the counterflow heat exchanger is significantly more efficient and more economical than the parallel-flow heat exchanger. In both cases the temperature difference between the two fluids is best approximated by the *Logrithimetic Mean Temperature Difference, or LMTD*. The equation is shown below:

$$LMTD = [\Delta T_{in} - \Delta T_{out}]/[\ln\ (\Delta T_{in}/\Delta T_{out}]$$

The nice thing about this form of the equation, it does not matter what end is called "in" and what end is called "out". One merely has to put the temperature difference of the fluids in the same order for the numerator as well as for the denominator.

Example 4.8

A heat exchanger is needed to operate under the following conditions: the hot fluid ranges from 300°C to 200°C and the cold fluid ranges from 30°C to 70°C. If the other conditions such as the required heat transfer and the heat transfer coefficients regarding the heat exchangers are similar, determine the size ratio needed, A_c/A_p.

Solution

The general equation for the heat transfer in a heat exchanger is:

$$Q = UA(LMTD)$$

So for the above conditions one would write:

$$Q = UA_{\text{counterflow}}\,(LMTD) = UA_{\text{parallel flow}}\,(LMTD)$$

or

$$A_c/A_p = [U_p(LMTD)_p]/[U_c(LMTD)_c]$$

and

$$U_p = U_c$$

Therefore:

$$\frac{A_c}{A_p} = \frac{\left(\dfrac{(300-20)-(200-70)}{\ln\left(\dfrac{300-20}{200-70}\right)}\right)}{\left(\dfrac{(300-70)-(200-20)}{\ln\left(\dfrac{300-70}{200-20}\right)}\right)}$$

and

$$A_c/A_p = (195.5)/203.9 = 0.958$$

Therefore, the heat transfer area required of the counterflow heat exchanger is 95.8% that of the parallel-flow heat exchanger.

PROBLEMS

The following information may be used for questions 4.1 and 4.2.

A 2-cm-diameter power wire used in a transmission line transmits 14,000 volts and 600 amps. The resistance of the wire is 0.0005 ohms/m and is wrapped with 0.01 cm insulation of $k = 0.07$ W/m·K. The convection coefficient is an average of 25 W/m^2·K and the outside temperature is a –10°C.

4.1 The power in kW dissipated in the wire per km is most nearly:

a. 50 kW c. 130 kW
b. 100 kW d. 180 kW

4.2 The temperature in °C of the outside radius of the wire is most nearly:

a. 154 c. 186
b. 175 d. 121

The following information may be used for questions 4.3 and 4.4.

A liquid metal is flowing inside of a 3-cm-diameter pipe, 5 meters long with an entering temperature of 400°C. The wall temperature is maintained at 500°C, and the mass flow rate of the metal is 50 kg/min. $\rho = 5500$ kg/m^3, $P_r = 0.0025$, $\mu = 0.0025$ N·s/m^2, $k = 6.2$ W/m·K, $c_p = 0.33$ kJ/kg·K

4.3 The convection coefficient of the liquid metal in W/m^2·K is most nearly:

a. 560 c. 895
b. 760 d. 1,290

4.4 The exit temperature of the liquid metal is most nearly:

a. 430 c. 465
b. 445 d. 500

The following information may be used for questions 4.5 and 4.6.

Oil at 300 K is flowing over a flat plate maintained at 400 K. The plate is 0.5 m long and 2 m wide. The velocity of flow is 0.2 m/s. The properties are evaluated at the mean temperature of 350°C. Properties of the oil are $v = 41.7 \times 10^{-6}$m^2/s, $k = 0.138$ W/m·K, $P_r = 546$, $\rho = 853$ kg/m^3.

4.5 The Nusselt number is most nearly:

a. 260 c. 230
b. 295 d. 180

4.6 The heat transfer in kW from the plate to the oil is most nearly:

a. 4.5 c. 6.4
b. 5.3 d. 7.1

The following information may be used for question 4.7 through 4.9.

A constant cross-section, rectangular fin 6 cm long, with a base of 1cm × 2mm is used to dissipate heat from a 400 °C surface. The convection coefficient is $h = 20\ \mathrm{W/m^2 \cdot K}$ and the conductivity is 40 W/m · K. The temperature of the air surrounding the fin is 35 °C.

4.7 The heat transfer in watts from the fin is most nearly:

a. 6.5
b. 5.8
c. 4.9
d. 7.1

4.8 The effectiveness of the fin is most nearly:

a. 21
b. 28
c. 32
d. 44

4.9 The efficiency of the fin is most nearly:

a. 32%
b. 61%
c. 45%
d. 58%

SOLUTIONS

4.1 **d.** The best approach to this problem is to write an energy balance equation.

Heat generated = Heat convected away from the wire

$$Q = I^2R = (600)^2(0.0005) = 180 \text{ W/m}$$

$$\text{Total } Q = (180)(1000)/1000 = 180 \text{ kW/km}$$

4.2 **a.** The temperature of the outside wall of the wire is obtained by using the composite wall equation for a pipe with insulation.

$$Q = \frac{T_w - T_\infty}{\frac{\ln(r_o/r_i)}{2\pi kL} + \frac{1}{hA}}$$

Rearranging and substituting in the values:

$$180 \text{ W}[\ln(2/1)]/[(2\pi)(0.07 \text{ W/m}\cdot\text{K})(1 \text{ m}) + 1/\{(25 \text{ W/m}^2\cdot\text{K})(2\pi)(0.02 \text{ m})(1\text{m})\}]$$
$$= (T_w - (-10))\,°\text{C}$$
$$T_w = 154.6°\text{C}$$

4.3 **d.** Writing an energy balance equation we get:

Heat transfer = Convection into the fluid from the pipe
= Internal energy change of fluid

$$Q = hA(T_w - T_{bave}) = mc_p(T_o - T_i)$$

To find the convection coefficient one has to obtain the Nusselt number, which has to be obtained using the liquid-metal equation for constant-wall temperature listed in the *FE Handbook*.

$$\text{Nu} = 7.0 + 0.025\,\text{Re}^{0.8}\,Pr^{0.8} = 7.0 + 0.025[(\rho DV)/\mu]^{0.8}\,Pr^{0.8}$$

where

$$V = m/\rho A = (50 \text{ kg/min})/(5500 \text{ kg/m}^3)(\pi)(0.03)^2/4 \text{ m}^2$$
$$V = 12.86 \text{ m/min} = 0.214 \text{ m/s}$$

$$\text{Nu} = 7.0 + 0.025[(5500 \text{ kg/m}^3)(0.03 \text{ m})(0.214 \text{ m/s})/0.0025 \text{ N}\cdot\text{s/m}^2]^{0.8}(0.0025)^{0.8}$$
$$\text{Nu} = 7.433 = hD/k$$

Solving for h:

$$h = (\text{Nu}k)/D = [(7.433)(6.2 \text{ W/m}\cdot\text{K})]/0.03 \text{ m} = 1288 \text{ W/m}^2\cdot\text{K}$$

4.4 **d.** Go back to the energy balance equation:

$$Q = mc_p(T_o - T_i) = hA_s(T_w - T_{bave})$$

where

$T_{bave} = (T_o - T_i)/2$ Assume mean temperature = 450°C

$$(1000 \text{ J/kJ})(50 \text{ kg/min})(0.35 \text{ kJ/kg}\cdot\text{K})(T_o - 400)°\text{C}$$
$$= (1288 \text{ W/m}^2\cdot\text{K})(\pi)(0.03\text{m})(5\text{m})(500 - 450)°\text{C}$$
$$(17{,}500)\text{J/min}(T_o - 400) = 30{,}348 \text{ J/s}$$
$$T_o = 504°\text{C}$$

4.5 **a.** In order to determine which equation must be used, the Re must first be calculated.

$$\text{Re} = (LV)/v = [(0.5\ \text{m})(0.2\ \text{m/s})]/[41.7 \times 10^{-6}\ \text{m}^2/\text{s}] = 2398$$

The equation for the Nusselt number may now be selected from the *FE Handbook*:

$$\text{Nu} = 0.648\ \text{Re}^{0.5} Pr^{0.33} = (0.648)(2398)^{0.5}(546)^{0.33} = 258.8$$

4.6 **d.** Therefore the heat transfer is:

$$Q = hA(T_p - T_{oil})$$

and

$$\text{Nu} = (hX)/k$$

or

$$h = (\text{Nu}k)/X$$

$$h = [(258.8)(0.138\ \text{W/m}\cdot\text{K})]/0.5\ \text{m} = 71.4\ \text{W/m}^2\cdot\text{K}$$

$$Q = (71.4\ \text{W/m}^2\cdot\text{K})(0.5 \times 2)\text{m}^2(400 - 300)°\text{C} = 7140\ \text{watts or } 7.14\ \text{kW}$$

Note that when k and/or h are used the temperature may be either °C or K because the problem involves a change in temperature.

4.7 **a.** Select the heat transfer equation for a fin from the *FE Handbook*.

$$Q = \sqrt{hpka_c}\,(T_b - T_\infty)\tanh mL_c$$

where

h = Convection coefficient $\text{W/m}^2\cdot\text{K}$
p = Perimeter m
k = Conductivity $\text{W/m}\cdot\text{K}$
A_c = Cross-sectional area
T_b = Temperature at the base of the fin
T_∞ = Temperature of the fluid
m = $[hp/kA_c]^{.5}$
L_c = Corrected length = $L + A_c/p$

$$Q = [(20\ \text{W/m}^2\cdot\text{K})(0.024\ \text{m})(40\ \text{W/m}\cdot\text{K})(0.01 \times 0.002)\text{m}^2]^{0.5}$$
$$(400 - 35)°\text{C} \tanh mL_c$$

$$Q = (0.0196\ \text{W/K})(365)\text{K(or C)}\tanh[(20)(0.024)/(40)(2 \times 10^{-5})]^{0.5}$$
$$[(0.06) + (2 \times 10^{-5})/0.024]$$

$$Q = 7.154 \tanh 1.49 = 6.47\ \text{W}$$

4.8 **d.** Effectiveness of the fin is equal to the heat transfer dissipated by the fin divided by the heat transfer from the area of the base of the fin if there were no fin.

$$Eff = Q_f/[(hA)(T_b - T_\infty)]$$
$$= (6.47\ \text{W})/[20\ \text{W/m}^2\cdot\text{K}(0.01 \times 0.002)\text{m}^2(400 - 35)]$$
$$Eff = 44.3$$

4.9 **b.** Efficiency of the fin is equal to the heat transfer dissipated by the fin divided by the heat transfer that the fin would dissipate if the fin were at the base temperature.

$$\eta_f = Q_{act}/Q_{ideal} = (6.47)/[hpL_c(T_b - T_f)]$$

$$\eta_f = (6.47\text{W})/[(20\text{W/m}^2 \cdot \text{K})(0.024 \text{ m})(0.0608)(400 - 35)] = 60.7\%$$

CHAPTER 5

Fluid Mechanics

OUTLINE

Fluid mechanics is the branch of engineering that deals with the actions of forces on and by fluids. The study of fluids may be divided into two branches: hydrostatics—fluids at rest,—and hydrodynamics—fluids in motion.

HYDROSTATICS

Fluid mechanics involves the study of liquids, vapors, and gasses (highly superheated vapors). Vapors and gasses usually are assumed to exert constant pressure throughout the contained volume. This of course does not hold true for our atmosphere because, as we know, the barometric pressure decreases as we go higher into the atmosphere, hence, pressurization of the inside of aircraft. Hydrostatics, however, deals mainly with the action of liquids. This is when the total pressure at any given point in the liquid is equal to the pressure at the surface plus the pressure exerted by the head of the liquid.

Manometry

Manometers are often used to measure pressures and differential pressures. They use a column of liquid as the indication of pressure.

$$\text{Pressure due to liquid column} = \rho g h$$

where

ρ = Density kg/m^3
g = Acceleration due to gravity 9.81 m/s^2
h = Vertical height of column of liquid m

Therefore pressure is

$$P = [(\rho \text{ kg/m}^3)(g \text{ m/s}^2)(h \text{ m})][\text{m}^2/\text{m}^2] = \text{N/m}^2 = \text{Pa}$$

Example 5.1

A U-tube manometer is used to measure the pressure in an air line. The 15 cm of liquid in the U-tube manometer is mercury, which has a specific gravity of 13.6 and is open to an atmosphere that is 1.2 bar. Determine the absolute pressure in the line. Refer to Exhibit 1.

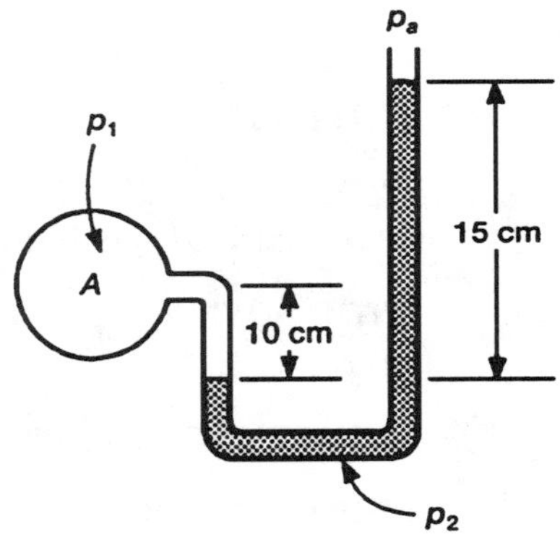

Exhibit 1

Solution

The generic equation for determining pressure may be written as:

$$P_{\text{absolute}} = P_{\text{gage}} + P_{\text{atmosphere}}$$

Keep in mind that the absolute pressure is the total pressure, and is equal to the pressure indicated by the gage added to the pressure of the atmosphere surrounding the gage. The generic equation may now be written as

$$P_{\text{abs}} = \rho gh + 1.2 \text{ bar} = 13.6(1000 \text{ kg/m}^3)(9.81 \text{ m/s}^2)(0.15 \text{ m}) + 1.2 \text{ bar}$$
$$P_{\text{abs}} = (20{,}012 \text{ kg/m}\cdot\text{s}^2)(\text{m}^2/\text{m}^2) + 1.2 \text{ bar}$$
$$P_{\text{abs}} = 20{,}012 \text{ N/m}^2 + 1.2 \times 10^5 = 1.4 \times 10^5 \text{ N/m}^2 = 1.4 \times 10^5 \text{ Pa} = 1.4 \text{ bar}$$

(Note that the density of mercury is equal to the specific gravity of mercury times the density of water where the density of water is taken as 1000 kg/m^3.)

Example 5.2

A company uses a number of multi-pass heat exchangers that use water as the cooling fluid. It is necessary to determine the power needs for the pumping of the water through the heat exchangers. A differential manometer is used to determine the pressure differential between the incoming and outgoing water. The manometer system is shown in Exhibit 2 with the distances shown. The second liquid used is mercury, with a specific gravity of 13.6. Determine the pressure drop in the heat exchanger in Pa.

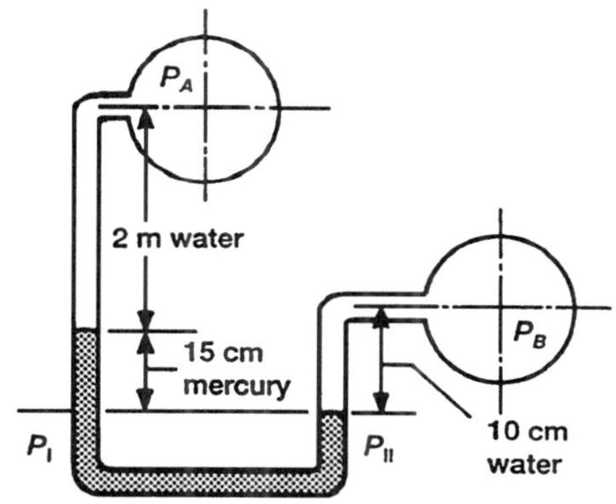

Exhibit 2

Solution

This heat exchanger is a little more complex than the former one. First, draw a horizontal line at the lowest interface between the two fluids. Then you may eliminate everything below this line. Next, equate the pressures at the slice of the interface section.

$$P_I = (\rho gh)_{hg} + (\rho gh)_{water} + P_A$$
$$P_{II} = (\rho gh)_{water} + P_B$$

Combining the two equations

$$(13.6)(1000 \text{ kg/m}^3)(9.81 \text{ m/s}^2)(0.15 \text{ m}) + (1000 \text{ kg/m}^3)(9.81 \text{ m/s}^2)(2 \text{ m}) + P_A$$
$$= (1000 \text{ kg/m}^3)(9.81 \text{ m/s}^2)(0.10 \text{ m}) + P_B$$

Solving for the differential pressure

$$P_A - P_B = 981 - 19{,}620 - 20{,}012 = -38{,}651 \text{ Pa}$$

This indicates that the pressure drop is 38.6 kPa.

Forces on Submerged Areas

Pressure is a non-directional force and acts perpendicularly on a surface (refer to Figure 5.1). It is also noted that this force varies directly with the depth or the distance down from the surface. The vertical distance from the surface of the liquid to the centroid of the body is labeled h_c. The force acting on the body is equal to

$$F = \gamma h_c A$$

where

F = Total force acting on the body
γ = Specific weight of the liquid
h_c = Distance from the surface of the liquid to the centroid of the body
A = Area of the body

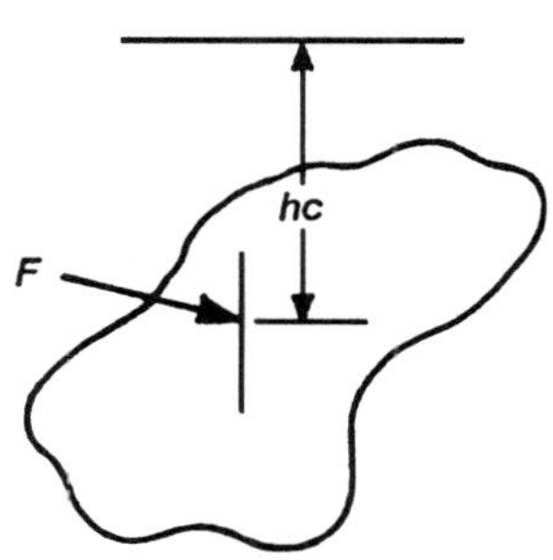

Figure 5.1

Example 5.3

A 6-m × 6-m plate is submerged in water as shown in Exhibit 3 with the top edge of the plate

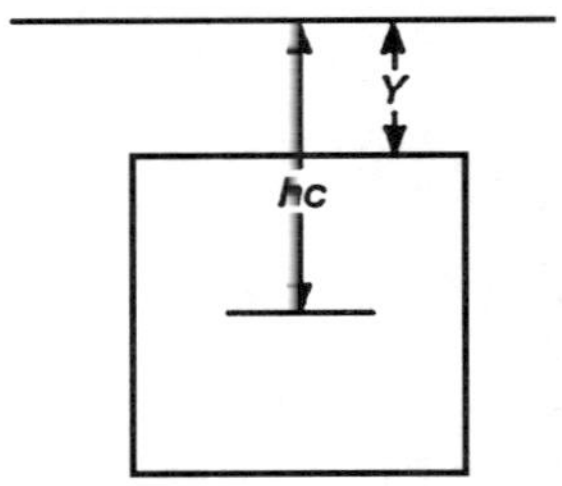

Exhibit 3

Determine the force, in kN, acting on the plate for the cases indicated

a) At the surface

b) 5 meters down

c) 100 meters down

Solution

a) $F = \gamma h_c A = (1000\ \text{kg/m}^3)(9.81\ \text{m/s}^2)(3\ \text{m})(6\ \text{m} \times 6\text{m})$
$= 1{,}059{,}480\ \text{kg}\cdot\text{m/s}^2$ or, force in kN = 1,059 kN

b) $F = (1000\ \text{kg/m}^3)(9.81\ \text{m/s}^2)(5 + 3)\text{m}\ (6\ \text{m} \times 6\ \text{m}) = 2{,}825\ \text{kN}$

c) $F = (1000\ \text{kg/m}^3)(9.81\ \text{m/s}^2)(100 + 3)\text{m}\ (6\ \text{m} \times 6\ \text{m}) = 36{,}375\ \text{kN}$

The center of pressure of a submerged body is that point where there would be no moment due to the pressure forces on the body. The equation for the calculation to the center of pressure is

$$h_p = h_c + (I_c / h_c A)$$

where
h_p = Distance to *center of pressure* of the body
h_c = Distance to *center of gravity* of body
I_c = *moment of inertia* of body
A = Area of the body

Determine the distance to the center of pressure for each of cases above: Calculating the moment of inertia and the area

$$I_c = bh^3/12 = [(6)(6)^3]/12 = 108\ \text{m}^4 \qquad A = bh = (6)(6) = 36\ \text{m}^2$$

Therefore

a) $h_p = 3 + (108)/(3)(36) = 4\ \text{m}$

b) $h_p = (3 + 5) + 1 = 9\ \text{m}$

c) $h_p = (3 + 100) + 1 = 104\ \text{m}$

Buoyancy

Buoyancy is the ability of a fluid to support a body in the fluid. A body has the buoyant force acting on its center of gravity equal to the displaced fluid. The buoyant force is equal to the mass of the fluid times the gravitational attraction.

Example 5.4

A 20-m × 30-m flat-bottom car ferry that is used to transport automobiles across the Mississippi has a mass of 35,000 kg and can carry sixty 5,000-kg automobiles. How far will the ferry sink into the water?

Solution

The buoyant force is equal to the displaced liquid

$$F = mg = \text{Displaced volume} \times \text{Specific weight}$$
$$F = [35{,}000 + 60(5000)]\text{kg}\ (9.81\ \text{m/s}^2) = 3.286 \times 10^6\ \text{kg}\cdot\text{m/s}^2$$

Equating to displacement volume times specific weight

$$F = 3.286 \times 10^6\ \text{N} = (20\ \text{m})(30\ \text{m})(\text{Vertical displacement } Y\ \text{m}) \times (1000\ \text{kg/m}^3)(9.81 \times \text{m/s}^2)$$
$$Y = 0.56\ \text{m}$$

Example 5.5

A 1.5-m-diameter and 4-m-tall cylinder with a mass of 2000 kg is closed at one end (Exhibit 4). It is placed in water such that the open end is down into the water. How far above the water will the tank float? Assume that the air and water are at the same temperature of 20°C.

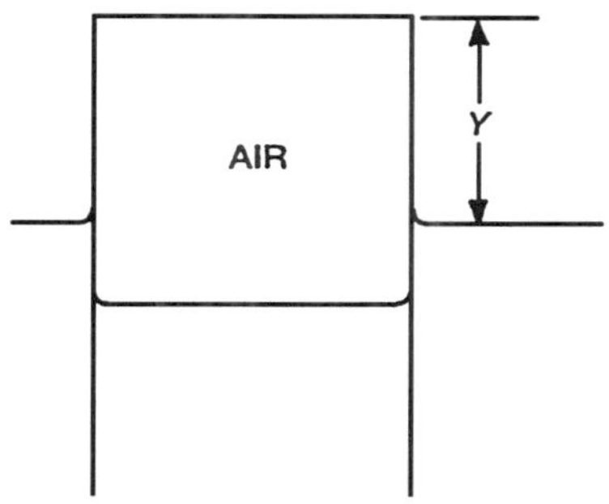

Exhibit 4

Solution

Since the tank is open downward the tank contains both air and water. Assuming, first, that the tank itself has insignificant displacement of water, the buoyant force is equal to the volume of water displaced by the entrapped air in the tank. The buoyant force has to be equal to the weight of the container. We can assume that the air in the tank has negligible weight.

$$F = (\text{mass of tank})(\text{gravity}) = (\text{Volume of displaced water})(\text{weight of water})$$
$$F = (2000\ \text{kg})(9.81\ \text{m/s}^2) = (\pi D^2 m^2/4)(\text{displacement m})(1000\ \text{kg/m}^3)(9.81\ \text{m/s}^2)$$
$$19{,}620\ \text{N} = \pi(1.5\ \text{m})^2/4(Y)\ \text{m}\ (9{,}810\ \text{kg/m}^2\cdot\text{s}^2)$$
$Y = 1.13$ m air in the tank will displace enough water for tank to float.

The pressure in the tank times the area of the opening of the tank is equal to the buoyant force, or

$$F = 19{,}620\ \text{N} = P_g = P_g\ \pi D^2/4 = P_g\ (3.14)(1.5)^2\text{m}^2/4$$
$$P_g = 11{,}103\ \text{Pa}$$
$$P_{tot} = P_g + P_{atm} = 11{,}103 + 101{,}300 = 112{,}403\ \text{Pa}$$

The volume of air in the tank may be obtained by using Charles' and Boyle's laws:

$$[P_1V_1]/T_1 = [P_2V_2]/T_2$$

and, for $T_1 = T_2$

$$P_1V_1 = P_2V_2$$

And solving for V_2

$$V_2 = V_1P_1/P_2 = V_1(101{,}300)/112{,}403 = 0.901V_1$$

This means that the column of air for the constant-area drum will be equal to 0.901×4 m, or equal to 3.60 m. The height of the tank sticking out of the water shall then be equal to 3.60 m minus the length of air needed to displace the equivalent amount of water

$$h_{\text{out of water}} = 3.60\text{ m} - 1.13\text{ m} = 2.47\text{ m}$$

HYDRODYNAMICS

Hydrodynamics is a study of liquids in motion. Since this study is about liquids, which are incompressible, two general conditions must be satisfied. The first condition that must be satisfied is the continuity equation. This simply means that mass flow in must equal to the mass flow out.

For incompressible, i.e., constant, density

$$A_1V_1 = A_2V_2$$

The second condition is that the summation of energy entering and leaving a system must be equal. This summation of energy is a simplification of the *first law of thermodynamics* and is known as the *Bernoulli equation.* This equation may be written as

$$P_1/\rho g + V_1^2/2g + Z_1 + h_A = P_2/\rho g + V_2^2/2g + Z_2 + h_L$$

where

P =	Pressure	Pa
ρ =	Density	kg/m^3
g =	Acceleration due to gravity	m/s^2
V =	Velocity of the fluid	m/s
Z =	Elevation	m
h_A =	Energy added to fluid	m
h_L =	Energy lost from the fluid	m
A =	Crossectional area of pipe	m2

Example 5.6

A water tank is used to supply the water for a village. The tank level is maintained at an elevation of 520 meters above sea level and is allowed to flow 200 m through a 4-cm pipe to a 1-cm-diameter nozzle at an elevation of 450 meters above sea level. If there is no friction, what is the velocity of flow at the exit of the nozzle?

Solution

Write the continuity equation relating the flow in the pipe to the flow in the nozzle

$$(\rho AV)_2 = (\rho AV)_3$$
$$\pi D_2^2 V_2 = \pi D_3^2 V_3, \text{ therefore:}$$
$$V_3 = [D_2/D_3]^2 V_2$$
$$V_3 = (4/1)^2 V_2 = 16 V_2$$

Write the Bernoulli equation for the pipe, noting that at the water tank end there is no velocity, and at both ends there is no pressure. Take the datum to be 450 meters above sea level. Write the equation, assuming that the flow is out of the full pipe, then make the continuity equation substitution to solve for the velocity

$$P_1/\rho g + V_1^2/2g + Z_1 = P_2/\rho g + V_2^2/2g + Z_2$$
$$70 \text{ m} = V_2^2/9.81 \text{ m/s}^2 \quad \text{Or} \quad V_2 = 26.2 \text{ m/s}$$

Therefore, the velocity out of the nozzle is

$$V_3 = (16)(26.2) = 419.2 \text{ m/s}$$

Example 5.7

Gasoline with a specific gravity of 0.80 is flowing through a 0.050-m-diameter pipe that has a venturi throat of 0.025 m (see Exhibit 5). The differential pressure between the throat and the pipe as shown in the Figure is 38 cm. The c_v of the venturi meter is 0.97, and the meter is on a 30° positive angle.

Determine

a) Difference in pressure between pipe and venturi throat, in meters of gasoline

b) Velocity at the throat ideal

c) Flow rate of gasoline, in l/min

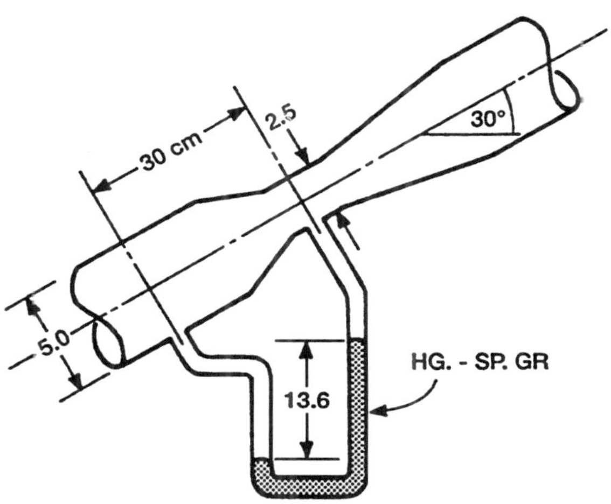

Exhibit 5

Solution

Write Bernoulli's equation for the venturi meter

$$P_1/\rho g + V_1^2/2g + Z_1 = P_2/\rho g + V_2^2/2g + Z_1 + \text{head loss due to friction}$$

Since we are only looking at a very short device there is not going to be any head loss due to friction. We do have to consider the venturi velocity coefficient, however.

Density of gasoline = Specific gravity of gasoline times density of water = 800 kg/m^3
Density of mercury = Specific gravity of mercury times density of water = 13,600 kg/m^3

a) The pressure difference is therefore due to the difference in the columns

$$[P_1 - P_2]/(800\ \text{kg/m}^3)(9.81\ \text{m/s}^2) = [(0.38\ \text{m})(13{,}600 - 800)\text{kg/m}^3] \times [9.81\ \text{m/s}^2]/(800\ \text{kg/m}^3)(9.81\ \text{m/s}^2) = 6.08 \text{ meters of gasoline}$$

b) The difference due to elevation change

$$Z_1 - Z_2 = -0.30\ \text{m} \times \sin 30° = -0.15 \text{ meters of gasoline}$$

Substituting these values into the Bernoulli equation

I)
$$6.08\ \text{m} - 0.15\ \text{m} = \left[V_2^2 - V_1^2\right]/(2)(9.81\ \text{m/s}^2)$$
$$\left(V_2^2 - V_1^2\right) = 116.3\ \text{m}^2/\text{s}^2$$

Using the continuity equation to determine the relation between the velocity in the pipe and the velocity in the throat of the venturi

II)
$$(\rho AV)_1 = (\rho AV)_2, \text{ with } \rho_1 = \rho_2$$
$$\left(\pi D_1^2/4\right)(V_1) = \left(\pi D_2^2/4\right)(V_2)$$
$$V_1 = V_2(0.025/0.050)^2 = V_2/4$$

Substituting equation II into equation I and solving for V_2

$$V_2^2 - (V_2/4)^2 = 116.3\ \text{m}^2/\text{s}^2$$
$$V_2 = 11.14\ \text{m/s}$$

c) To find the flow rate of gasoline we must multiply the c_v times the velocity in the pipe to get the pipe velocity.

$$V_1 = (V_2\text{m/s})/4 = 11.14/4 = 2.785\ \text{m/s ideal}$$
$$V_{1\ \text{actual}} = 0.97(2.785) = 2.70\ \text{m/s}$$
$$\text{Volume flow rate} = VA = (2.70\ \text{m/s})(\pi)(D_1)^2/4\ \text{m}^2 = (2.70)(\pi)(0.050)^2/4\ \text{m}^3/\text{s}$$
$$= 0.00530\ \text{m}^3/\text{s}(100\ \text{cm/m})^3(60\ \text{s/min})/(1000\ \text{cm}^3/\text{l}) = 318\ \text{l/min}$$

Dimensional Analysis

Dimensional analysis is a method by which one may predict physical parameters that affect fluid flow phenomena. By Newton's law we may operate in two different, yet related, systems. These systems are the MLT system and the FLT system.

$$F = ma = \text{ML/T}^2$$

where

F is force with F as the primary dimension
m is mass with M as the primary dimension
a is acceleration with L/T^2 as the primary dimensions

Example 5.8

Determine the generic equation for the drag on a body, considering the following factors that may influence drag

$$F_D = f(D, V, \rho, \mu)$$

where

F_D = Drag force	F
D = Diameter of pipe	L
V = Velocity of the fluid	L/T
ρ = Density of the fluid	M/L^3
μ = Viscosity of the fluid	M/LT

Solution

After choosing the dimensions from the FLT or MLT systems, assign the dimensions and set the exponents.

$$ML/T^2 = F_D = CL^a(L/T)^b(M/L^3)^c(M/LT)^d$$

Set up the table of exponents for

M:	$1 = c + d$	$c = 1 - d$
L:	$1 = a + b - 3c - d$	$a = 1 - (2 - d) + 3(1 - d) + d = 2 - d$
T:	$-2 = -b - d$	$b = 2 - d$

$$F_D = CD^{2-d}\, V^{2-d}\, r^{1-d}\, \mu^d$$

Grouping in terms of like exponents

$$F_D = C(DV)^2[1/(\rho DV)/\mu](\rho)$$

Rearranging the terms, noting that $(\rho DV)/\mu = \text{Re}$

$$\text{Drag force } F_D = f(\text{Re})D^2V^2\rho$$

Momentum Force

When a fluid flows through a pipe that rounds a bend, forces are exerted by the fluid upon whatever resists the bend from movement. The forces are the summation of momentum in the various directions. The generic equation is

$$mV_1 + \int F dt = mV_2$$

Example 5.9

A jet of water that is 5 cm in diameter and is moving to the right at 30 m/s hits a stationary flat plate held normal to the flow. What force is required to hold the plate stationary?

Solution

A summation of momentum equation in the x direction should be used. Note that when the flow hits the stationary plate all x-direction velocity is gone because the water fans out into the y- and z-directions.

Writing the momentum equation in the x direction

$$mV_{x1} + F_x dt = mV_{x2}$$
$$[1000 \text{ kg/m}^3][(\pi D^2)/4] \text{ m}^2[30 \text{ m/s}]^2 + F_x = 0$$
$$F_x = -1767 \text{ N} \quad \text{or} \quad 1.77 \text{ kN to the left}$$

Example 5.10

Water flows in a 6-cm-diameter stream at 50 m/s to the right. The stream of water hits a stationary, curved plate, and the stream is deflected 60° upward and to the right. Determine the x and y forces required to hold the plate stationary.

Solution

Again, the momentum equation is to be used; however, this time it is to be written in the x direction and in the y direction.

$$mV_{x1} + F_x dt = mV_{x2}$$
$$(1000 \text{ kg/m}^3)[(\pi D^2)/4] \text{ m}^2(50 \text{ m/s})^2 + F_x = (1000)[(\pi D^2)/4][(50 \text{ m/s})(\cos 60°)]^2$$
$$F_x = -7.069 + 1.767 = -5.30 \text{ kN to the left.}$$
$$mV_{y1} + F_y dt = mV_{y2}$$
$$0 + F_y = (1000)[(\pi D^2)/4](50 \text{ m/s} \sin 60°)^2$$
$$F_y = 5.301 \text{ kN} \quad \text{Upward}$$

FLOW THROUGH PIPES

If we go back to the Bernoulli equation that was introduced and defined in the section on *hydrodynamics*, we notice that h_L is the energy lost by the fluid between the two points in question, 1 and 2. This energy loss is due to fittings and friction in the line, such as elbows, meters, valves, and normal flow. The losses are calculated using the Darcy–Weisbach equation, which is shown below.

$$h_L = f\,(L/D)(V^2/2g)$$

where

h_L = Head loss	meters
f = Friction factor	no units
L = Length of the pipe	meters
D = Diameter of the pipe	meters
V = Velocity of flow in pipe	m/s
g = Acceleration due to gravity	m/s^2

Note that the value D is given as the diameter of the pipe. If the pipe is really a duct then the value of D is a characteristic dimension and, for flow in open channels, amounts to four times the *hydraulic radius*. The *hydraulic radius* is equal to the *cross-sectional area of flow* divided by the *wetted perimeter*.

The term f is called the friction factor and is dependent upon the *Reynolds number* and the roughness of the pipe. If the Reynolds number is less than 2,000 the flow is called *laminar flow*, and the friction factor $f = 64/\text{Re}$. When the Reynolds number is above 10,000 the flow is considered to be *turbulent*, and the value of f is taken from the Moody diagram. Between the two numbers the flow is considered to be in transition.

Example 5.11

How much power, in kW, does it require to pump 250 m^3/hr of oil with a specific gravity of 0.85 and a viscosity of 0.011 kg/m·s through 5 km of 30 cm diameter commercial steel pipe. The pipe rises some 150 m and has 5 elbows and two gate valves with coefficients of 0.85 and 1.25, respectively.

Solution

Starting with the Bernoulli equation

$$P_1/\rho g + V_1^2/2g + Z_1 + h_A = P_2/\rho g + V_2^2/2g + Z_2 + h_L$$

Then, noting that since this is a pump power problem, everything on the left hand side of the equation may be simplified down to the energy added (h_A) to the oil. Assuming that the flow is open to the atmosphere, or that the flow just makes it to the end of the pipe, we can also drop P_2.

$$h_A = V_2^2/2g + Z_2 + h_L$$

Solving for the terms in the Bernoulli equation

$$V_2^2/2g = \{[(250\ \text{m}^3/\text{hr})/(3600\ \text{s/hr})]/[(\pi D^2/4)\ \text{m}^2]\}^2/[2(9.81\ \text{m/s}^2)]$$
$$V_2^2/2g = \{(0.0694\ \text{m}^3/\text{s})/[\pi(0.30)^2/4]\ \text{m}^2\}^2/[19.62\ \text{m/s}^2] = 0.050\ \text{m}$$

where the velocity is equal to 0.982 m/s

Z_2 = 150 m

h_L = Head loss due to elbows, gate valves, and the friction flow in the pipe.

$$h_{L(\text{elbows})} = C_{\text{elbows}}\ (\text{Number of elbows})(V^2/2g) = (0.85)(5)(0.50) = 2.125\ \text{m}$$
$$h_{L(\text{Gate valves})} = C_{\text{valves}}\ (\text{Number of valves})(V^2/2g) = (1.25)(2)(0.50) = 1.25\ \text{m}$$
$$h_{L(\text{Due to friction})} = f\,(L/D)(V^2/2g)$$

Where f for commercial steel pipe has to be determined after the Reynolds number has been calculated.

$$\text{Re} = \rho DV/\mu = (1000\ \text{kg/m}^3)(0.85)(0.30\ \text{m})(0.982\ \text{m/s})/(0.011\ \text{kg/m}\cdot\text{s}) = 22{,}764$$

The Reynolds number is greater than 2000, and, as a consequence, we have turbulent flow. We then must go to the Moody diagram in the *FE Supplied-Reference Handbook* and select the value of f for commercial steel pipe. The handbook lists the roughness as e = 0.046 mm. Calculating the relative roughness as e/D we get

$$e/D = (0.046\ \text{mm})/(300\ \text{mm}) = 0.0001533$$

The Moody diagram lists the friction factor for an e/D of 0.0001533 and Re = 22,764 to be 0.0265. Now we can calculate the friction head loss in the pipe.

$$h_L = [(0.0265)(5000\ \text{m})/(0.30\ \text{m})](0.050\ \text{m}) = 22.08\ \text{m}$$

The values may then be put in the simplified Bernoulli equation

$$h_A = 0.050 \text{ m} + 150 \text{ m} + 2.125 \text{ m} + 1.25 \text{ m} + 22.08 = 175.5 \text{ m oil}$$

Therefore, the work added to the oil is

$$\text{Power} = P(AV) = (\rho gh)(250 \text{ m}^3/\text{hr})/(3600 \text{ s/hr}) = (\rho gh)(0.0694 \text{ m}^3/\text{s})$$
$$\text{Power} = (1000 \text{ kg/m}^3)(0.85)(9.81 \text{ m/s}^2)(175.5 \text{ m}) \times (0.0694 \text{ m}^3/\text{s})[(\text{N})/(\text{kg}\cdot\text{m/s}^2)](\text{kJ}/10^3 \text{ N}\cdot\text{m})$$
$$\text{Power} = 101.56 \text{ kW}$$

OPEN-CHANNEL FLOW

Open-channel flow is the type of flow that has a free surface in contact with atmospheric pressure. The flow is caused by a sloping of the channel. Accurate solutions are very difficult to obtain because of the varied conditions under which the data is obtained.

Example 5.12

What flow can be expected in a 3-m-wide rectangular, cement-lined channel with a slope of 5 m in a distance of 15,000 m. The water is to flow 0.5 m deep.

Solution

Take the Manning equation from the *Handbook* to solve for the velocity. Then take the velocity and multiply by the area of flow to obtain the flow rate.

$$V = (1/n)R^{2/3}S^{1/2}$$

where

V = velocity	m/s
n = roughness coefficient	0.015
R = Hydraulic radius	m
S = Slope of channel	m/m

$$V = (1/0.015)[(3 \times 0.5 \text{ m}^2)/(0.5 + 3 + 0.5 \text{ m})]^{2/3}(5 \text{ m}/15{,}000 \text{ m})^{1/2} = 0.633 \text{ m/s}$$
$$\text{Flow rate} = AV = (3 \times 0.5 \text{ m}^2)(0.633 \text{ m/s}) = 0.95 \text{ m}^3/\text{s}$$

PROBLEMS

5.1 It is necessary to have a relationship regarding the shear stress of a fluid flowing in a pipe. Shear stress is thought to be a function of density, viscosity, and velocity of the fluid, as well as a function of the diameter and the roughness of the pipe containing the flowing fluid. The dimensionless equation relating these variables is

a. $C\rho V^2/\mathrm{Re}^b$
b. $C\rho D/V^2$
c. $CD^2V/L\mu\rho$
d. $Cd\rho\mu/V$

Use the following information for Problems 5.2 & 5.3.

Given a 10-m × 10-m submerged gate used for flood control. The gate is hinged at the top and held in place by a force at the bottom. Consider that the top of the hinged gate is 10 meters below the surface of the water.

5.2 The center of pressure of the hinged gate is most nearly

a. 10 m
b. 12.67 m
c. 15.56 m
d. 18.43 m

5.3 The force, in kN, required to hold the gate in position is most nearly

a. 10,000
b. 12,500
c. 6,500
d. 8,200

Use the following information for Problems 5.4 through 5.7

A 40-cm-diameter pipeline has a flow rate of 1,000 m^3/hr of water at standard temperature. The pipeline is 5 km long, is made out of galvanized iron, and has a lift of 75 m. The water has a viscosity of 1.005×10^{-3} kg/m·s and the pump has an efficiency of 70%.

5.4 The friction factor for the above pipeline is most nearly

a. 0.014
b. 0.018
c. 0.022
d. 0.025

5.5 The pump work required, in meters of water added, is most nearly

a. 80 m
b. 90 m
c. 100 m
d. 120 m

5.6 The hp required for the pump is most nearly

a. 215
b. 330
c. 450
d. 620

5.7 If the flow rate were decreased to 300 m^3/hr how many hp is needed?

a. 75 hp
b. 125 hp
c. 105 hp
d. 135 hp

5.8 A heavy fuel oil with a specific gravity of 0.94 and a viscosity of 2.01 poise is to be pumped at the rate of 3000 m^3/day through an old 30 cm *ID* pipe that is badly corroded on the inside. What would be the friction factor?

a. 0.033
b. 0.055
c. 0.077
d. 0.093

5.9 An ore carrier 120 m long × 10 m wide displaces 8500 m^3 of fresh water. It is moved into a lock in fresh water that is 140 m long × 15 m wide and then is loaded with 3500 metric tons of ore. What will be the increase in the depth of the water in the lock after the ship is loaded?

a. 0.83 m	c. 2.67 m
b. 1.89 m	d. 3.89 m

5.10 A careless hunter shoots a 7.00-mm-diameter hole in the vertical side of a water tank one meter above the ground. The level of the water in the tank is 10.0 m above the ground. How far from the base of the tank will the water strike the ground?
Assume $C = 1.0$

a. 2.67 m	c. 6.00 m
b. 4.75 m	d. 7.88 m

5.11 The water level of a reservoir is 150 m above a power plant. A 30.0 cm *ID* pipe 200 m long connects the reservoir to the plant. If the friction factor for flow through the pipe is 0.02, how much water will flow to the plant?

a. 0.80 m^3/s	c. 1.2 m^3/s
b. 1.0 m^3/s	d. 1.4 m^3/s

5.12 A piece of glass weighs 125 g in air, 75 g in water, and 92 g in gasoline. What is the specific gravity of the gasoline?

a. 0.88	c. 0.72
b. 0.80	d. 0.66

5.13 A 30- × 30-cm timber 3.6 meters long floats level in fresh water with 11.0 cm above the water surface. One end is just touching a rock, which prevents that end from sinking any deeper. How far out from the supported end can a 68-kg man walk before the free end submerges? See Exhibit 5.13.

a. 1.0 m	c. 2.1 m
b. 1.5 m	d. 2.6 m

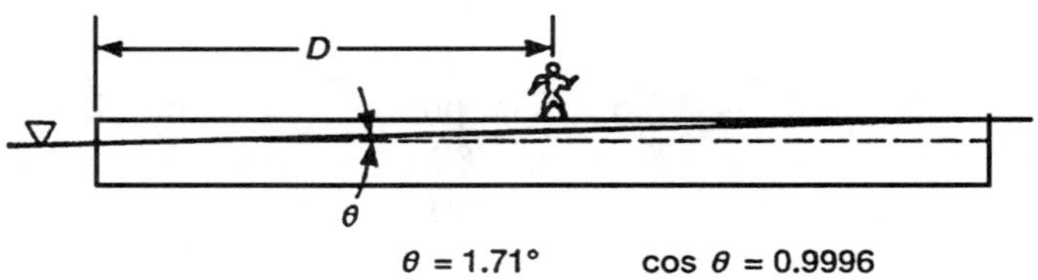

Exhibit 5.13

5.14 An 20 cm *ID* pipeline, 1070 meters long, conveys water from a pump to a reservoir whose water surface is 450 ft above the pump which is pumping at the rate of 1.0 m^3/sec. Use constants for cast-iron pipe, disregard velocity head and minor losses, and determine the gauge pressure in pascals at the discharge end of the pump. The absolute viscosity of water is 1.0 centipoise.

a. 1.70 MPa	c. 2.75 MPa
b. 2.04 MPa	d. 3.01 MPa

5.15 Water is flowing, horizontally, from a reservoir through a circular orifice under a head of 50 ft. The coefficient of discharge is 0.7, and the diameter of the orifice is 10 cm. How many cubic meters per second are being discharged from the orifice?

a. $0.07 \text{ m}^3/\text{sec}$ c. $0.11 \text{ m}^3/\text{sec}$
b. $0.09 \text{ m}^3/\text{sec}$ d. $0.15 \text{ m}^3/\text{sec}$

5.16 A swimming pool is being cleaned with a brush fixture on the end of a rubber hose which is 5.0 cm inside diameter, through which the water is drawn by a vacuum pump. The hose is 15 meters long, leading to the pump 3.6 meters above the bottom of the pool. If the water in the pool is 2.4 meters deep, what suction (in Pascals) would be required at the pump to draw 2.25 m^3 per minute through the hose? (Assume a loss coefficient at the brush of 0.5 $v^2/2g$, where v is the velocity in the hose. Assume the absolute viscosity of water = 1.0 centipoise.

a. −14 kPa c. −21 kPa
b. −17 kPa d. −24 kPa

5.17 Water flows from a supply tank through 25 meters of welded steel pipe 15 cm in diameter to a hydraulic mining nozzle 5.0 cm in diameter. The coefficient of the nozzle may be assumed to be 0.90, and the temperature of the water may be assumed to be 75F with a kinematic viscosity of 1.0 centipoise. If 28 l/s of water is flowing, how far must the nozzle lie below the elevation of the water in the tank?

a. 9 m c. 15 m
b. 12 m d. 18 m

5.18 Water falling from a height of 30 meters at the rate of 57 m^3/min drives a water turbine connected to an electric generator at 120 rpm. If the total resisting torque due to friction is 1.8 kN at 30 cm radius and the water leaves the turbine blades with a velocity of 3.70 m/sec find the power developed by the generator.

a. 308 kW c. 275 kW
b. 289 kW d. 266 kW

5.19 A venturi meter with a throat 5 cm in diameter is placed in a pipeline 10 cm in diameter carrying fuel oil. If the differential pressure between the upstream section and the throat is 25 kPa, how much oil is flowing in liters per minute? The discharge coefficient of the venturi meter is 0.97. The specific gravity of the oil is 0.94.

a. 790 *l*/min c. 890 *l*/min
b. 840 *l*/min d. 925 *l*/min

5.20 A rectangular sluice gate, 122 cm wide and 183 cm deep, hangs in a vertical plane. It is hinged along the top (122 cm) edge. If there are 6.4 meters of water above the top of the gate, what horizontal force applied at the bottom of the gate will be necessary to open it? Refer to Exhibit 5.20.

a. 75 kN c. 92 kN
b. 83 kN d. 98 kN

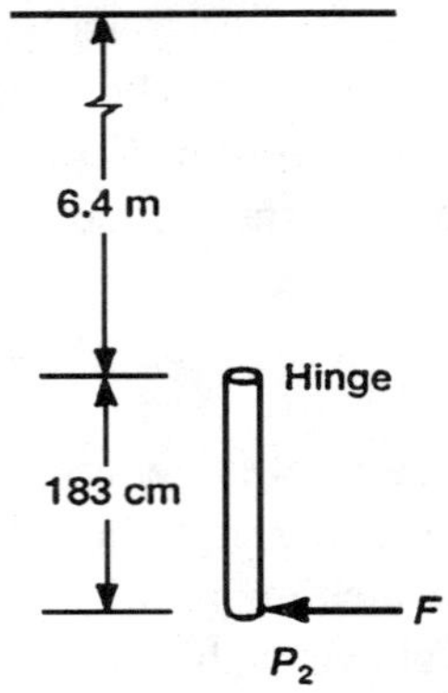

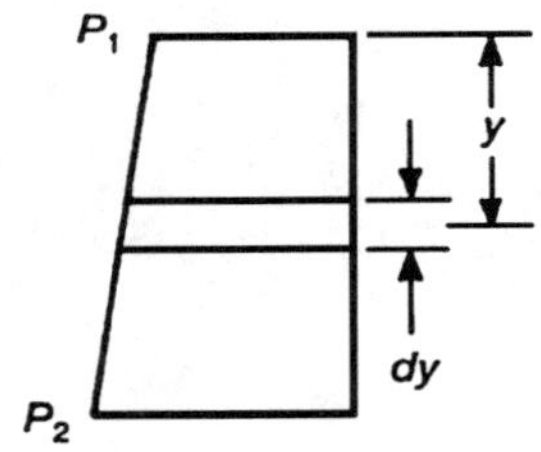

Exhibit 5.20

5.21 If a pipe is corroded on the *ID* the corrosion will affect the flow through it in the following manner
a. For a constant pressure supply it will reduce the flow if the flow is laminar.
b. If the Reynolds number is above 10,000 the corrosion will not have any effect on the flow.
c. For a Reynolds number of 100,000 it will cause a higher friction factor than for an uncorroded pipe.
d. For a given flow rate the Reynolds number will be higher than for an un-corroded pipe.

5.22 A dam 15 meters high has water behind it to a depth of ten meters. What is the overturning-force per meter of width?
a. 1.6 MJ c. 2.3 MJ
b. 2.0 MJ d. 1.8 MJ

5.23 A two-meter *ID* pipe is flowing half full. What is the hydraulic radius?
a. 1.00 m c. 1.50 m
b. 0.50 m d. 1.75 m

5.24 Estimate the total time required to empty a tank which is a paraboloid of revolution with every horizontal section a circle whose radius equals the square root of the height above the bottom, through which is cut a 25-mm-round sharp-edged orifice. The depth of water at the start of time is 3.0 meters. Refer to Exhibit 5.24.
a. 2 hr 20 min c. 1 hr 50 min
b. 2 hr 5 min d. 1 hr 38 min

5.25 A 30-cm *ID* cast-iron pipe, 1220 meters in length, conveys water from a pump to a reservoir whose water surface is 76.0 meters above the pump, which is pumping at a rate of 85 liters per sec. Assuming the efficiency of the pump to be 89 percent, determine the power input. The viscosity of water is 1 centipoise.
a. 65 kW c. 77 kW
b. 70 kW d. 82 kW

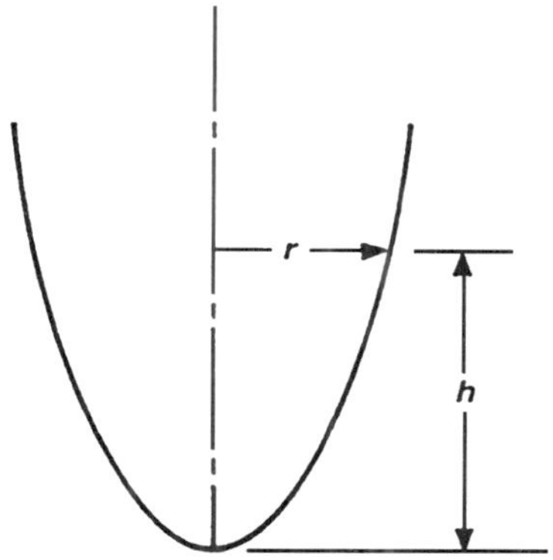

Exhibit 5.24

5.26 A pitot tube having a coefficient of unity is inserted in the exact center of a long, smooth tube of 25.0 mm *ID* in which crude oil (sp. gr. = 0.9 and $\mu = 16.7$ centipoises) is flowing. Determine the average velocity in the tube if the velocity pressure is 100 mm of water.

a. 0.56 m/sec c. 0.74 m/sec
b. 0.65 m/sec d. 0.82 m/sec

5.27 Water flows through 915 linear meters of 90-cm-diameter pipe that branches into 610 linear meters of 45-cm-diameter pipe and 730 linear meters of 60-cm-diameter pipe. These rejoin, and the water continues through 460 meters of 75-cm-diameter pipe. All pipes are horizontal, and the friction factors are 0.016 for the 90-cm diameter pipe, 0.017 for the 60-cm and 75-cm diameter pipes, and 0.019 for the 45-cm pipe. Find the pressure drop in pascals between the beginning and the end of the system if the steady flow is 1.70 m^3/sec in the 90-cm pipe. Disregard minor losses. Refer to Exhibit 5.27.

a. 280 kPa c. 320 kPa
b. 300 kPa d. 335 kPa

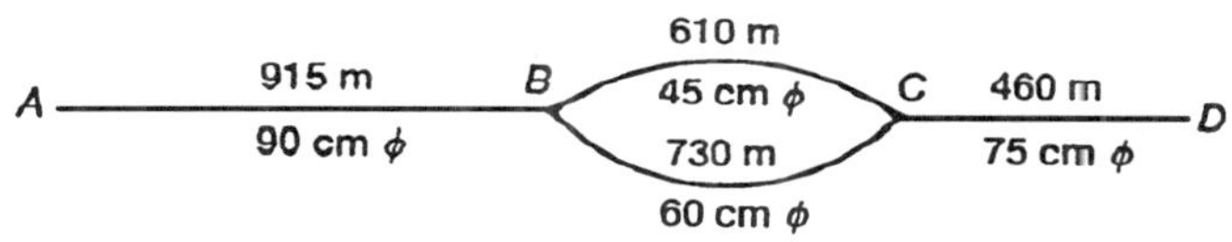

Exhibit 5.27

5.28 A piece of lead (sp. gr. 11.3) is attached to 40 cm^3 of cork (sp. gr. 0.25). When fully submerged the combination will just float in water. What is the mass of the lead?

a. 25 gm c. 30 gm
b. 27 gm d. 33 gm

5.29 A loaded timber 30 cm square and 1.83 meters long floats upright in fresh water with 60 cm of its length exposed. What will be the length of the timber projecting above the surface if the water is covered with a layer of oil 30 cm thick? The specific gravity of the oil is 0.8. Refer to Exhibit 5.29.

a. 54 cm c. 59 cm
b. 56 cm d. 61 cm

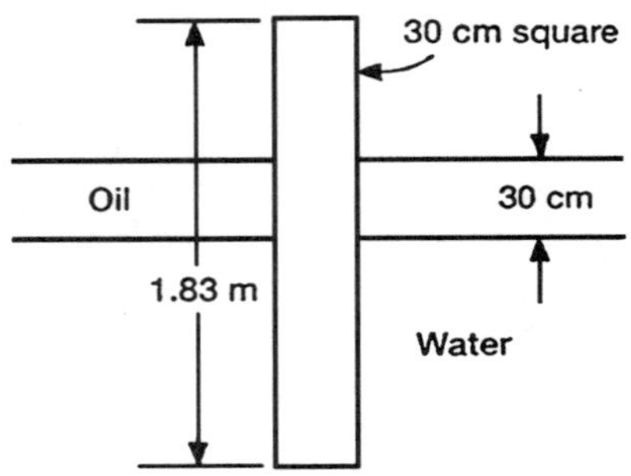

Exhibit 5.29

5.30 The outlet and inlet of a Venturi meter are each 10 cm in diameter, and the throat is 76 mm in diameter. The inlet velocity is 3.0 m/sec. The inlet has a static head of 3.05 meters of water. It there is no loss due to friction between the inlet and the throat of the meter, what will be the head in the throat?

a. 1.7 m c. 2.6 m
b. 2.1 m d. 2.9 m

5.31 A conduit having a cross section of an equilateral triangle of sides *b* has water flowing through it at a depth *b*/2. Find the "hydraulic radius."

a. 0.39*b* c. 0.24*b*
b. 0.32*b* d. 0.17*b*

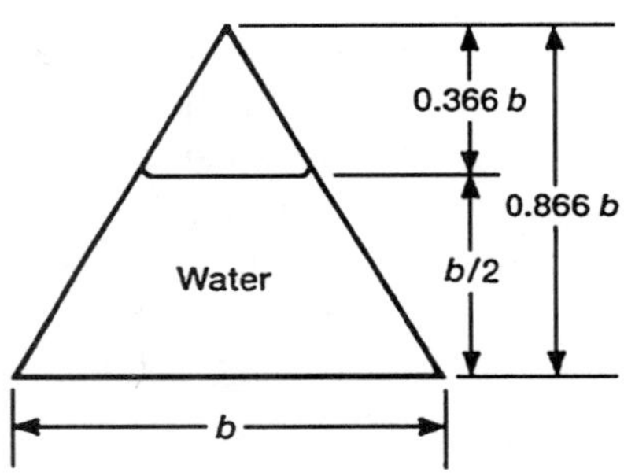

Exhibit 5.31

5.32 A mass of copper, suspected of being hollow, weighs 523 g in air and 447.5 g in water. If the specific gravity of copper is 8.92, what is the volume of the cavity, if any?

a. 11 cm^3 c. 17 cm^3
b. 14 cm^3 d. 21 cm^3

5.33 A 27-cm *OD* galvanized iron pipe 4.6 m long, having a mass of 47.6 kg/m, is closed at one end by a pipe cap with a mass attached so that the pipe will float upright. The cap and attached mass total 11.3 kg and displace 14.151 of water. How much of the pipe will show above the surface of seawater of 1.025 sp. gr.?

a. 0.69 m c. 0.79 m
b. 0.74 m d. 0.84 m

5.34 An oil pipeline with a 25-cm *ID* is 160 km long, and is made of welded steel. The discharge end is 150 meters above the intake. The rate of flow is 122 cm/sec. What is the total pressure in P pascals, gauge, at the intake when the whole length is full of gasoline at 16 °C and 0.68 sp. gr.? The absolute viscosity of the gasoline is 0.30 centipoise.

a. 5.1 MPa
b. 5.4 MPa
c. 5.9 MPa
d. 6.2 MPa

SOLUTIONS

5.1 **a.** Starting with a generalized grouping of the possible variables in terms of generic equation

$$\tau = f(\rho,\mu,V,D,R)$$

Selecting the characteristic dimensions for the above generic equation based on Newton's 2nd law we obtain

τ = Shearing stress	F/L^2
ρ = Density	M/L^3
μ = Viscosity	M/LT
V = Velocity	L/T
D = Diameter	L
R = Roughness	L/L

Using Newton's second law relationship

$$F = ML/T^2$$

Our equation becomes

$$(ML/T^2)/L^2 = M/LT^2 = C(M/L^3)^a(M/LT)^b(L/T)^c(L)^d(L/L)^e$$

Since the roughness relation is non dimensional it will drop out of the relationship. The only way that it will again appear is through experimentation.

Set up table for like exponents

M:	$1 = a + b$	$a = 1 - b$
L:	$-1 = -3a - b + c + d + e - e$	
T:	$-2 = -b - c$	$c = 2 - b$

Solving for d

$$d = -1 + 3(1 - b) + b - (2 - b) = -1 + 3 - 3b + b - 2 + b = -b$$

Thus

$$\tau = C(\rho)^{1-b}(\mu)^b(V)^{2-b}(D)^{-b}(R)^0$$

Grouping into terms of like exponents

$$\tau = C(\rho i^2)(\mu/\rho VD)^b = C(\text{Re})^{-b}\rho V^2$$

5.2 **c.** Selecting the equation from the *Handbook* for the coordinates of the center of pressure

$$Y^* = (\gamma I \sin\alpha)/(p_c A) + Y_c$$

For a vertical plate $\sin \alpha = 1.0$, and the equation may be rearranged to be

$$Y_p = Y_c + I/(Y_c A) = 15 \text{ m} + [(1/12)(bh^3)\text{m}^4/[(15)\text{m}(10 \times 10)\text{m}^2]$$
$$Y_p = 15 \text{ m} + [(1/12)(10)(10^3)]/1500 = 15.56 \text{ m}$$

5.3 **d.** Taking summation of moments about the hinge (A) we find

$$\Sigma M_A = Y_p' F - R(10) = (5.56 \text{ m})(\gamma A Y_c) - 10\,R = 0$$
$$(5.56 \text{ m})(1000 \text{ kg/m}^3)(9.81 \text{ m/s}^2)(10 \times 10 \text{ m}^2)(15\ Y_c m)(\text{N/kg}\cdot\text{m/s}^2) = (10\text{m})R$$
$$R = 8{,}182 \text{ kN}$$

5.4 **a.** To determine the friction factor of the pipe we need first to determine the velocity of flow using the continuity equation, calculate the Reynolds number, calculate the e/D ratio, and obtain the friction factor off the Moody diagram.

$$(AV)_1 = (AV)_2 = (1000 \text{ m}^3/\text{hr}) = (\pi D^2/4 \text{ m}^2)(V \text{ m/s})$$
$$V = [(1000 \text{ m}^3/\text{hr})/(3600 \text{ s/hr})]/[\pi(0.40)^2/4)\text{m}^2] = 2.21 \text{ m/s}$$

Therefore

$$\text{Re} = \rho DV/\mu = [(1000 \text{ kg/m}^3)(0.40 \text{ m})(2.21 \text{ m/s})]/1.005 \times 10^{-3} \text{ kg/m}\cdot\text{s}$$
$$\text{Re} = 8.796 \times 10^5$$

Turning to the Moody diagram in the *Handbook* we find that a galvanized iron pipe has a roughness of 0.15 mm; therefore, $e/D = 0.15 \text{ mm}/400 \text{ mm} = 0.000375$.

The friction factor from the Moody diagram is:

$$f = 0.014$$

5.5 **d.** The work input needed to cause the above flow rate may be determined by using the Bernoulli equation simplified for this problem.

$$\text{Pump Work} = Z_2 - Z_1 + (f\,L/D)V^2/2g$$
$$= 75 \text{ m} + [(0.014)(5000 \text{ m})/0.4 \text{ m}][(2.21)^2 \text{ m}^2/\text{s}^2]/(2)(9.81 \text{ m/s}^2)$$
$$= 118.6 \text{ m}$$

5.6 **d.** The power required may be calculated using the pump power equation in the *Handbook*.

$$W = (Q\gamma h)/\eta = [(1000 \text{ m}^3/\text{hr})/(3600 \text{ s/hr})](1000 \text{ kg/m}^3) \times (9.81 \text{ m/s}^2)(118.6 \text{ m})/0.70$$
$$W = (461.7 \text{ kW})(1.341 \text{ hp/kW}) = 619 \text{ hp.}$$

5.7 **b.** It is simplest to make ratios using the previous calculations to determine the new hp needed.

$$V = (300/1000)(2.21 \text{ m/s}) = 0.663 \text{ m/s}$$

Therefore

$$\text{Re} = 0.3(8.796 \times 10^5) = 2.639 \times 10^5$$

Using the same e/D ratio and the new Re, a new friction factor can be selected

$$f = 0.0175$$

Calculating the new pump work in head of water added to the system

$$W = 75 + (0.0175)(5000/0.40)(0.663)^2/2(9.81) = 79.9 \text{ m}$$
$$\text{Power} = [(300/3600)(1000)(9.81)(79.9)/0.70](1.341)/1000 = 125 \text{ hp}$$

5.8 **d.** The friction will be a function of the Reynolds number

$$\text{Re} = \rho D v/\mu$$

where

$$\rho = 0.94 \times 1{,}000 = 940 \text{ kg/m}^3$$
$$D = 0.300 \text{ m}$$
$$A = 0.0707 \text{ m}^2$$
$$Q = 3{,}000/(24 \times 3{,}600) = 0.0347 \text{ m}^3/\text{sec}$$
$$v = 0.0347/0.0707 = 0.491 \text{ m/s}$$
$$\mu = 201 \times 0.001 = 0.201 \text{ Pa}\cdot\text{s}$$
$$\text{Re} = 940 \times 0.300 \times 0.0471/(0.201/g) = 648$$

This indicates that the flow will be well into the laminar-flow range so $f = 64/\text{Re} = 64/648 = 0.0988$

5.9 **d.** A floating body displaces a mass of fluid equal to its own mass, so the loaded ship would displace an additional mass of water equal to 3500 metric tons, which would equal—

$$(3500 \times 1000 \text{ kg})/(1000 \text{ kg/m}^3) = 3500 \text{ m}^3$$

The area of the lock = $140 \times 15 = 2100 \text{ m}^2$

But the area of the ore carrier is equal to $120 \times 10 = 1200 \text{ m}^2$, so the net open area of water is only equal to $2100 - 1200 = 900 \text{ m}^2$, assuming the ore carrier's area remains constant as it sinks farther into the water. The water would thus rise 3500/900 = 3.889 meters.

Another way of looking at this problem is that the ore carrier would sink an amount equal to 3500/1200 = 2.917 meters, which would displace a volume of water in the lock equal to $2.917 \times 1200 = 3500 \text{ m}^3$. The total volume of water in the lock would remain constant, so, as the ore carrier sank, the level of the water in the lock would rise. The total volume of water plus the submerged portion of the ore carrier would increase by 300 m^3. The water would then rise in the area outside the ore carrier but inside the walls of the lock. The water would rise 3500/(2100 – 1200) = 3.889 meters.

5.10 **c.** The velocity of the water out of the tank will be equal to

$$v = C\sqrt{(2gh)} = \sqrt{(2)(9.8066)(9.00)} = 13.286 \text{ m/s}$$

The jet is 1.00 m above the ground.

$$h = v_0t + {}^1\!/_2at^2$$

where a is the acceleration of gravity.

For this case, the vertical velocity at time zero is zero. The time for the water to fall 1.00 m is

$$t = (2h/g) = 0.452$$

The distance from the base of the tank the water will strike is

$$s = 0.452 \times 13.286 = 6.00 \text{ m}$$

5.11 **b.** Apply Bernoulli's equation

$$P_1/\rho g + v_1^2/2g + z_1 + h_A = P_2/\rho g + v_2^2/2g + z_2 + h_L$$

between the surface of the reservoir and the discharge from the pipe. There is no energy added, and the velocity of the surface of the water is zero. The pressures at the surface and at the pipe discharge are both atmospheric, so they cancel. The only term remaining on the left side of the equation is the elevation of the surface of the reservoir. The difference in elevation between the surface of the reservoir and the outlet of the pipe, is equal to $Z_1 - Z_2 = 150$ m. On the right hand side of the equation would be the velocity term, $v^2/2g$, and the work done in overcoming pipe friction

$$h_L = fL/D \times v^2/2g$$
$$fL/D = 0.02 \times 200/0.30 \times v^2/2g = 13.33\ v^2/2g$$

so,

$$150 = (1 + 13.33)v^2/2g$$
$$v^2/2g = 10.47 \text{ m, and}$$
$$v = 14.33 \text{ m/s}$$
$$Q = (14.33 \text{ m/s})(0.0707 \text{ m}^2) = 1.013 \text{ m}^3\text{/s}$$

5.12 **d.** The glass displaces 125 – 75 = 50 grams = 50 cm^3 of water. 50 cm^3 of gasoline has a mass of 125 – 92 = 33 g; therefore, the specific gravity of the gasoline is equal to 33/50 = 0.66.

5.13 **c.** See Exhibit 5.13.

$$\text{Volume of timber submerged} = (0.3 - 0.11)(0.3)(3.60) = 0.205 \text{ m}^3$$
$$\text{Mass of timber} = (0.205)(1000) = 205 \text{ kg}$$

Take moments about the point touching rock

$$\Sigma M_R = [(-205)(3.6/2) - 68D] + [(3.6/2)(205)] + [(0.667 \times 3.6)(^1/_2)(0.30)(0.11)(3.6)(1000)] = 0$$
$$369 + 68D = 369 + 142.6$$
$$D = 142.6/68 = 2.10 \text{ meters}$$

The mass of the timber and the mass of the man would produce moments in the clockwise direction. These would be opposed by counterclockwise moments due to the buoyancy of the timber and the triangular section shown by the dotted line. The moment arm of the buoyancy of the triangular section would be equal to 2/3 of the distance from the apex to the base, 2.4 meters. The buoyancy would be equal to the volume of water displaced times the density of water, $0.0594 \times 1000 = 59.4$ kg. This buoyancy times the moment arm would be equal to 142.6 kg·m. The small amount of difference in the length of the moment arm introduced by the cosine of angle θ (0.9995) would not be enough to markedly affect the accuracy of the answer.

5.14 **a.** This problem can best be solved with the aid of Bernoulli's equation

$$P_1/\rho g + v_1^2/2g + z_1 + h_A = P_2/\rho g + v_2^2/2g + z_2 + h_L$$

where

P_1 and P_2 are zero
z_1 is equal to zero
z_2 is equal to 137 meters
v_1 is equal to zero
h_L is a function of the Reynolds number and the type of pipe.

$$A = 0.0314 \text{ m}^2$$
$$v = 0.085/0.0314 = 2.707 \text{ m/sec}$$
$$v^2/2g = 0.3736 \text{ m}$$
$$\text{Re} = \rho D v/\mu$$
$$\mu = 0.01 \text{ poise, or } \mu = 0.001 \text{ kg/sec·m}$$
$$\rho = 1000 \text{ kg/m}^3$$
$$D = 0.20 \text{ m}$$
$$\text{Re} = (1000)(0.20)(2.707/0.001) = 541{,}400$$
$$\text{Head loss} = f(L/d)(v^2/2g)$$

where f is the the friction factor for Re = 5.4×10^5

From the *FE Handbook,* for cast iron pipe

$$e/D = 0.25/200 = 0.00125$$
$$f = 0.0205$$
$$h_L = 0.0205(1{,}070/0.200)(0.3736) = 40.97 \text{ m}$$

The head to be supplied by the pump (ignoring the velocity head and minor losses) is equal to

$$40.97 + 137 = 177.97, \text{ or } 178 \text{ m}$$

The required output pressure is thus equal to

$$178\rho g = 178(1000)(9.807) = 1{,}745{,}000 \text{ Pa, or } 1.75 \text{ MPa}$$

5.15 **b.**

Orifice area, $A = 0.10^2(\pi/4) = 0.007854 \text{ m}^2$ The head is 15 meters.

$$Q = 0.70(\sqrt{2gh})A$$
$$Q = 0.70A\sqrt{(2)(9.807)(15)} = 0.094 \text{ m}^3\text{/s}$$

5.16 **d.** Again, we should look at Bernoulli's equation

$$P_1/\rho g + v_1^2/2g + z_1 + h_A = P_2/\rho g + v_2^2/2g + z_2 + h_L$$

The differential head is equal to 3.6 – 2.4 = 1.2 meters, which must be supplied by the pump. The pump must also supply the frictional head loss in the hose, plus one-half of the velocity head at the entrance to the brush, plus one velocity head. Flow area is equal to

$$0.05^2(\pi/4) = 0.001964 \text{ m}^2$$
$$Q = (225/60) = 3.75 \text{ l/s} = 0.00375 \text{ m}^3\text{/s}$$
$$v = 1.909 \text{ m/s}$$
$$v^2/2g = 0.186 \text{ m}$$

To calculate the frictional head loss it is necessary to first calculate the Reynolds number

$$\text{Re} = \rho D v/\mu$$

where

$D = 0.05$ m

$\mu = 0.01$ poise $= 0.001$ kg/s · m

$$\text{Re} = 1000(0.05)(1.909/0.001) = 95{,}450$$

The rubber hose would have a smooth *ID* similar to drawn tubing, so *e* can be taken as 0.0015 mm. The relative roughness would be equal to 0.0015/50 = 0.00003 and $f = 0.018$ from the chart in the *Handbook* for this Re and *e*/*D*. The frictional head loss would be equal to

$$h_L = f(L/D)(v^2/2g) = 0.018(15/0.05)(0.186) = 5.40(0.186) = 1.004 \text{ m}$$

Adding 1.5 velocity heads, 0.279 m, and the increase in elevation of 1.2 m yields a total suction requirement of 1.00 + 0.28 + 1.20 = 2 48 m of water.

$$P_1/\rho g - P_2/\rho g = -2.48 \text{ m}$$
$$P_1 - P_2 = -1000(9.807)(2.48) = -24.3 \text{ kPa}$$

5.17 **b.** Again, check Bernoulli's equation

$$qP_1/\rho g + v_1^2/2g + z_1 + h_A = P_2/\rho g + v_2^2/2g + v_2^2/2g + z_2 + h_L$$

In this case there will be frictional line losses, one velocity head in the line, and a loss of 0.10 velocity head through the nozzle.

Area of the pipe = $0.15^2(\pi/4) = 0.0177\ m^2$
Area of nozzle = $0.00196\ m^2$

$$\text{Velocity through pipe, } v = 0.028/0.0177 = 1.582\ \text{m/s}$$
$$v^2/2g = 0.1276\ \text{m}$$
$$\text{Re} = \rho Dv/\mu = 1000(0.15)(1.582/0.001) = 237{,}300$$
$$e/D = 0.046/150 = 0.0003$$

From the curve in the *Handbook*, $f = 0.017$ (welded steel pipe is the same as commercial steel pipe).

Frictional head loss $= f(L/D)(v^2/2g) = 0.017(25/0.15)(0.1276) = 0.362$ m

For the nozzle

$$v = 0.028/0.00196 = 14.286\ \text{m/s}$$
$$v^2/2g = 10.405\ \text{m}$$

The nozzle has a discharge coefficient of 0.9, so the head required at the nozzle would be equal to

$$10.405/0.9 = 11.561\ \text{m}$$

The total head required to offset the total head loss would thus be equal to

$$11.561 + 0.362 + 0.1276 = 12.051\ \text{meters}$$

5.18 **d.** The potential energy of the water

$$(57/60)(1000)(30) = 28{,}500\ \text{kg}\cdot\text{m/s}$$
$$28{,}500(9.807) = 279{,}500\ \text{kg}\cdot\text{m}^2/\text{s}^3 = 279{,}500\ \text{N}\cdot\text{m/s, or } 279\ \text{kW}$$

Energy contained in discharge

$$KE = {}^1/_2\ mv^2$$

For each kg

$KE = {}^1/_2(1)(3.70^2) = 6.845(\text{N}\cdot\text{sec}^2/\text{m}) \times \text{m}^2/\text{sec}^2 = 6.845$ J energy leaving the turbine

For (57/60)(1000) kg/s, this becomes 6.50 kW of energy lost in the discharged water.

Power to overcome friction = $(1.8\ \text{kN})(0.30)(2\pi)(120/60)/\text{s} = 6.79$ kN/s, or 6.79 kW

So, the power developed by the generator is

$$P = 279 - 6.50 - 6.8 = 265.7\ \text{kW}$$

5.19 **c.** Again, we resort to Bernoulli's equation

$$P_1/\rho g + v_1^2/2g + z_1 + h_A = P_2/\rho g + v_2^2/2g + z_2 + h_L$$

The pressure difference between the throat and the line is 25 kPa = 25 kN/m^2.

The density of the oil is 940 kg/m^3, so

$$\rho g = (940)(9.807) = 9219 \text{ kg/m}^3$$
$$25{,}000/9219 = 2.712 \text{ meters of oil pressure difference.}$$

So Bernoulli's equation reduces to—

$$2.712 + v_1^2/2g = v_2^2/2g$$
$$v_2/v_1 = (10/5)^2$$

or

$$v_2 = 4v_1$$
$$(16 - 1)(v_1^2/2g) = 2.712$$
$$v_1 = \sqrt{[(2)(9.807)(2.712/15)]} = 1.883 \text{ m/s}$$
$$Q = (1.883)(0.10^2)(\pi/4) = 0.0148 \text{ m}^3/\text{s}$$

or

$$Q = (14.8)(60) = 888 \text{ l/min}$$

5.20 **b.** See Figure 5.8

$$\text{Moment about hinge, } M = \int y dF$$
$$dF = (1000)(9.807)(1.22)(6.4 + y)dy$$
$$dF = 76{,}573 + 11{,}965y$$
$$M = \int(76{,}573 \cdot y + 11{,}965 \cdot y^2)dy, \text{ between } y = 0 \text{ and } y = 1.83 \text{ m}$$
$$M = (76{,}573/2)(1.83^2) + (11{,}965/3)(1.83^3) = 128{,}217 + 24{,}442$$
$$= 152{,}659 \text{ N}\cdot\text{m moment about the hinge}$$

The resulting horizontal force at the bottom of the gate is then

$$152{,}659/1.83 = 83{,}420 \text{ N, or } 83.4 \text{ kN}$$

5.21 **c.** From the Moody Diagram in the *Handbook* it can be seen that the internal roughness of a pipe has no effect on the friction factor for laminar flow. For Reynolds numbers of 10,000 and above the friction factor is definitely affected by the relative roughness of the inside of the pipe. The curves for relative roughness show that for high Reynolds numbers the internal roughness of a conduit can have an appreciable effect on the friction factor, and for a Reynolds number of 100,000 internal roughness will cause a higher friction factor than would be the case for an uncorroded pipe. The internal roughness of the pipe does not enter into the determination of the Reynolds number.

5.22 **a.** The vertical pressure distribution behind the dam acting on its face will be triangular, ranging from zero at the top to

$$P = (1000\ \text{kg/m}^3)(10\ \text{m deep})(9.807) = 98.07\ \text{kPa}$$

at the base. This will produce a triangle of force acting on the face of the dam equal to

$$F = (\tfrac{1}{2})(10)(98.07 \times 10^6)\ \text{N}$$

which will produce a moment of

$$M = (490.35)(10/3) = 1.635\ \text{MN/m}$$

of width, or 1.635 MJ/m. (The centroid of a triangle is one-third the distance from the base to the apex, and this is the point at which the total force can be considered to act.)

5.23 **b.** The hydraulic radius, R_H, is equal to the cross-sectional area of the liquid divided by the wetted perimeter.

$$\text{Area} = (\tfrac{1}{2})(D^2)(\pi/4) = 1.5708\ \text{m}^2$$
$$\text{Wetted perimeter} = \pi D/2 = 3.1416\ \text{m}$$
$$R = 0.500\ \text{m}$$

5.24 **a.** See Exhibit 5.24

$$dV = \pi r^2 dh = \pi h dh,\ \text{since}\ r = \sqrt{h}$$
$$dV/dt = C_D a\sqrt{(2gh)}$$
$$a = (0.025^2)\pi/4 = 0.000491$$
$$dV/dt = (0.60)(0.000491)(4.429)\sqrt{h} = 0.00130\sqrt{h}\ \text{m}^3/\text{s}$$

Volume of water in tank (paraboloid of revolution), $V = (\tfrac{1}{2})\pi r^2 h = (\tfrac{1}{2})\pi h^2$

$$dV = \pi h\ dh$$
$$dV/dt = \pi h\ dh/dt = 0.00130 h^{1/2}$$

Combining terms gives

$$h^{1/2}\ dh = 0.000414t$$

Integrating both sides gives

$$(\tfrac{2}{3})\ h^{3/2} = 0.000414t + \text{constant} = \tfrac{2}{3}$$

Calculating from $h = 3$ to $h = 0$ gives:

$t = (0.667)(5.196/0.000414) = 8371$ s, or 140 min = 2 hr and 20 min

5.25 **c.** Check the values in Bernoulli's equation

$$P_1/\rho g + v_1^2/2g + z_1 + h_A = P_2/\rho g + v_2^2/2g + z_2 + h_L$$
$$P_1/\rho g = P_2/\rho g = 0$$
$$z_1 - z_2 = -76.0\ \text{m}$$
$$h_A = v_2^2/2g + z_2 + h_L = \text{the head added to the fluid}$$

$$\text{Flow area} = (0.30^2)(\pi/4) = 0.0707 \text{ m}^2$$
$$v = 0.085/0.0707 = 1.202 \text{ m/s}$$
$$v^2/2g = 0.0737 \text{ m}$$
$$\mu = 0.01 \text{ poise} = 0.001 \text{ kg/s} \cdot \text{m}$$
$$\text{Re} = \rho Dv/\mu = (1000)(0.30)(1.202/0.0010) = 360{,}600$$

From the chart in the *Handbook*

$$e/D = 0.25/300 = 0.00083$$

For this relative roughness and the calculated Reynolds number, a friction factor of $f = 0.0195$ is obtained from the Moody diagram in the *FE Handbook*, The head loss is thus equal to

$$h_L = (0.0195)(1220/0.30)(0.0737) = 5.844 \text{ m}$$

The total head to be supplied to the water is equal to

$$0.0737 + 76.0 + 5.844 = 81.9 \text{ meters}$$

The power to be added to the water is equal to

$$(0.085)(1000)(9.807)(81.9) = 68{,}271 \text{ W or } 68.3 \text{ kW}$$

The power to be supplied to the pump would be equal to 68.3/0.89 = 76.7 kW.

5.26 **c.** The velocity at the center of the tube, $v = \sqrt{(2gh)}$

$$h = 0.10/0.9 = 0.111 \text{ m}$$
$$\text{Velocity at the center of the tube} = \sqrt{(2 \times 9.807 \times 0.111)} = 1.476 \text{ m/sec}$$
$$\text{Re} = \rho Dv/\mu$$

where μ is the viscosity = (0.10)(0.167) = 0.0167 kg/s·m.

$$\text{Re} = (900)(0.025)(1.476/0.0167) = 1989$$

so the flow is in the laminar range. Flow in the laminar range is parabolic, so the average velocity is one-half the maximum velocity at the center of the tube. The average velocity is thus equal to 1.476/2 = 0.738 m/s.

5.27 **b.** The problem of split-flow, or flow through two branches, is discussed in the *Handbook*. The pressure drop, or head loss, must be the same through each branch, so the drop through the 45-cm-diameter pipe and that through the 60-cm-diameter pipe will be the same. The instructions say to disregard minor losses, so the losses due to the fittings will be ignored, and only the frictional losses will be calculated. $Q = 1.70 \text{ m}^3\text{/s}$

See Exhibit 5.27

The velocity from A to B, $v = 1.70/0.636 = 2.672$ m/s

$$L/D = 915/0.90 = 1017$$
$$v^2/2g = 0.364 \text{ m}$$
$$h_L = (0.016)(1017)(0.364) = 5.923 \text{ m}$$

From B to C, top branch

$$h_L = (0.019)(610/0.45)(v^2/2g) = 1.313\ v_T^2$$

From B to C, bottom branch

$$h_L = (0.017)(730/0.60)(v^2/2g) = 1.055 v_B^2$$

The flow splits between the top and bottom branches so

$$1.70 = v_T(0.159\ m^2) + v_B(0.283\ m^2)$$

which gives

$$v_T = 10.692 - 1.780\ v_B$$

The head loss through the top branch is equal to the head loss through the bottom branch, so

$$1.313\ v_T^2 = 1.055\ v_B^2$$

Taking square roots of both sides gives

$$v_T = 0.896 v_B$$

so,

$$0.896\ v_B = 10.692 - 1.780\ v_B$$
$$v_B = 3.996\ m/s$$

and

$$v_T = 3.580\ m/s$$
$$\text{Head loss, top branch} = (1.313)(12.816) = 16.83\ m$$
$$\text{Head loss, bottom branch} = (1.055)(15.968) = 16.85\ m$$
$$\text{Head loss from C to D} = (0.017)(460/0.75)(v^2/2g) = 0.532v^2$$
$$A = (0.563)(\pi/4) = 0.442\ m^2$$
$$v = 1.70/0.441 = 3.848\ m$$
$$v^2 = 14.807$$
$$h_L = 7.88\ m$$
$$\text{Total head loss} = 5.92 + 16.84 + 7.88 = 30.64 \text{ meters of water}$$
$$\text{Pressure drop} = (30.64)(1000)(9.807) = 300{,}486\ Pa, \text{ or } 300\ kPa$$

5.28 **d.** Assume that the lead and the cork, together, displace $(40 + C)$ cm^3 of water. The cork has a mass of 10 g, so its mass will account for (40)(0.25) g = 10 g of the displacement.

The lead would add C cm^3 to the volume displaced

$$(C\ cm^3)(11.3) = 11.3C\ cm^3$$
$$(40 + C)\ cm^3 = (10 + 11.3C)\ gram$$

Since one cc of water has a mass of one gram,

$$40 - 10 = (11.3 - 1)C$$
$$C = 30/10.3 = 2.913\ cm^3$$
$$\text{Mass of lead} = (2.913)(11.3) = 32.92\ g$$

5.29 **a.** See Exhibit 5.29

$$1.83 - 0.60 = 1.23 \text{ m submerged}$$

$$\text{Volume submerged} = (0.30)(0.30)(1.23) = 0.111 \text{ m}^3$$

$$\text{Mass of timber} = (0.111)(1000) = 111 \text{ kg} = \text{mass of water displaced.}$$

The same mass of liquid would be displaced when the oil layer was added.

$$[(0.30)(800) + L(1000)](0.090) = 111 \text{ kg}$$
$$L = 0.993 \text{ m}$$

1.293 meters are immersed, and 1.83 – 1.293 = 0.537 m, or 54 cm, of timber projects above the surface of the oil.

5.30 **b.** Check Bernoulli's equation

$$P_1/\rho g + v_1^2/2g + z_1 + h_A = P_2/\rho g + v_2^2/2g + z_2 + h_L$$

Reviewing the factors in the equation we see that it reduces to

$$P_1/\rho g + v_1^2/2g = P_2/\rho g + v_2^2/2g$$

The difference in the pressures at the two points is equal to the difference in the velocity heads.

$$v_1/v_2 = (7.6/10.0)^2 = 0.578$$
$$v_2^2 = 2.993v_1^2$$
$$v_1 = 3.0 \text{ m/s}$$
$$v_2 = 3.0/0.578 = 5.190 \text{ m/s}$$

The difference in the velocity heads is then equal to

$$(26.936 - 9.00)/2g = 0.914 \text{ meters}$$

The head in the throat is equal to 3.05 – 0.914 = 2.14 meters

Or, the difference in the two heads is

$$\Delta h = v_2^2/2g - v_1^2/2g$$
$$= (2.993 - 1.00)(9.00)/2g = 0.914 \text{ m,}$$

The head in the throat is equal to 2.14 m, as before.

5.31 **d.** The hydraulic radius is equal to the area of flow divided by the wetted perimeter.

Since this is an equilateral triangle, all of the internal angles are 60°. The height of the triangular duct would be

$$H = b(\sin 60°) = 0.866b$$

The width of the flow surface would be

$$W = (0.366/0.866)b = 0.423b$$

The area of the trapezoidal cross-section would be

$$A = [(0.423b + b)/2](b/2) = 0.356b^2$$

The wetted perimeter would be

$$P = b + 2(b/2)/\sin 60° = 2.155b$$
$$R_H = 0.356b^2/2.155b = 0.165b$$

5.32 **c.** The volume of water displaced is

$$V_W = 523 - 447.5 = 75.5 \text{ cm}^3$$

If it had been solid copper the volume of the copper would have been

$$V_C = 523/8.92 = 58.6 \text{ cm}^3$$

The volume of the void is, therefore

$$V_V = 75.5 - 58.6 = 16.9 \text{ cm}^3$$

5.33 **a.** The mass of the length of pipe is

$$M_P = (4.6)(47.6) = 218.96 \text{ kg}$$

Adding the mass of the cap and the attached mass gives a total mass of

$$M_T = 218.96 + 11.3 = 230.26 \text{ kg}$$

The pipe assembly would therefore displace a 230.26 kg mass of seawater, or 230,260/1.025 = 223,644 cm^3.

The area of the *OD* of the pipe is

$$A = \pi(27/2)^2 = 572.6 \text{ cm}^2$$

The pipe would therefore sink 223,644/572.6 = 390.6 cm into the seawater, leaving 4.6 − 3.91 = 0.69 m or 69 cm of the pipe extending above the surface.

5.34 **c.** Apply Bernoulli's equation

$$P_1/\rho g + v_1^2/2g + z_1 + h_A = P_2/\rho g + v_2^2/2g + z_2 + h_L$$

Simplifying the equation for the pertinent factors

$$P_1/\rho g = v^2/2g + z_2 + h_L$$

The head loss is

$$h_L = f(L/D)(v^2/2g)$$
$$v^2/2g = (1.22^2)/2g = 0.0759 \text{ m}$$
$$L/D = 160{,}000/0.25 = 640{,}000$$

The friction factor is a function of both the Reynolds number and the type of pipe.

$$\mathrm{Re} = \rho D v/\mu$$

$\mu = 0.30$ centipoises or 0.0030 poise, which is equal to 0.00030 kg/s · m

$\rho = 680\ \mathrm{kg/m^3}$

$\mathrm{Re} = (680)(0.25)(1.22/0.00030) = 691{,}333$

For welded-steel pipe the roughness would be the same as that for commercial steel pipe, which is equal to 0.046 mm (from the *Handbook*).

$$e/D = 0.0046/25 = 0.000184$$

From the Moody Diagram in the *Handbook*, $f = 0.15$.

$$h_L = (0.015)(640{,}000)(0.0759) = 728.6\ \mathrm{m}$$

Adding one velocity head gives a total head H of 729 + 150 = 879 m ($z_2 = 150$ m).

$$P_1 = \rho g H = (680)(9.807)(879) = 5{,}861{,}840\ \mathrm{Pa,\ or\ 5.86\ MPa}$$

CHAPTER 6

Fans, Pumps, and Compressors

OUTLINE

FANS

A fan is a device that moves gasses or vapors from one location to another. Since fans are usually low-velocity devices, the gas, which usually is air, can be considered to be incompressible for the majority of engineering calculations. The general characteristics of fan operation are the following:

a) Volumetric fan output varies directly with the fan speed of rotation for a given fan.

b) The pressure, or head, of the fan varies directly with the square of the speed of the fan.

c) The power required to run a given fan varies directly with the cube of the speed.

d) For a given installation and constant fan speed, the pressure output and the operating power required will be proportional to the density of the gas.

e) For a constant mass flow rate of gas, the fan speed, the volumetric output, and the pressure vary inversely with the density of the gas. In addition, the power required varies inversely with the square of the density of the gas.

f) At a constant pressure, the speed, volumetric output, and power vary inversely with the square root of the density of the gas.

Example 6.1

A coal-fired boiler requires 1700 m^3/min of air at a pressure of 15 cm of water for the combustion process. The fan has a mechanical efficiency of 58% at the given conditions. How large of a motor, in kW, is required to power this fan?

Solution

Generally, we may conclude that fans are low-pressure devices and, as a consequence, that the air is incompressible. The power output, we learned back in thermodynamics, is equal to pressure times the volume rate of flow.

$$\text{Power} = \text{Pressure} \times \text{Volume rate of flow}$$

where

$$\text{Pressure} = \rho g h = (1000 \text{ kg/m}^3)(9.81 \text{ m/s}^2)(0.15 \text{ m}) = 1471 \text{ N/m}^2$$

Therefore

$$\text{Power} = (1471 \text{ N/m}^2)(1700 \text{ m}^3\text{/min})/(60 \text{ s/min}) = 41{,}678 \text{ N}\cdot\text{m/s}$$
$$\text{Power} = 41.7 \text{ kW}$$

and

$$\text{Motor kW} = (\text{Air power})/\eta = 41.7/0.58 = 71.9 \text{ kW motor needed.}$$

Example 6.2

An existing ventilating fan delivers 400 m^3/min of air against a back pressure of 5 cm water when the fan is operating at 450 rpm. The plant wishes to increase the air flow to 500 m^3/min.

Determine

a) Air power produced by the fan

b) New speed to obtain the increased air flow

c) The new back pressure

d) The new air power at the increased flow rate

Solution

a) Knowing that power is equal to pressure times volume rate of flow we get

$$\text{Power} = (\rho g h)(\text{Volume flow rate}) = (1000 \text{ kg/m}^3)(9.81 \text{ m/s}^3)(0.05 \text{ m})(400 \text{ m}^3\text{/min})$$
$$\text{Power} = 196{,}200 \text{ kg}\cdot\text{m}^2\text{/s}^2\cdot\text{min} = 196{,}200 \text{ N}\cdot\text{m/min}$$
$$\text{Power} = (196{,}200 \text{ N}\cdot\text{m/min})/60 \text{ s/min} = 3270 \text{ N}\cdot\text{m/s or J/s}$$

Therefore

$$\text{Power} = 3.27 \text{ kW}$$

b) Knowing that the volumetric air flow rate is directly proportional to the fan speed we get

$$(\text{Speed})_{\text{new}}/(\text{Speed})_{\text{old}} = (\text{Flow rate})_{\text{new}}/(\text{Flow rate})_{\text{old}}$$
$$(\text{Speed})_{\text{new}} = [(500 \text{ m}^3\text{/min})/(400 \text{ m}^3\text{/min})](450 \text{ rpm})$$
$$(\text{Speed})_{\text{new}} = 563 \text{ rpm}$$

c) Knowing that the back pressure is directly proportional to the square of the speed of the fan

$$(\text{Pressure})_{new}/(\text{Pressure})_{old} = [(\text{Speed})_{new}/(\text{Speed})_{old}]^2$$
$$(\text{Pressure})_{new} = [(563 \text{ rpm})/(450 \text{ rpm})]^2(5 \text{ cm}) = 7.83 \text{ cm}$$

d) Using the equation for power, Power = Pressure × Volume flow rate, at the new conditions

$$\text{Power} = (1000 \text{ kg/m}^3)(9.81 \text{ m/s}^2)(0.0783 \text{ m})(500 \text{ m}^3/\text{min})$$
$$= 384{,}062 \text{ N}\cdot\text{m/min} = 6.4 \text{ kW}$$

A second method is to use the power law, which is

$$\text{Power}_2 = \text{Power}_1\left(V_2/V_1\right)^3 = 3.27(563/450)^3 = 6.4 \text{ kW}$$

which is the same value as calculated by the first method.

Example 6.3

A blower delivers 350 m^3/min of air at standard atmosphere and 20°C when rotating at 500 rpm, with a back pressure of 5 cm of water gage pressure. If the air temperature rises to 90°C and the speed of the fan remains constant, determine:

a) New back pressure in cm of water

b) Ratio of the new power required to the original power required

Solution

We should examine these changes in light of the fan laws. The volumetric flow rate remains constant because the speed of rotation is unchanged. We find, however, that the pressure head is proportional to the density of the gas and that the operating power is also proportional to the density of the gas.

a) Using the perfect gas law, written in terms of density

$$\rho = P/RT$$

Letting the atmospheric pressure remain the same along with the gas constant

$$\rho_1/\rho_2 = [P/RT]_1/[P/RT]_2 = T_2/T_1 = (90 + 273)/(20 + 273) = 1.24$$

The new density is

$$\rho_2 = \rho_1/1.24$$

Therefore, using the fan laws

$$P_2 = P_1/(\rho_2/\rho_1) = 5 \text{ cm}(\rho_1/1.24)(1/\rho_1) = 4.03 \text{ cm } H_2O$$

b) Using the density ratio calculated above

$$[\text{Power}_2/\text{Power}_1] = \rho_2/\rho_1 = 1/1.24 = 0.806$$

This means that the power requirement of the 90°C air is only 80.6% of the required power needed if the air were at 20°C.

Example 6.4

Using the same fan installation as given in Example 6.3, it is desired to maintain the same mass flow rate of air at 90 °C as the mass flow rate at 20 °C. Determine

a) New fan speed

b) New volumetric flow rate

c) New back pressure

d) New power required

Solution

Going back to the fan laws it is noted that, for a constant mass flow rate the fan speed, the fan volumetric flow rate and the back pressure vary inversely with the density of the gas. In addition, the power required varies inversely with the square of the density of the gas.

a) New fan speed = (Old fan speed)(ρ_1/ρ_2) = (500 rpm)(1.24) = 620 rpm

b)
$$\text{New volumetric flow rate} = (\text{Old volumetric flow rate})(\rho_1/\rho_2) = (350\ \text{m}^3/\text{min})(1.24) = 434\ \text{m}^3/\text{min}$$

c)
$$(\text{New back pressure}) = (\text{Old back pressure})(\rho_1/\rho_2) = (5\ \text{cm})(1.24) = 6.2\ \text{cm}$$

d)
$$\begin{aligned}\text{New power required} &= \text{Old power required } (\rho_1/\rho_2)^2 \\ &= [(\rho g h_1)(\text{Volumetric flow rate})](1.24)^2 \\ &= (1000\ \text{kg/m}^3)(9.81\ \text{m/s}^2)(0.05\ \text{m})(350\ \text{m}^3/\text{min})(1.24)^2 \\ &= 263{,}967\ \text{N}\cdot\text{m/min} = 4.4\ \text{kW}\end{aligned}$$

PUMPS

A pump is a device that is used to increase the pressure of a liquid. Pumps fall into different categories, such as axial-flow pumps, centrifugal-flow pumps, piston pumps, and other types of positive-displacement pumps. A typical centrifugal-flow pump has the liquid entering at the central portion of the pump, or hub, and then the liquid is accelerated outward by means of the rotating impeller. Some of the kinetic energy imparted to the liquid results in a pressure rise of the liquid. The efficiency of a typical pump is equal to

$$\eta_p = (\text{Fluid power out of the pump})/(\text{Power input to the pump})$$

Example 6.5

A 12-kW pump running at 1500 rpm has an inlet diameter of 25 cm, and the discharge line is 15 cm in diameter. The output from the pump is 3100 l/min of 20 °C water at 110 kPa. The pressure at the suction side is a negative 12 cm of mercury and the centerline of the pump discharge is 1 meter above the centerline of the intake pipe. Determine the efficiency of the pump.

Solution

Write the Bernoulli equation in order to find the energy added to the water. The efficiency may then be obtained using the above mentioned efficiency equation.

$$(P/\rho g)_1 + Z_1 + V_1^2/2g + E = (P/\rho g)_2 + Z_2 + V_2^2/2g + F$$

where, in addition to the normal terms for the Bernoulli equation covered in the *Fluid Mechanics* chapter, E = Energy added to the system F = Friction loss in the system, which is equal to zero in this case

Solving for each term of the Bernoulli equation

$$\begin{aligned}(P/\rho g)_1 &= [(-12 \text{ cm Hg})(13.6 \text{ cm } H_2O/\text{cm Hg})(1000 \text{ kg/m}^2)\\ &\quad \times (9.81 \text{ m/s}^2)]/[(1000 \text{ kg/m}^2)(9.81 \text{ m/s}^2)]\\ &= -1.632 \text{ m } H_2O\end{aligned}$$

$$\begin{aligned}(P/\rho g)_2 &= [(110{,}000 \text{ Pa})(1 \text{ N/m}_2/\text{Pa})(1 \text{ kg}\cdot\text{m/s}^2)/1 \text{ N}]/[(1000 \text{ kg/m}^2)(9.81 \text{ m/s}^2)]\\ &= 11.2 \text{ m } H_2O\\ Z_2 - Z_1 &= 1.0 \text{ m } H_2O\end{aligned}$$

To determine the velocity head we must use both the continuity equation and the kinetic energy of the fluid.

$$A_1V_1 = A_2V_2 = 3100 \text{ l/min} = 3.1 \text{ m}^3/\text{min} = 0.05167 \text{ m}^3/\text{s}$$

$$A_1V_1 = \left(\pi D_1^2/4\right)V_1 = [\pi(0.25 \text{ m})^2]/4V_1 = 0.05167 \text{ m}^3/\text{s}, \quad \text{or} \quad V_1 = 1.053 \text{ m/s}$$

Therefore, the velocity head entering is

$$V_1^2/2g = (1.053 \text{ m/s})^2/[2(9.81 \text{ m/s}^2)] = 0.0565 \text{ m } H_2O$$

Now, examining the fluid leaving the pump

$$A_1V_1 = A_2V_2$$

Therefore

$$\begin{aligned}V_2 &= (A_1/A_2)V_1\\ &= \left[\left(\pi D_1^2/4\right)/\left(\pi D_2^2/4\right)\right]V_1 = (0.25/0.15)^2(1.053) = 2.925 \text{ m/s}\end{aligned}$$

Therefore, the velocity head leaving is

$$V_2^2/2g = (2.925)^2/[2(9.81)] = 0.436 \text{ m } H_2O$$

Now that all of the terms have been calculated we may put them together into the Bernoulli equation to determine the amount of energy added

$$-1.632 + Z_1 + 0.0565 + E = 11.2 + (Z_1 + 1.0) + 0.436$$

Therefore

$$E = 14.21 \text{ m } H_2O \text{ is added to the system}$$

Now we need to calculate the power of the pump if the system were 100% efficient.

$$\begin{aligned}\text{Power} &= \text{Pressure} \times \text{Volume flow rate}\\ &= (14.21 \text{ m})(1000 \text{ kg/m}^3)(9.81 \text{ m/s}^2)(3.1/60)\text{m}^3/\text{s}\\ &= 7202 \text{ kg}\cdot\text{m}^2/\text{s}^3 = 7202 \text{ N}\cdot\text{m/s} = 7202 \text{ J/s}, \quad \text{or} \quad 7.202 \text{ kW}\end{aligned}$$

Therefore, the efficiency is

$$\eta = (7.202 \text{ kW})/(12 \text{ kW}) = 60\%$$

PUMP AFFINITY LAWS

The pump affinity laws give approximate results regarding the operation of centrifugal pumps. For pumps of constant speed

a) Capacity varies directly with the impeller diameter.

b) Head varies directly with the square of the impeller diameter.

c) Horsepower varies directly with the cube of the impeller diameter.

In addition, the laws for the operation of fans also hold true. Specifically:

a) The capacity of the flow varies directly with the speed of the impeller.

b) The pressure head varies directly with the square of the speed.

c) The power required varies directly with the cube of the speed.

Example 6.6

Using the basic information given in Example 6.5, replace the 1500-rpm motor with a 1760-rpm motor and determine

a) New flow rate

b) New output pressure

c) New power requirement

Assume that the efficiency of the unit remains constant.

Solution

We may use the affinity laws for the solution of the problem.

a) Capacity varies directly with impeller rpm.

$$\text{Capacity}_2 = \text{Capacity}_1[(\text{rpm}_2)/(\text{rpm}_1)] = 3.1 \text{ m}^3/\text{min } (1760/1500)$$
$$= 3.63 \text{ m}^3/\text{min}$$

b) Pressure varies directly with the square of the impeller rpm.

$$P_2 = P_1[(\text{rpm}_2)/(\text{rpm}_1)]^2 = (110 \text{ kPa})[1760/1500]^2 = 151.4 \text{ kPa}$$

c) Power required varies directly with cube of the impeller rpm.

$$\text{Power}_2 = \text{Power}_1[(\text{rpm}_2)/(\text{rpm}_1)]^3 = 12 \text{ kW}(1760/1500)^3 = 19.38 \text{ kW}$$

CAVITATION

When the local pressure of a liquid falls below its vapor pressure the liquid will vaporize, and, by vaporization, it will form a bubble of vapor that will push the liquid away from that location. This bubble formed is called a cavity or a void. When the outer, cooler liquid absorbs the heat from the cavity the vapor in the cavity returns to a liquid form. In general, it will occupy about 1/1000 of the space, and the collapse is essentially instantaneous and can actually be quite violent. This cavitation can cause extensive damage in impellers and even in other parts on the pump. In addition, the collapse may be quite noisy and cause a sever reduction in the efficiency of the pump. Consequently, care should be made to keep the suction side of the pump at a pressure higher than the vapor pressure of the liquid.

Example 6.7

A pump is used to draw 20°C water out of a mine shaft. The velocity in the intake line to the pump is an average of 3 m/s. At a particular point in the intake line the pressure is 10.0 kPa. How far above this point would cavitation in the line start?

Solution

We may determine the vapor pressure of the water from the saturated water tables in the *Handbook.* At 20°C the saturation pressure, i.e., vapor pressure, is equal to 2.239 kPa absolute. In the same table, at the same conditions, the specific volume ($v = 1/\rho$) is equal to 0.001002 m^3/kg. Therefore, density, ρ, is equal to 998 kg/m^3.

The total pressure in the line at the reference point is equal to the gage pressure plus the atmospheric pressure. Therefore, assuming that the atmospheric pressure is standard, the total pressure is

$$P_{total} = P_{gage} + P_{atmospheric} = 10 \text{ kPa} + 101.3 \text{ kPa} = 111.3 \text{ kPa}$$

In order for the water to flash into vapor, the total pressure would have to be equal to the vapor pressure, or 2.239 kPa. This means we would have to have a drop in pressure created by a negative head of water equivalent to 111.3 kPa – 2.239 kPa = 109.06 kPa. Using the manometer equation

$$P = \rho g h$$

or

$$\begin{aligned} h &= P/\rho g \\ &= [(111.3 \text{ kPa})(1000 \text{ N/m}^2)/\text{kPa}]/[(998 \text{ kg/m}^3)(9.81 \text{ m/s}^2)(\text{N/kg}\cdot\text{m/s}^2)] \\ &= 11.37 \text{ m} \end{aligned}$$

Therefore, the pump intake cannot be located any higher than this particular point in order to prevent cavitation in the line.

NPSH (NET POSITIVE SUCTION HEAD)

Cavitation is also able to occur in the impeller of a centrifugal pump due to the rotating impeller. To prevent this situation from occurring, a pump manufacturer will test the pump to determine how much pressure is needed at the intake to prevent cavitation during pump operation. This pressure is called *Net Positive Suction Head (NPSH).* The equation defining the NPSH is equal to

$$\text{NPSH} = P_i/\rho g + V_i^2/2g - P_v/\rho g$$

where

P_i = Inlet pressure to the pump (absolute pressure/total pressure)
ρ = Density at inlet conditions
V_i = Velocity at the inlet to the pump
P_v = Vapor pressure of the liquid being pumped

Example 6.8

Liquid propane is contained in a storage tank at a pressure of 1.4 MPa, which is the equilibrium vapor pressure of the propane at the pumping conditions. The propane has a specific gravity of 58. If the level of the liquid propane is 4.00 meters above the centerline of the pump, and if the frictional losses between the supply and the pump is 2.00 meters, determine the NPSH when the velocity in the suction line is 1.5 m/s.

Solution

Starting with the generic NPSH equation given in the *Handbook* and calculating each of the terms we get

$$\begin{aligned} \text{NPSH} &= P_i/\rho g + V_i^2/2g - P_v/\rho g \\ &= [1.4\text{ MPa} + 4.00\text{ m} - 2.00\text{ m}] + (1.5\text{ m/s})^2/2g - 1.4\text{ MPa} \\ &= 2.00\text{ m} + (1.5\text{ m/s})^2/2(9.81\text{ m/s}^2) = 2.11\text{ m} \quad \text{available} \end{aligned}$$

Example 6.9

A pump discharges 7500 l/min of a certain brine solution (s.g. of 1.20) to an evaporating pond. The intake line is 30 cm in diameter and is at the same level as the pump discharge line, which is 20 cm in diameter. The pressure at the inlet of the pump is –150 mm of mercury. The pressure gage connected to the pipe discharge reads 140 kPa, and its center is 1.50 m above the center of the discharge flange. The vapor pressure of the brine solution is 35 mm of Hg.

a) If the pump efficiency is 82%, how much power is required?

b) If vapor pressure of the brine is 35.3 mm Hg, what is the NPSH?

c) If the flow rate is increased to 10,000 l/min, what is the new NPSH?

Solution

a) Starting with the Bernoulli equation to determine the energy added to the brine we get

$$P_1/\rho g + V_1^2/2g + Z_1 + E = P_2/\rho g + V_2^2/\rho g + Z_2 + F$$

Calculating the values for each of the above terms (noting that F = 0)

$$\begin{aligned} P_1/\rho g &= -[(150/760)(101.3\text{ kPa})(10^3\text{ N/m}^2)/\text{kPa}]/ \\ &\qquad [(1.20)(1000\text{ kg/m}^3)(9.81\text{ m/s}^2)\text{N/kg}\cdot\text{m/s}^2] \\ &= -1.698\text{ m} \\ V_1^2/2g &= [(AV)/A]^2/2g \\ &= \left[(7.5\text{ m}^3/\text{min})/\left(\pi D_1^2/4\text{ m}^2\right)(60\text{ s/min})\right]^2/2(9.81\text{ m/s}^2) \\ &= [(0.125\text{m}^3/\text{s})/\pi(0.30)^2/4\text{ m}^2]^2/(19.81\text{ m/s}^2) \\ &= (1.77\text{ m/s})^2/19.81\text{ m/s}^2 = 0.158\text{ m} \\ Z_2 - Z_1 &= 0 \end{aligned}$$

The pressure head developed by the pump will be a function of the gage pressure at its location plus the effect of the location of the pressure gage.

$$P_2/\rho g = P_{gage} + \text{Elevation of the gage}$$
$$= [(140{,}000 \text{ Pa})(\text{N/m}^2)/(\text{Pa})]/[(1000 \text{ kg/m}^3)(1.2)(9.81 \text{ m/s}^2)\text{N}/(\text{kg}\cdot\text{m/s}^2)] + 1.5 \text{ m} = 13.39 \text{ m}$$

It is safe to assume that the density of the brine is constant across the pump; therefore, the velocity is merely a function of the area.

$$A_1V_1 = A_2V_2$$

or,

$$V_2 = V_1(A_1/A_2)$$
$$= (1.77 \text{ m/s})(D_1/D_2)^2$$
$$V_2 = (1.77 \text{ m/s})(30/20)^2 = 3.98 \text{ m/s}$$

Calculating the pressure head on the pump outlet side

$$V_2^2/2g = (3.98 \text{ m/s})^2/[2(9.81 \text{ m/s}^2)] = 0.807 \text{ m}$$

Therefore

$$E = V_2^2/2g - V_1^2/2g + P_2/\rho g - P_1/\rho g$$
$$E = 0.807 - 0.158 + 13.39 - (-1.698) = 15.737 \text{ m brine}$$

The power required to drive the pump may be obtained using the equation in the Fluid Mechanics section of the *Handbook*

$$W = Q\gamma h/\eta$$
$$= \{[(7.5 \text{ m}^3/\text{min})/60 \text{ s/min}](1000 \text{ kg/m}^3)(9.81 \text{ m/s}^2)(1.2)(15.737 \text{ m})\}/0.82$$
$$= (28{,}240 \text{ kg}\cdot\text{m}^2/\text{s}^2)(\text{N/kg}\cdot\text{m/s}^2)(\text{J/N}\cdot\text{m}) = 28.24 \text{ kW}$$

b)

$$\text{NPSH} = P_{absolute}/\gamma + V_1^2/2g - h_v$$

Assuming that the atmospheric pressure is standard and equal to 1.013 Bar

$$\text{NPSH} = [P_{\text{pump inlet}} + P_{barometric}]/\gamma + V_1^2/2g - P_v/\gamma$$

Solving for each of the terms

$$P_v = (35/760)(1.013 \text{ bar}) = 0.04665 \text{ bar}$$
$$\gamma_{brine} = \rho g = [(1000 \text{ kg/m}^3)(1.2)](9.81 \text{ m/s}^2) = 11{,}772 \text{ kg/m}^2\cdot\text{s}^2$$
$$h_v = [(0.04665 \text{ bar})(10^5\text{N/m}^2)/\text{bar}][(\text{kg}\cdot\text{m/s}^2)/\text{N}]/11{,}772 \text{ kg/m}^2\cdot\text{s}^2 = 0.396 \text{ m}$$

$h_{\text{pump inlet}}$ has been previously calculated and is equal to -1.698 m

$$h_{barometric} = [(1.013 \text{ bar})(10^5 \text{ N/m}^2)/(\text{bar})][(\text{kg}\cdot\text{m/s}^2)/(\text{N})]/(11{,}772 \text{ kg/m}^2/\text{s}^2)$$
$$h_{barometric} = 8.61 \text{ m}$$

$h_{\text{velocity head}}$ previously calculated as 0.159 m

$$\text{NPSH} = 8.61 - 1.698 + 0.159 - 0.396 = 6.675 \text{ m}$$

c) The flow rate is increased to 10 m^3/min; determine the new NPSH.

Knowing that the vapor pressure remains constant and that the velocity head is increased as the ratio of the square of the flow rates

$$h_{\text{velocity head}} = (10/7.5)^2(0.159) = 0.283 \text{ m}$$
$$h_v = 0.396 \text{ m}$$

According to the affinity laws regarding pump operation, the head varies directly with the square of the flow rate. Therefore

$$h_{\text{pump inlet}} = (10/7.5)^2\ (-1.698) = -3.02 \text{ m}$$
$$\text{NPSH}_{\text{new}} = 8.61 - 3.02 + 0.283 - 0.396 = 5.477 \text{ m}$$

COMPRESSORS

Oftentimes compressors are positive-displacement gas/vapor "pumps" using reciprocating pistons. The capacity of the compressor is usually stated as the quantity of gas handled at intake conditions.

Example 6.10

Assuming isothermal compression with a compressor having no clearance, determine the work required to compress 150 m^3/hr of *standard air* to 700 kPa gage in a single-stage compressor?

Solution

One may take the isothermal work equation from the *Handbook*

$$W = mRT\ln(P_2/P_1)$$

Using the perfect gas relation, $PV = mRT$ and $P_1 = 101{,}300$ Pa as standard pressure, work may be calculated as

$$\begin{aligned} W &= P_1V_1\ln(P_2/P_1) \\ &= [(101{,}300 \text{ Pa})(\text{N/m}^2)/\text{Pa}][(150 \text{ m}^3/\text{hr})/(3600 \text{ s/hr})]\ln(101.3/801.3) \\ &= -8729.3 \text{ N}\cdot\text{m/s} = 8.73 \text{ kW of work done on the gas.} \end{aligned}$$

PROBLEMS

The following information may be used for Problems 6.1 and 6.2.

A fan delivers 225 m^3/min of standard air against a back pressure of 10 cm of water while it is operating at 500 rpm. The efficiency of the fan is 85%.

6.1 The power, in kW, required to operate the fan is most nearly
a. 3.7
c. 5.6
b. 4.3
d. 6.8

6.2 If the speed is increased to 700 rpm how much power is required, and what is the new flow rate?
a. 7.6
c. 10.1
b. 8.5
d. 11.9

The following information may be used for problems 6.3 to 6.6.

A pump uses a 1200 rpm motor to pump 100,000 kg/min of standard temperature water that has a density of 1000 kg/m^3 (which may be assumed constant through the pump), and has a pressure rise from 1 bar to 40 bar using a 10 cm diameter impeller. The inlet diameter is 35 cm, and the exit diameter is 25 cm. The outlet from the pump is at the same elevation as the inlet to the pump. The pump has an efficiency of 75%.

6.3 Power, in kW, required to operate the pump is most nearly
a. 7300
c. 8200
b. 5600
d. 9600

6.4 If a similar pump is used but has an impeller of 12 cm in diameter and an rpm of 1760, what is the power input in kW?
a. 75,500
c. 160,000
b. 140,000
d. 195,000

6.5 For the new pump, determine the new pressure rise in bar.
a. 120
c. 85
b. 100
d. 230

6.6 For the new pump, what is the new flow rate in kg/min?
a. 150,000
c. 170,000
b. 193,000
d. 211,000

The following information may be used for Problems 6.7 and 6.8.

A water-cooled air compressor operating on the polytropic process with n = 1.2 is used to compress 500 m^3/min of air at 1 bar and 20°C to a pressure of 15 bar. The compressor has an isentropic efficiency of 75%.

6.7 Power input required, in kW, is most nearly

a. 2500
b. 3800
c. 3100
d. 4500

6.8 Heat transfer out of the compressor, in kJ/s, is most nearly

a. 1500
b. 1800
c. 2100
d. 2400

SOLUTIONS

6.1 **b.** Knowing that the equation for fan power is P = Pressure × Volume rate of flow, we may write the equation as

$$\begin{aligned}\text{Power} &= \rho g h \times \text{Volume rate of flow}\\ &= [(1000\ \text{kg/m}^3)(9.81\ \text{m/s}^2)(0.10\ \text{m})(225\ \text{m}^3/\text{min})]\\ &= 3678.8\ \text{kg}\cdot\text{m}^2/\text{s}^3\end{aligned}$$

Or, power may be written in more conventional units as

$$\begin{aligned}\text{Air Power} &= (3678.8\ \text{kg}\cdot\text{m}^2/\text{s}^3)(\text{N})/[(\text{kg}\cdot\text{m})/\text{s}^2]\\ &= 3.678\ \text{kN}\cdot\text{m/s} = 3.678\ \text{kW}\end{aligned}$$

The power to operate the fan is then

$$\text{Power} = 3.678/0.85 = 4.32\ \text{kW}$$

6.2 **d.** Since we are looking at a fan in which the basic conditions are unchanged, with the exception of the fan rpm we must make use of the fan laws.

$$\begin{aligned}(\text{Volumetric flow rate})_2 &= (\text{rpm}_2/\text{rpm}_1)(\text{Volumetric flow rate})_1\\ (\text{Volumetric flow rate})_2 &= (700/500)(225) = 315\ \text{m}^3/\text{min}\\ \text{Power}_2 &= (700/500)^3(4.32\ \text{kW}) = 11.85\ \text{kW}\end{aligned}$$

6.3 **d.** The Bernoulli equation must be written to determine the energy added to the water and thus the energy needed to power the pump.

$$(P_1/\rho gh) + Z_1 + V_1^2/2g + E = (P_2/\rho gh) + Z_2 + V_2^2/2g + F$$

Noting that ΔZ and the friction loss F are equal to zero, and solving first for the velocity using the continuity relationship

$$m = (\rho AV)_1 = (\rho AV_2)$$

Assuming that ρ is constant

$$\begin{aligned}AV &= m/\rho = (100{,}000\ \text{kg/min})/(1000\ \text{kg/m}^3) = 100\ \text{m}^3/\text{min}\\ V_1 &= [(100\ \text{m}^3/\text{min})/(60\ \text{s/min})]/\left(\pi D_1^2/4\ \text{m}^2\right)\\ &= (1.67\ \text{m}^3/\text{s})/[(\pi)(0.35)^2/4\ \text{m}^2] = 17.36\ \text{m/s}\\ V_2 &= V_1(A_1/A_2) = V_1(D_1/D_2)^2 = (17.36\ \text{m/s})(35/25)^2 = 34.02\ \text{m/s}\end{aligned}$$

Therefore

$$\begin{aligned}E &= \left(V_2^2 - V_1^2\right)/2g + (P_2 - P_2)/\rho g\\ &= \{[(34.02)^2 - (17.36)^2]\ \text{m}^2/\text{s}^2\}/[2(9.81\ \text{m/s}^2)]\\ &\quad + [(40 - 1)\ \text{bar}]/(1000\ \text{kg/m}^3)(9.81\ \text{m/s}^2)\\ &= 43.63\ \text{m} + \{[(39\ \text{bar})(10^5\ \text{N}\cdot\text{m/bar})]\\ &\quad (\text{kg}\cdot\text{m/s}^2)/\text{N}\}/(9810\ \text{kg/m}^2\cdot\text{s}^2) = 441.2\ \text{m}\end{aligned}$$

Solving now for power

Power = Pressure × Volume flow rate

$= (441.2\ \text{m})(1000\ \text{kg/m}^3)(9.81\ \text{m/s}^2)\ (100\ \text{m}^3/\text{min})/(60\ \text{s/min})$

$= 7{,}213{,}600\ \text{kg}\cdot\text{m}^2/\text{s}^2 = 7{,}213\ \text{kJ/s} = 7{,}213.6\ \text{kW}$

power to the water

Power to the pump = 7,213.6/0.75 = 9618 kW

6.4 **a.** Using the *scaling laws* in the *Handbook* we may calculate the new power required.

$$[W/(\rho N^3 D^5)]_1 = [W/(\rho N^3 D^5)]_2$$
$$W_2 = W_1\left(\rho N_2^3 D_2^5\right)/\left(\rho N_1^3 D_1^5\right)$$
$$= (9{,}618)[(1760)^3(12)^5]/[(1200)^3(10)^5]$$
$$= 75{,}500\ \text{kW} \quad \textit{New Power}$$

6.5 **a.** To calculate the pressure rise, again use the scaling law in the *Handbook* for pressure.

$$[P/(\rho N^2 D^2)]_1 = [P/(\rho N^2 D^2)]_2$$
$$P_2 = P_1\left(\rho N_2^2 D_2^2\right)/\left(\rho N_1^2 D_1^2\right)$$
$$= 39[(1760)^2(12)^2]/[(1200)^2(10)^2]$$
$$= 120.8\ \text{bar} \quad \textit{New Pressure Rise}$$

6.6 **d.** The scaling law for flow shall be used

$$[m/(\rho N D^3)]_1 = [m/(\rho N D^3]_2$$
$$m_2 = m_1\left(\rho N_2 D_2^3\right)/\left(\rho N_1 D_1^3\right)$$
$$= [100{,}000\ \text{kg/min}\ (1760)(12)^3]/[(1200)(10^3)]$$
$$= 211{,}200\ \text{kg/min} \quad \textit{New Mass Flow Rate}$$

6.7 **b.** Select the work equation from the *Handbook* for the steady-flow polytropic compression system.

$$w = n(P_2V_2 - P_1V_1)$$
$$= [n/(n-1)]RT_1\left[1-(P_2/P_1)^{(n-1)/n}\right]$$
$$= 1.2/(1.2-1)[(8.314/29)\ \text{kJ/kg}\cdot\text{K}]293\ \text{K}\left[1-(15/1)^{(1.2-1)/1.2}\right]$$
$$= -287\ \text{kJ/kg}$$

To find the total work we must find the mass flow, which can be obtained using the perfect gas equation.

$$m = PV/RT$$
$$= [(1\ \text{bar})(10^5\ \text{N/m}^2/\text{bar})][(500\ \text{m}^3/\text{min})]/\{[(8.314/29)\ \text{kJ/kg}\cdot\text{K}] \times (10^3\ \text{N}\cdot\text{m/kJ})(293\ \text{K})\}$$
$$= 595.2\ \text{kg/min} = 9.92\ \text{kg/s}$$

Therefore

$$W = (9.92 \text{ kg/s})(-287 \text{ kJ/kg}) = 2847 \text{ kJ/s Input to the air}$$

Total work supplied to the compressor

$$W_c = W/\eta = 2847/0.75 = 3796 \text{ kW}$$

6.8 **c.** Writing the first law of thermodynamics as a basis for the solution

$$Q/m + h_1 + V_1^2/2g + Z_1 = h_2 + V_2^2/2g + Z_2 + W/m$$

For a compressor the effect of the elevation is essentially equal to zero ($\Delta Z = 0$).

In addition, many compressors are sized such that $\Delta KE = 0$

Therefore

$$Q = m(h_2 - h_1) + W$$

To solve for heat transfer, the final temperature needs to be found using the polytropic relation found in the *Handbook.*

$$T_2 = T_1(P_1/P_2)^{(1-n)/n} = 293 \text{ K } (1/15)^{(1-1.2)/1.2} = 460 \text{ K}$$

Therefore, the heat transfer is

$$Q = mc_p\Delta T + W$$

where c_p is found in the *Handbook.*

$$Q = (9.92 \text{ kg/s})(1 \text{ kJ/kg·K})(460 - 293) - 3796 \text{ kJ/s}$$
$$Q = 1656.6 - 3796 = 2139 \text{ kJ/s out of the compressor.}$$

CHAPTER 7

Stress Analysis

OUTLINE

The subject of stress analysis is usually studied in courses entitled "Mechanics of Materials" or "Strength of Materials" and is applied in problems of machine design to determine whether the device will withstand the loads to which it may be subjected. Stress analysis is actually an expansion of the more general subject of mechanics and includes the study of the effects of the elastic properties and strengths of the structural materials being considered.

Problems of stress analysis include both static and dynamic loading conditions. They may also include environmental conditions such as temperature, corrosion, radiation, and other factors that can influence material properties in the short or long term.

The simplest type of a stress problem is a pair of static forces acting in opposite directions along the longitudinal axis at the ends of a straight bar of constant cross section. In this case the stress is equal to the force divided by the cross-sectional area of the bar.

Example 7.1

A mass of 100 kg is suspended vertically on the end of a 5.00-mm-diameter wire. What is the stress in the wire?

Solution

A 100-kg mass exerts a gravitational force of $100g = 980.66$ N. The cross-sectional area of a 5.00-mm-diameter wire is equal to 19.64×10^{-6} m^2 the stress in the wire is

$$\sigma = 980.66/(19.64 \times 10^{-6}) = 49.93 \text{ MPa}$$

STRAIN

Stress in a member is always accompanied by strain in the ratio

$$\varepsilon = \sigma/E$$

where

ε = strain, m/m
σ = stress, Pa
E = Modulus of Elasticity, Pa

Example 7.2

In Example 7.1, if the supporting wire is 10.00-m long, how much will it stretch (elongate) when the mass is attached to it?

Solution

The modulus of elasticity of steel is given in the *FE Handbook* as 2.1×10^{11} Pa

$$\varepsilon = (49.93 \times 10^{6})/(2.1 \times 10^{11}) = 2.378 \times 10^{-4} \text{ m/m}$$

The total strain (elongation) is then

$$\Delta L = \varepsilon \times L = (10)(2.378 \times 10^{-4}) = 2.378 \times 10^{-3} \text{ m, or } 2.378 \text{ mm}$$

THERMAL STRESS

Most metals expand when their temperature is increased. If a member is constrained from elongating, it will be subject to strain. Since stress and strain are proportional, a strain will result in a stress, or will be evidence of a stress. The stress in a constrained member is calculated as if the member is allowed to expand and then is physically compressed the amount of the expansion.

$$\sigma = E \times \varepsilon$$

The thermal expansion, or compression, of a member is determined by multiplying the thermal coefficient of expansion by the temperature differential. The thermal coefficient of expansion for steel is $\alpha = 11.7 \times 10^{-6}$ m/m·°C. In most cases it will simplify the calculations if the strain due to stress and the strain due to temperature are calculated separately and are then added algebraically to obtain the net, or final, result. This can best be illustrated by means of examples.

Example 7.3

A 1-m-long copper bar having a circular cross section of 2.5 cm in diameter is arranged as shown in Exhibit 1, with a 0.025 mm gap between its end and a rigid wall at room temperature. If the temperature increase 35 °C, find the stress in the bar if its thermal coefficient of expansion is 16.7×10^{-6} m/m·°C and its modulus of elasticity is 1.2×10^{11} Pa.

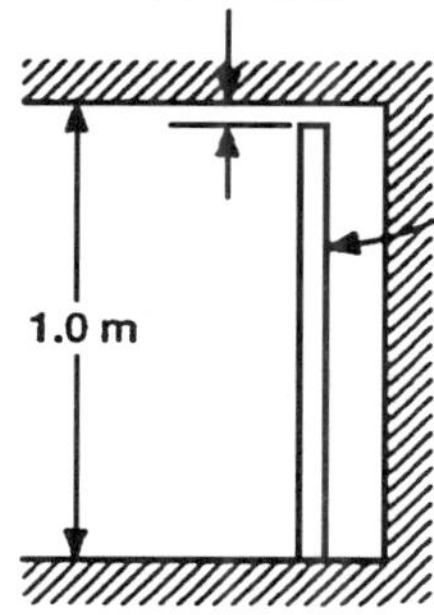

Exhibit 1

Solution

The unrestrained increase in length due to the increase in temperature would have been

$$\Delta L = (1.0)(35)(16.7 \times 10^{-6} \text{ m/m}\cdot{}^\circ\text{C}) = 0.000585 \text{ m, or } 0.585 \text{ mm}$$

The bar can expand 0.025 mm and then will be restrained. This is the same (stresswise) as if it had been allowed to increase 0.585 mm and was then compressed.

$$0.585 - 0.025 = 0.560 \text{ mm}$$

The resulting stress would be

$$S = E \times \varepsilon = (1.2 \times 10^{11} \text{ Pa})(0.560/1000) = 67.2 \text{ MPa}$$

There are a few assumptions that would have to be made, such as

1. The walls must be perfectly rigid, i.e., undergo neither expansion nor deformation.
2. The bar (column) must remain straight.
3. The yield point of the copper must not be exceeded.

Example 7.4

An exhaust manifold on a diesel engine is held rigidly at both ends. Under extended operation at high power output the manifold may become cherry red in color, indicating that it has reached a very high temperature, say 705 °C. If the manifold were bolted in place at a temperature of 25 °C, what would be the apparent stress in the manifold at the higher temperature? Assume the manifold is held rigidly and the rest of the engine assembly does not change its dimensions.

Solution

The easiest way to look at a problem like this is to assume that the heated member is free to expand to the higher temperature and then is physically forced to its former dimension. The theoretical longitudinal strain of the steel manifold would equal be

$$\varepsilon = \Delta L/L = \Delta T \times \alpha$$
$$= (705\,°\text{C} - 25\,°\text{C})(11.7 \times 10^{-6}\ \text{m/m}\cdot°\text{C}) = 0.00796\ \text{m/m}$$

It would then be compressed an equal amount when the temperature of the manifold cooled to the ambient temperature. This gives an apparent stress of

$$\sigma = \varepsilon \times E = (0.00796)(2.1 \times 10^{11}) = 1.67\ \text{GPa}$$

This is well above the yield point of stainless steel, so the manifold would yield in compression as it was heated. Since the length of the manifold between the supports would now be shorter, it would yield in tension when the manifold cooled. After a number of such cycles the manifold would fail. This has been termed "low-cycle fatigue."

HOOP STRESS

If the wall thickness of a cylindrical vessel is smaller than one-twentieth of the radius it is considered to be a thin-walled vessel. The hoop stress can then be calculated with the equation

$$\sigma = PD/2t$$

where
P = internal pressure, Pa
D = internal diameter, meters
t = wall thickness, meters

Similarly, if the pressure vessel were a closed vessel, the longitudinal force acting on the cylindrical shell would be equal to the internal area times the pressure, and the longitudinal stress would be equal to

$$PD^2(\pi/4)/\pi Dt = PD/4t, \text{ or just half of the hoop stress}$$

Example 7.5

If a 750-mm-diameter cylinder with a 5.00-mm-thick wall is pressurized to 1.5 MPa, what is the hoop stress?

Solution

$$PD/2t = \sigma = (1.5 \times 10^6)(0.750)/2(0.005) = 112.5\ \text{MPa}$$

Similarly the longitudinal stress is equal to one half of the hoop stress, or

$$\sigma = 112.5/2 = 56.25\ \text{MPa.}$$

These constitute the two principal stresses in the wall of the vessel. The maximum shear stress, from the diagram of Mohr's circle in the *FE Handbook*, is equal to one half of the algebraic difference between the two principal stresses.

$$\tau_{max} = [(112.5 - 56.25) \times 10^6]/2 = 28.125\ \text{MPa}$$

MOHR'S CIRCLE

This brings up the topic of Mohr's circle and combined stresses. Mohr's circle was developed as a simplified method for determining the maximum stress resulting from a number of differently applied loads or combined stresses. We shall restrict ourselves here to a brief discussion of one method of determining the principal stresses and the maximum shearing stress for a condition of combined loading, a method that is based on Mohr's circle.

Example 7.6

Derive the equation for the maximum shearing stress in a member that is subjected to combined tension, torsion, and circumferential tension, specifically, a pipe under pressure with torsion and axial tension applied, using the diagram in Exhibit 2 (Mohr's circle). The equation is to be derived for the maximum shearing stress, from the information in the diagram, in terms of $S_{s,\max}$ = maximum shearing stress, S_s = stress due to torsion, S_x = stress due to pure tension, and S_y = stress due to circumferential tension. In addition, sketch a stress diagram for a unit area of the external pipe surface.

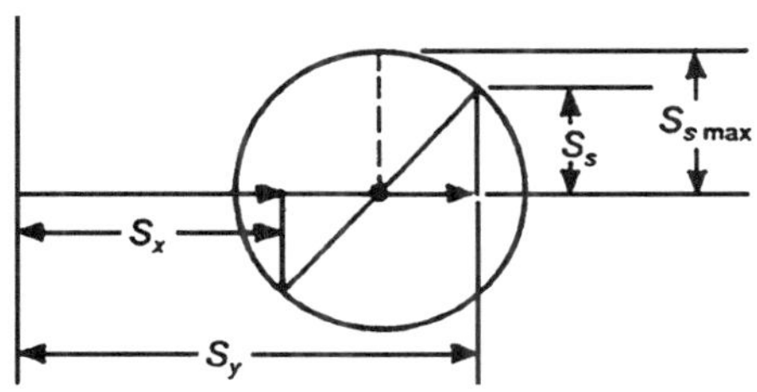

Exhibit 2

Solution

This example illustrates a relatively simple application of Mohr's circle. The subject can be examined in a more general sense to obtain maximum benefit from the example in describing the application of the principles of Mohr's circle.

First, determine the stresses acting at a section on the outer surface of the pipe (see Exhibit 3).

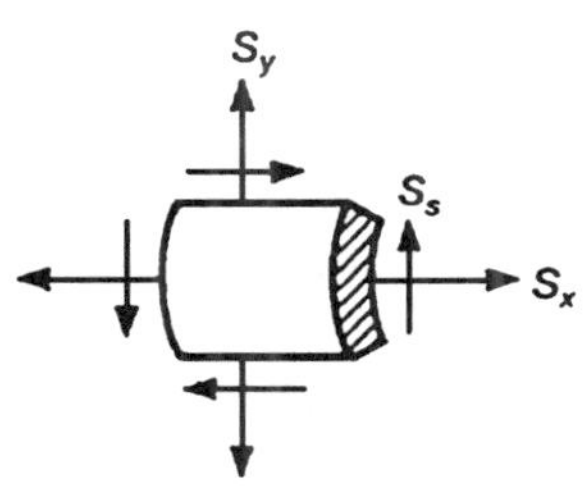

Exhibit 3

$S_x = F/A$ (axial load divided by the cross-sectional area)
$S_y = Pd/2t$ (assuming a thin-walled tube)
$S_s = TC/J$

Having determined these stresses, we can construct Mohr's circle for the given case. The general procedure is shown in Exhibit 4. First, lay out the coordinate axes. From the origin, point 0, lay off a distance equal to $(S_x + S_y)/2$ along the abscissa.

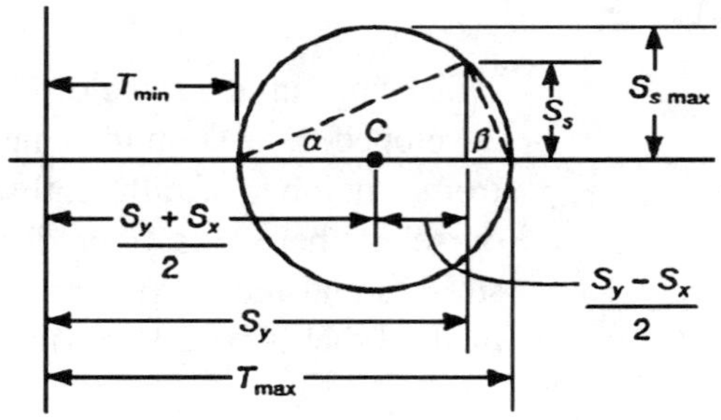

Exhibit 4

This gives the location of C, the center of the circle. Note here that $S_x + S_y$ is the algebraic sum of the two perpendicular stresses; if one of these stresses had been compressive, we would have had the arithmetic difference instead of the arithmetic sum. At a distance from the origin equal to S_y draw a vertical line, and at a distance equal to S_s above the abscissa mark a point on this line. This point lies on the circumference of the circle. Now, draw the circle.

The maximum shearing stress is equal to the radius of the circle. From the figure, the trigonometric relationship is

$$S_{s,\max} = \sqrt{\left[S_s^2 + \{(S_y - S_x)/2\}^2\right]}$$

The principal stresses can also be quickly obtained and are shown in Exhibit 5, which shows a section of the tube surface with the applied stresses. The minimum and maximum stresses, $T_{\min}$ and $T_{\max}$ are shown, as well as the directions of the stresses by the angles α and β.

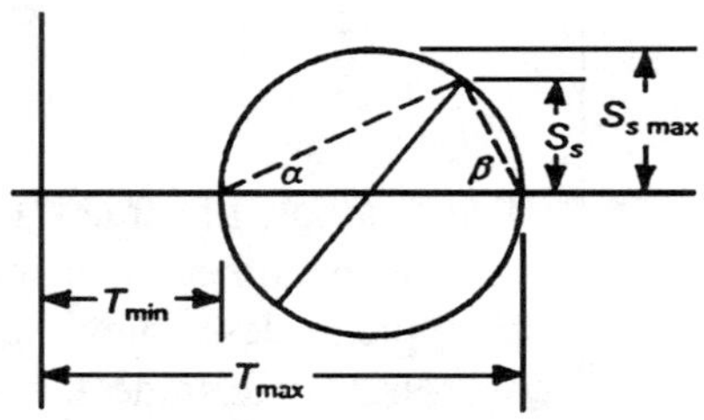

Exhibit 5

The lines drawn from the ends of the horizontal diameter to the construction point on the circumference give these angles, and the principal stresses act in directions that are perpendicular to these lines. That is, $T_{\min}$ acts at an angle of β, and $T_{\max}$ acts at an angle of α, to the direction of S_y. The maximum shearing stress acts on a plane at an angle of 45° to these two principal planes.

A stress diagram for a unit area of the external tube surface, Exhibit 6, shows the directions of the different stresses and would ordinarily be drawn before Mohr's circle was constructed.

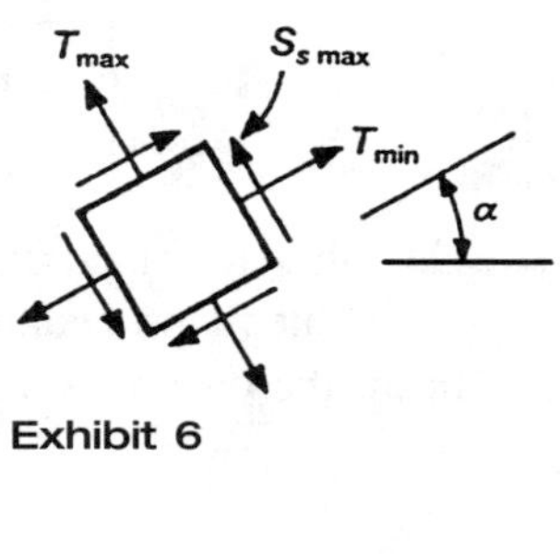

Exhibit 6

To sum up, the equations for determining the maximum stresses can be listed as follows

$$\text{Principal stress (1)} = (S_x + S_y)/2 + \sqrt{\left[S_x^2 + \{(S_y - S_x)/2\}^2\right]}$$
$$\text{Principal stress (2)} = (S_x + S_y)/2 - \sqrt{\left[S_s^2 + \{(S_y - S_x)/2\}^2\right]}$$
$$\text{Maximum sheer stress} = \sqrt{\left[S_s^2 + \{(S_y - S_x)/2\}^2\right]}$$

The equation for the principal stress can be used to check the maximum stress obtained in Example 7.5.

From the example, S_y = 56.25 MPa and S_x = 112.5 MPa. Since there is no applied shear stress, the maximum principal stress would be

$$S_{s,\max} = (S_x + S_y)/2 + \sqrt{\{(S_y - S_x)/2\}^2} = 84.375 + 28.125 = 112.5 \text{ MPa}$$

SHRINK FIT

Parts are often assembled by heating the outer member, and thus expanding it, while simultaneously cooling the inner member and shrinking it, if necessary. This puts the outer member in tension and the inner member in compression.

Example **7.7**

A 1.00-m cast-iron wheel has a 1.00-mm-thick steel ring shrunk onto it as a tire. The interference fit is 0.1000 mm. What is the stress in the ring? What is the contact pressure?

Solution

The diametral strain is 0.100/1000 = 0.000100 m/m. The circumferential strain would be the same, though the circumferential elongation would be equal to π times the diametral elongation.

$$\text{Stress, } \sigma = E \times \varepsilon = (2.1 \times 10^{11})(1 \times 10^{-4}) = 21.0 \text{ MPa stress in the ring}$$
$$\text{Hoop stress } \sigma = PD/2t\text{, so}$$
$$P = 2t\sigma/D = 2(0.001)(21.0 \times 10^6)/1.0 = 42{,}000 \text{ Pa}$$

The contact pressure would thus be equal to 42.0 kPa.

It is assumed that the cast iron wheel does not shrink since the ring is very thin, by comparison, and would exert minimal force on the wheel.

Example **7.8**

A steel liner is assembled in an aluminum pump housing by heating the housing and cooling the liner in dry ice. At 20 °C, before assembly, the outside diameter of the liner is 8.910 cm, and the inside diameter of the housing is 8.890 cm. After assembly and inspection, several units were rejected because of poor liners. It is desired to salvage the housing by heating the complete unit to a temperature that would cause a difference (clearance) in diameter of 0.050 mm between the liner and the housing and permit free removal of the liner. Determine the temperature at which this may be possible. The thermal coefficient of linear expansion for steel, α_S, is equal to 11.7×10^{-6} m/m·°C, and that for aluminum, α_A, is equal to 28.8×10^{-6} m/m·°C.

Solution

At 20°C the interference is 8.910 – 8.890 = 0.020 cm. A clearance of 0.050 mm is desired, so the expansion of the aluminum housing must be 0.025 cm more than the expansion of the steel liner

$$\Delta D = 0.020 \text{ cm (interference)} + 0.005 \text{ cm (clearance)} = 0.025 \text{ cm}$$

The unit expansion of the aluminum must be 0.025/8.890 = 0.00281 m/m greater than that of the steel. The differential expansion is equal to

$$\alpha_A - \alpha_S = 0.0000288 - 0.0000117 = 17.1 \times 10^{-6} \text{ m/m·°C}$$

The required temperature increase would then be

$$\Delta T = (0.00281 \text{ m/m})/[17.1 \times 10^{-6} \text{ m/m·°C}] = 164\text{°C}$$

Therefore, the assembly would have to be heated to a temperature of 184°C.

To check

$$164(11.7 \times 10^{-6} \text{ m/m·°C}) = 0.001919 \text{ m/m, so the liner would expand to } 8.910 + (8.910)(0.001919) = 8.927 \text{ cm}$$

The housing would have a thermal strain of

$$164(28.8 \times 10^{-6} \text{ m/m·°C}) = 0.004723 \text{ m/m, so it would expand to } 8.890 + (8.890 \times 0.004723) = 8.932, \text{ giving a clearance of } 8.932 - 8.927 = 0.005 \text{ cm, which checks.}$$

TORSION

Pure torsional loading (couple only, no bending) produces a shearing stress in a shaft with the magnitude of the stress, in any cross section, being proportional to the distance from the center of the shaft (Figure 7.1). This follows from the fact that the deformation at any point is equal to $\rho d\theta$, and, since stress is proportional to strain, the stress increases from zero at the center to a maximum at the outside. Let S_s be the maximum shearing stress at the outer fiber. Then S_s/r will be the stress one meter from the center and $(\rho/r) \times S_s$ will be the stress at a distance ρ from the center. Force is equal to stress times area, so the force resisting the applied torque, due to the area dA, is equal to $(\rho/r)S_s\, dA$; the resisting moment about the axis of

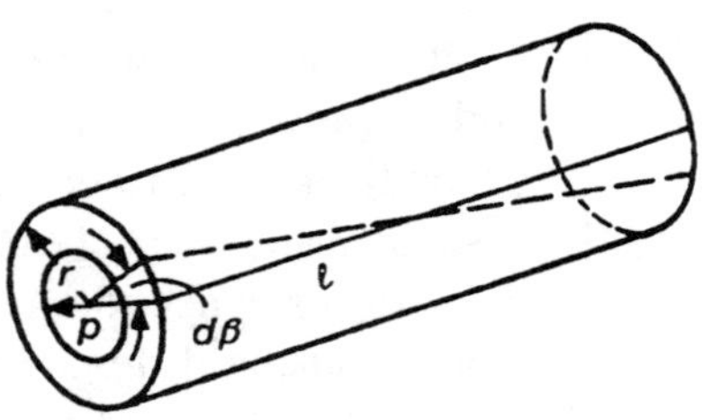

Figure 7.1

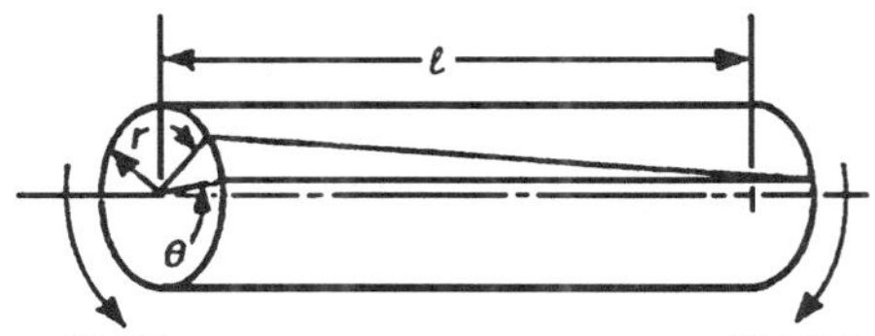

Figure 7.2

the bar due to this force is equal to $(\rho^2/r)S_s\,dA$, and $(S_s/r) \times \int \rho^2 dA$ is the total moment about the axis due to all of the internal shearing forces. The quantity $\int \rho^2 dA$ is, we recall from Mechanics (see *FE Handbook*), the moment of inertia of the cross-sectional area about the axis through its center or the *polar moment of inertia of the area*, which is ordinarily represented by the symbol J. This gives us the relationship that torque, T, is equal to $(S_s/r) \times J$, and, therefore, the maximum stress in the shaft due to the applied torque is $S_s = rT/J$. The stress at any other distance ρ from the center of the shaft is equal to

$$(\rho/r) \times (rT/J) = \rho T/J$$

Within the proportional limit, $\varepsilon_s = S_s/E_s$, which gives the relationship $S_s = E_s\varepsilon_s = (E_s r\theta)/l$, since, from Figure 7.2, $\varepsilon_s = \theta/L$. This also gives us $T = (E_s J\theta)/L$.

Example 7.9

A solid round shaft 9.0 cm in diameter transmits 75 kW of power at 200 rpm. What is the maximum torsional stress in the shaft? The appropriate equations are given in the *FE Handbook*.

Solution

The maximum stress will occur in the outer fibers. The power transmitted by a shaft is equal to $2\pi NTq$. In Exhibit 7, torque is equal to tension, T, times the radius, r, which gives Newton-meters or joules, and joules/sec is equal to watts. So N, in this equation must be equal to revolutions per second, or 3.333 rev/s. Then we have that the torque is

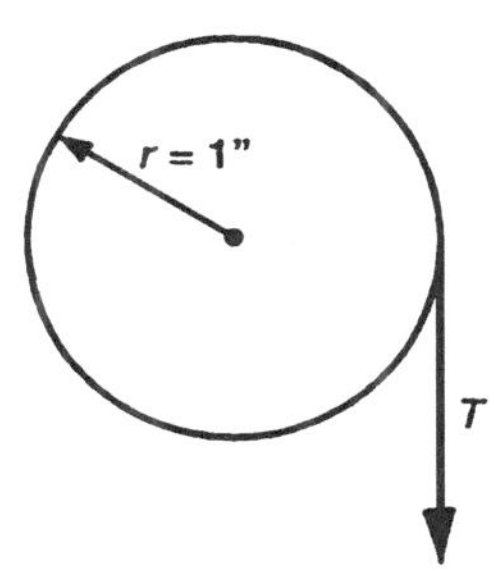

Exhibit 7

$$Tq = 75{,}000/2\pi(3.333) = 3581.3 \text{ J}$$

The polar moment of inertia of a circular section is

$$J = (\pi r^4)/2 = \pi(0.045)^4/2 = 0.00000644 \text{ m}^4$$

The maximum shear stress is thus

$$S_{s,\max} = (0.045)(3581.3)/0.00000644 = 25.02 \text{ MPa}$$

Example 7.10

For a hollow shaft of the same outer diameter, but with an inner diameter equal to half of the outer diameter, or 4.5 cm, what would be the increase in stress?

Solution

Again, the maximum stress will occur in the outermost fibers and will be equal to rT/J where r is equal to the distance from the neutral axis to the outermost fibers, T is equal to the torque, and J is equal to the polar moment of inertia

$$J = \pi/32 \times (OD^4 - ID^4) = (0.0982)[(65.61 - 4.10) \times 10^{-6}\ \text{J}] = 6.040 \times 10^{-6}\ \text{m}^4$$

which is equal to 6.040/6.44 = 0.938, or 93.8%, of that of a solid shaft. The maximum shear stress is equal to

$$S_{s,\max} = OD/2 \times T/J = (0.045)(3581.3)/(6.040 \times 10^{-6}) = 26.68\ \text{MPa}$$

which gives an increase of 1.66 MPa, or an increase of 6.6%.

Example 7.11

Referring to Exhibit 8, what would be the torque M_B such that the maximum unit shear stress would be the same in both parts. Disregard stress concentrations.

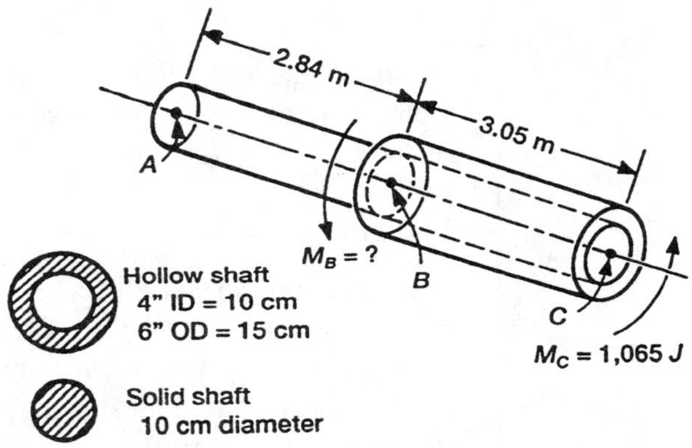

Exhibit 8

Solution

The maximum stress due to torsional loading is equal to $r \times T/J$, where J is equal to the polar moment of inertia. For the hollow shaft

$$J = \pi/32 \times (OD^4 - ID^4) = (0.09817)[(506.25 - 100.0) \times 10^{-6}] = 39.882 \times 10^{-6}\ \text{m}^4$$

The maximum shear stress in the hollow shaft is then

$$S_{s,\max} = (0.075)(1065)/(39.882 \times 10^{-6}\ \text{m}^4) = 2.00\ \text{MPa}$$

For the solid shaft

$$J = (0.09817)(0.10)^4 = 9.817 \times 10^{-6}\ \text{m}^4$$

For the shear stress in the solid shaft to be equal to 2.00 MPa, the applied torque would need to be

$$T = (2.00 \times 10^6)(9.817 \times 10^{-6}\ \text{m}^4)/0.05 = 392.7\ \text{J}$$

The required torque MB would then be

$$M_a = M_C - T = 1065 - 392.7 = 672.3\ \text{J}$$

Example 7.12

What would be the angle of twist in the overall length of the shaft in Example 7.11 if the shaft were made of steel, using the torques as defined and calculated above?

Solution

The angle of twist is given in the *FE Handbook* as

$$\phi = T \times L/(G \times J)$$

where
ϕ is in radians
T is in joules
L is in meters
G is the shear modulus, which is equal to 8.3×10^{10} Pa (see *FE Handbook*)
J is the polar moment of inertia, as above

From A to B

$$\phi = (392.3)(2.84)/(8.3 \times 10^{10})(9.817 \times 10^{-6}) = 0.00137 \text{ radians}$$

From B to C

$$\phi = (1065)(3.05)/(8.3 \times 10^{10})(39.882 \times 10^{-6}) = 0.000981 \text{ radians}$$

The total twist would be equal to 0.00235 radians, or $0.00235(180/\pi) =$ 0.135 degrees.

Example 7.13

A motor running at 1750 rpm transmits 10 kW of power through a 25.0-mm-diameter shaft that is 2.0 m long. What is the angle of twist in the shaft?

Solution

$$\text{Watts} = \text{J/s} = \text{N}\cdot\text{m/s} = 2\pi NT, \text{ where } N \text{ is rev/s} = 1750/60 = 29.167$$
$$T = \text{Watts}/2\pi N = 10{,}000/(2\pi)(29.167) = 54.57 \text{ N}\cdot\text{m}$$
$$J = \pi(0.025)^4/32 = 3.835 \times 10^{-8} \text{ m}^4$$
$$G = 8.3 \times 10^{10} \text{ Pa, from the } FE\ Handbook$$
$$\phi = (54.57)(2.00)/(3.835)(8.3)(100) = 0.0343 \text{ radians}$$
$$\text{Angle of twist} = (0.0343)(180/\pi) = 1.965°$$

Example 7.14

What would the shaft diameter in Example 7.13 have to be to reduce the angle of twist to 1.00°?

Solution

The polar moment of inertia would have to be increased in the ratio of 1.9647:1.000 so that the new shaft diameter would be

$$D = (25.0)(1.9647)^{0.25} = (25.0)(1.1839) = 29.60 \text{ mm}$$

Example 7.15

Given the bi-metal shaft shown in Exhibit 9, if a torque of 2.00 kJ is applied at the point of juncture, and both ends are restrained at their ends, what would be the reaction where the steel section is attached to the wall? What would be the stress in the aluminum section?

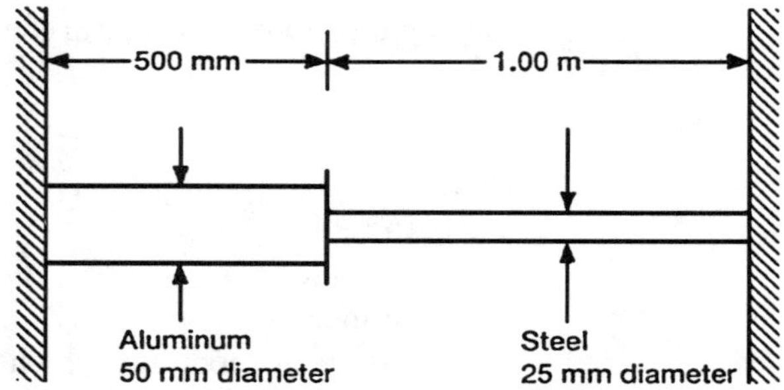

Exhibit 9

Solution

From the *FE Handbook* the total angle of twist of a shaft subjected to a torque

$$\phi = TL/GJ$$

where

T = torque, joules
L = length of shaft, meters
G = shear modulus, Pa
J = polar moment of inertia, m^4

The total angle of twist would be the same for both the steel and the aluminum sections.

For the aluminum section

$$J = \pi D^4/32 = \pi(0.050)^4/32$$
$$= 6.136 \times 10^{-7}\ m^4$$
$$L = 0.500\ m$$
$$G = 2.8 \times 10^{10}\ Pa$$
$$\phi = T_{Al}(2.910 \times 10^{-5})$$

For the steel section

$$J = 3.835 \times 10^{-4}\ m^4$$
$$L = 1.000\ m$$
$$G = 8.3 \times 10^{10}\ Pa$$
$$\phi = T_{Steel}(31.42 \times 10^{-5})$$

The total angles of twist are equal so,

$$T_{Steel}(31.42) = 2.910T_{Al} \text{ and}$$
$$T_{Steel} = 0.0926T_{Al}$$

The sum of the torques in the steel and the aluminum sections is equal to the total applied torque.

$$1.0926T_{Al} = 2000\ J$$
$$T_{Al} = 1831\ J$$
$$T_{Steel} = 2000 - 1831 = 169 \text{ joules}$$

The stress due to applied torque, $\sigma = Tc/J$.

For the aluminum section

$$\sigma = (1831)(0.025)/(6.136 \times 10^{-7}) = 74.6 \text{ MJ}$$

BEAMS

The general relationship for determining the tensile or compressive stress in a beam due to bending is

$$S = (Mc)/I$$

where

M = moment, J

I = moment of inertia of the cross section

c = distance from the neutral axis to the outermost fiber

The stress at any other point than the outer fiber is proportional to its distance from the center and would be equal to

$$S = (y/c)(Mc/I) = (My)/I$$

The values of c and I can be determined from the geometry of the beam. The moment, M, depends upon the loading of the beam. As can be deduced, the maximum stress will occur at the location of the greatest moment (for a beam of constant cross section).

The transverse shear stress at any point is equal to the magnitude of the shearing force divided by the cross-sectional area at that point. It is seldom possible to determine by a glance just where the maximum shear and the maximum moment will occur in a loaded beam, so it is usually desirable to construct both the shear diagram and the moment diagram for the beam under consideration; then the points of maximum shear and maximum moment will be readily apparent.

SHEAR DIAGRAM

The shear at any section of a beam is equal to the algebraic sum of all the external forces on *either* side of the section; it is considered positive if the segment of the beam on the left of the cross section tends to move up with respect to the segment on the right. The algebraic sum of all of the forces on both sides of the section will, of course, equal zero, since the system is in static equilibrium. It is usually simplest to add all the shearing forces to the left of the section being considered; this sum is equal to the shearing force acting at that point. If this process is repeated for different points over the length of the beam, a complete shear diagram may be constructed.

MOMENT DIAGRAM

The moment diagram may be constructed in a similar manner; one must remember that the bending moment at any section of a beam is the algebraic sum of the moments of all the external forces on either side of the section. When calculating the bending moment, use the segment (or side) for which the arithmetic will be the simplest. A positive moment is one that tends to cause the beam to be concave on the upper side (top fibers in compression). An easy method for arriving at the correct sign in calculating bending moment is to give plus signs to the moments

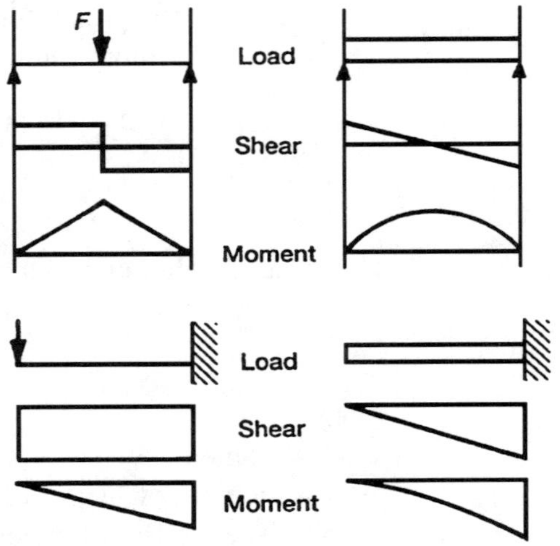

Figure 7.3

of upward forces and minus signs to the moments of downward forces. This method will give the correct sign to the algebraic sum of the moments, whether the forces to the right of the section or to the left of the section are used.

Examples of shear and moment diagrams for a few of the more common types of loading are shown in Figure 7.3.

For combined loading the shear and moment diagrams may be constructed by adding (algebraically) the individual shear and moment diagrams due to the individual loadings. This will give the total values for shear and moment for all parts of the beams. Each load will produce the same effect as if it had acted alone; it is unaffected by the other loads. Another method is to calculate the moments at different points on the beam and draw a smooth curve. Care must be taken, however, to make certain that the points of maximum total shear and maximum total moment are included. These points of maximum magnitude will be more easily determined by sketching the component shear and moment diagrams and determining the total shear and moment for all points of maximum magnitude (both positive and negative) on the component diagrams.

There are a few relationships that are of value in constructing and in checking shear and moment diagrams. One of these is that the derivative of the moment, M, with respect to distance, is equal to the shear V, or $V = (dM)/(dx)$. This means that the slope of the moment curve at any point is equal to the magnitude of the shear at that point. The relationship may also be written as

$$\int dM = \int V(x)\,dx, \qquad \text{from point 1 to point 2, or}$$

$$M_2 - M_1 = \int V(x)dx, \qquad \text{from point 1 to point 2,}$$

which means that the difference between the values of the moments at points 1 and 2 is equal to the area under the shear diagram between points 1 and 2. Similarly, we have

$$W = (dV)/(dx)$$

where W is the load at any point

$$V_2 - V_1 = \int W(x)dx$$

($W(x)$ means W as a function of x.)

Example 7.16

A wooden beam is made up of three timbers: two 10 by 15 cm and one 10 by 20 cm, as shown in Exhibit 10. If this section is used as a uniformly loaded beam in a simple span 10 meters long, what maximum total load could it sustain, assuming a maximum bending stress of 8.3 MPa?

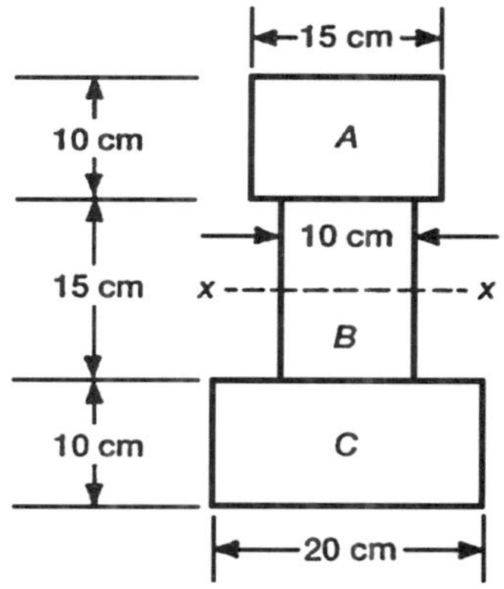

Exhibit 10

Solution

First, determine the location of the neutral axis. This can be obtained by taking moments about the lower edge.

$$\begin{aligned} y &= \text{(moments of individual areas)/(total area)} \\ &= [(10)(20)(5) + (10)(15)(17.5) + (10)(15)(30)]/(200 + 150 + 150) \\ &= 16.25 \text{ cm from the bottom edge} \end{aligned}$$

The moment of inertia, I, for a rectangular area is equal to $bh^3/12$, and the parallel axis theorem states that the moment of inertia of a section, I, about an axis other than that which passes through the centroid of the section is equal to the I about the centroidal axis plus the area of the section times the square of the distance between the axis through the centroid and the new axis. This is explained in the *FE Handbook*.

$$I_A = (0.15)(0.10)^3/12 = 12.5 \times 10^{-6} \text{ m}^4$$

Similarly,

$$I_B = (0.10)(0.15)^3/12 = 28.125 \times 10^{-6} \text{ m}^4\text{, and}$$
$$I_C = (0.20)(0.10)^3/12 = 16.667\ 10^{-6} \text{ m}^4$$

The moment of inertia about the beam centroidal axis is equal to

$$I = [(12.5 \times 10^{-6}) + (0.015)(0.1375^2)] + [(28.125 \times 10^{-6}) + (0.015)(0.0125^2)] + [(16.667 \times 10^{-6}) + (0.020)(0.125^2)]$$

$$I = [(12.5 + 284.59) + (25.125 + 2.34) + (16.667 + 312.4)] \times 10^{-6} = 656.6 \times 10^{-6} \text{ m}^4$$

$$\text{Stress} = Mc/I$$

where $M = (wL^2)/8$ (see *FE Handbook*)

$$c = 35 - 16.25 = 18.75 \text{ cm}, \quad \text{or} \quad 0.1875 \text{ m}$$

The total load, as determined by compression in the top portion of the beam, is equal to

$$P = wL = (SI/c) \times 8/L = (8.3 \times 10^6 \text{ Pa})(656.6 \times 10^{-6})/(0.1875)(8/10)$$
$$= 23{,}252 \text{ N} = 2371 \text{ kg}$$

If the load is determined by tension in the bottom fibers, the distance c would be equal to 0.165 meters, and the permissible load would be

$$P = [(8.3)(6565.6)/0.1625](8/10) = 26{,}830 \text{ N}, \quad \text{or} \quad 2736 \text{ kg}$$

Example 7.17

Construct the shear and moment diagrams for the beam shown in Exhibit 11, and show the positions and magnitude of the maximum points. What is the maximum moment in the beam?

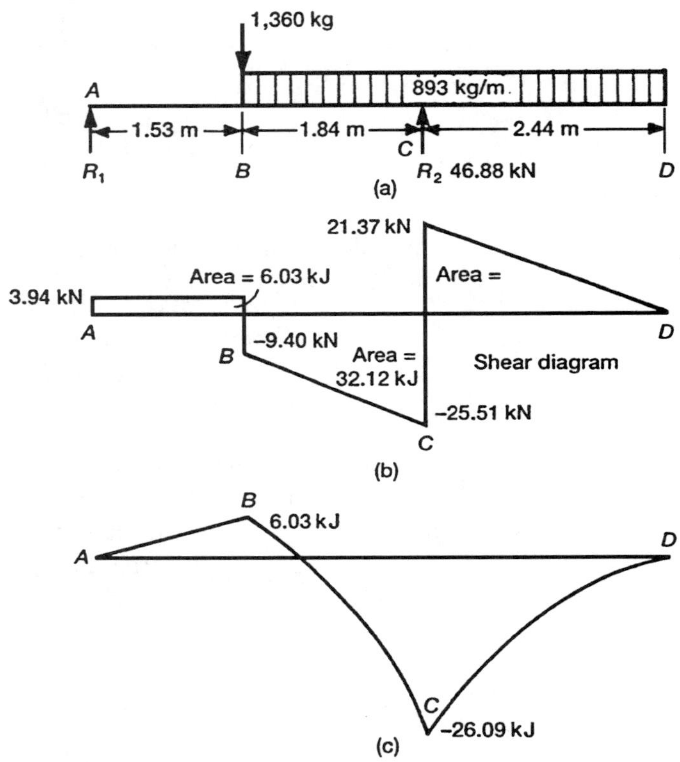

Exhibit 11

Solution

To make certain that the location and magnitude of the maximum moment are determined accurately, it is best to construct the shear and moment diagrams. The diagrams will be constructed just below the beam diagram so that locations of the important points will be shown in their proper positions. First it is necessary to determine the reactions at R_1 and R_2. The loads are given in kg, these must be converted to newtons. The uniform loading of 893 kg/m equals 893g or 8758 N/m and the concentrated mass loading of 1360 kg exerts a force of 13.34 kN.

Taking moments about R_1 gives

$$R_2 = [(1.53 \times 13{,}340) + (1.53 + 2.14) \times (8{,}758 \times 4.28)]/3.37 = 46.88 \text{ kN}$$

The total load on the beam is equal to

$$13{,}340 + 8{,}758 \times 4.28 = 50.82 \text{ k}$$

so the reaction at the left end is equal to

$$R_1 = 50.82 - 46.88 = 3.94 \text{ kN}$$

The force at point *A* is equal to 3.94 kN up, so this provides a positive shear force up which remains constant to point *B*. At point *B* there is an applied mass of 1360 kg which produces a force of 13,338 N down, so the sharing force becomes 3940 – 13,338 equals –9398 N, or a negative shearing force of 9,398 N. The downward, or negative, shear force increases uniformly by $893g = 8758$ N/m for 1.84 meters to point *C*. At point *C* the magnitude of the shearing force is equal to

$$V = -9{,}398 - 8{,}758 \times 1.84 = -25{,}513 \text{ N}$$

At point *C* there is an upward force of 46.88 kN, so the magnitude of the shear force becomes

$$V = -25.51 + 46.88 = 21.37 \text{ kN}$$

This shear force will reduce linearly by the applied load of 8.76 kN/m for 2.44 meters to point *D*, where it becomes zero, which checks.

From the previously given relationship

$$M_2 - M_1 = \int V(x)dx \text{ from Points 1 to 2 (see } \textit{FE Handbook}\text{)}$$

This means that the difference between the values of the moments at Points 1 and 2 is equal to the area under the shear diagram between Points 1 and 2.

Calculating the different areas gives

From *A* to *B*, A = 3.90 kN × 1.53 m = 6.03 kJ
From *B* to *C*, A = 1.84 × (9.40 + 25.51)/2 = 32.12 kJ
From *C* to *D*, A = 2.44 × 21.37/2 = 26.07 kJ

The moment at point B *is* equal to 6.03 kJ and the moment at point *C* is equal to 6.03 – 32.12 = –26.09 kJ. The moment at point *D*, the end of the beam, is equal to 26.09 – 26.07 = 0.02, which checks within the computational accuracy. The maximum moment is equal to –26.09 kJ and occurs at the location of the support, R_2.

Example 7.18

What would be the maximum stress in a cantilever beam holding a mass of 500 kg at its end if it is 3.00 m long and is made of a 200 mm wide by 300 mm high wooden beam?

$$\text{Stress, } \sigma = Mc/I$$

The maximum stress would occur where the moment was a maximum.

$$M_{max} = 500 \times 9.8066 \times 3.00 = 14{,}710 \text{ J}$$

The moment of inertia of the beam,

$$I = bh^3/12 = 0.200 \times 0.300^3/12 = 4.500 \times 10^{-4} \text{ m}^4$$

The distance from the neutral axis to the outermost fiber,

$$c = 0.150 \text{ m}$$

The maximum stress is thus equal to

$$\sigma = 4.903 \text{ MPa}$$

As noted in the *FE Handbook* the moment of inertia of an area about an axis that is parallel to, but does not pass through its centroid equals its moment of inertia about its centroidal axis plus its area times the square of the distance between the two axes.

Example 7.19

What is the moment of inertia about its centroidal axis of a beam composed of two 200 mm by 300 mm beams connected in the form of a tee, see Exhibit 12.

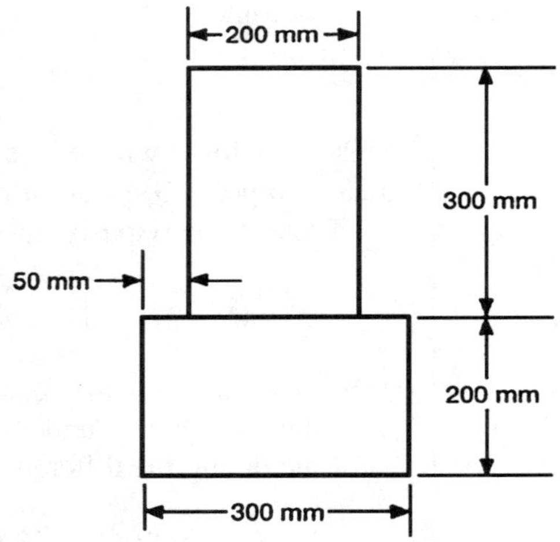

Exhibit 12

Solution

First it is necessary to determine the location of the centroidal axis. Take moments of the areas about a horizontal axis through the top edge of the composite beam and divide by the total cross-sectional area.

$$c = (0.150 \times 0.060 + 0.400 \times 0.060)/(0.060 + 0.060) = 0.275 \text{ m}$$

The centroid of the composite beam is 0.275 m down from the top.

The moment of inertia of the vertical member about its horizontal centroidal axis, parallel to the axis through the centroid of the composite beam is

$$I = bh^3/12 = 450 \times 10^{-6} \text{ m}^4$$

and about the new axis,

$$I = 0.060 \times (0.275 - 0.150)^2 + 450 \times 10^{-6} = 1387.5 \times 10^{-6} \text{ m}^4$$

The moment of inertia of the cross member about its centroidal axis,

$$I = 0.300 \times 0.200^3/12 = 200 \times 10^{-6} \text{ m}^4$$

and about the new axis,

$$I = (0.125^2 \times 0.060) + 200 \times 10^{-6} = 1137.5\ 10^{-6}\ \text{m}^4$$

The total moment of inertia of the composite beam about its centroidal axis,

$$I = (1387.5 \times 10^{-6}\ \text{m}^4) + (1137.5 \times 10^{-6}\ \text{m}^4) = 2525 \times 10^{-6}\ \text{m}^4$$

Example 7.20

What would be the moment of inertia of an aluminum box-beam shown in Exhibit 13? What would be the maximum deflection of such a beam if it were uniformly loaded cantilever beam 3.00 m long, with a distributed loading of 100 kg/m?

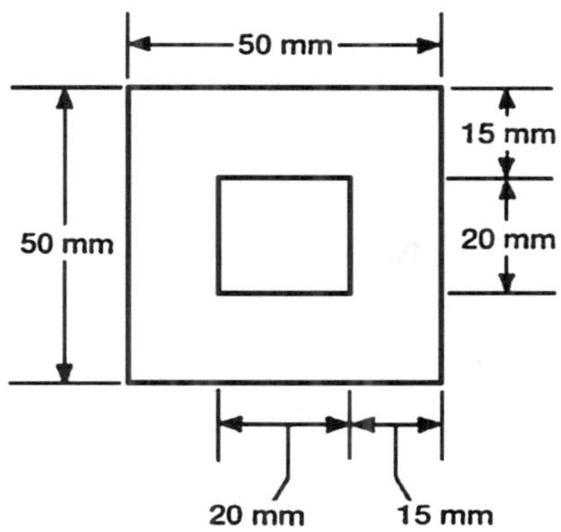

Exhibit 13

Solution

I = $bh^3/12$, which reduces to $b^4/12$ for a square cross section. For this case, the moment of inertia of the section would equal the I of the 50-mm square minus the I of the 20-mm core section.

$$I = (0.050^4 - 0.020^4)/12 = 5.075 \times 10^{-7}\ \text{m}^4$$

The loading would be equal to 100 × 9.8066 = 980.66 N/m.

The formula for the maximum deflection of a uniformly loaded cantilever beam as given in the *FE Handbook* is

$$\delta_{max} = w_0 \times L^4/(8EI)$$
$$\delta_{max} = (\text{N/m}) \cdot L^4/8EI = 980.66 \times 81/(8 \times 5.075 \times 10^{-7} \times 6.9 \times 10^{10})$$

where

$w_0 = 980.66$ N/m
$\delta = 0.284$ m, or 28.4 cm

What would be the maximum stress in the beam?

$$\delta = Mc/I = [L \times L/2]c/I$$
$$M = 980.66\ L^2/2$$
$$\delta = (4413)(0.025)/(5.075 \times 10^{-7}) = 217.4\ \text{Pa}$$

Example 7.21

What concentrated load at the end of the cantilever beam in Example 7.20 would produce the same deflection?

Solution

From the *FE Handbook* the deflection of the end of an end-loaded cantilever beam equals

$$\delta = PL^3/(3EI)$$

and for a uniformly loaded beam

$$\delta = L^4/8EI, \text{ so}$$
$$P = (3EI/L^3)(L^4/8EI) = (3/8)L = (0.375)(980.66)(3.0) = 1103 \text{ N}$$

or a concentrated load of 113 kg as opposed to a uniformly distributed load of 300 kg.

SHEAR STRESSES IN BEAMS

A loaded beam will strain in some fashion. When it does, it flexes and the longitudinal fibers are placed in tension or compression every place except at the neutral axis. The effect of flexure on a simple beam is shown in Figure 7.4. If the beam were made of two parallel pieces, originally of the same length which were placed in contact, but not fixed to one another, the ends of the top member would overlap the ends of the bottom member as shown in Figure 7.5. The bottom-most fibers of the top member would strain in tension and elongate. The uppermost fibers of the bottom member would strain in compression and reduce in length. If the two members were joined together in such a manner that the differential strain were prevented, there would be a definite longitudinal shearing stress at the junction, the neutral axis, of the beam shown in Figure 7.5.

The longitudinal shear stress is related to the transverse (vertical in this case) shear stress by the relationship

$$S_s = VQ/(I \times t) \text{ Pa} \quad \text{or,}$$
$$\text{Shear flow} = VQ/I \text{ N/m}$$

where

V = vertical shear load at the point being considered.
$Q = \int fy\, dA = yA$

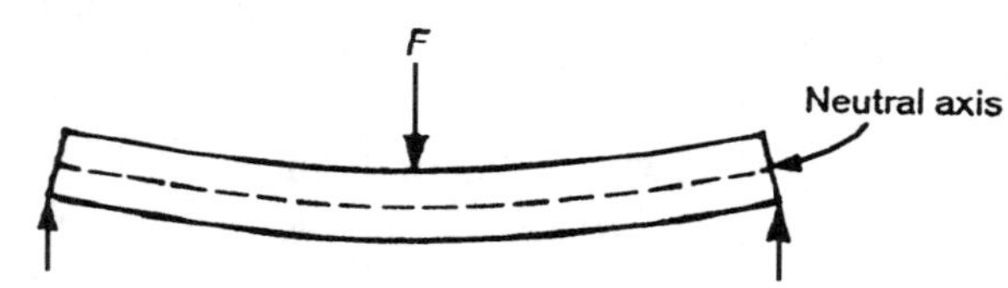

Figure 7.4

Figure 7.5

where

y = the distance from the neutral axis of the beam to the neutral axis of area A
A = the area of the section above the plane at which the shear stress acts
I = transverse moment of inertia
t = length (width) of section being considered.

Example 7.22

Take, for example, a rectangular cross-section beam as shown in Exhibit 14. What is the maximum longitudinal shear in the beam?

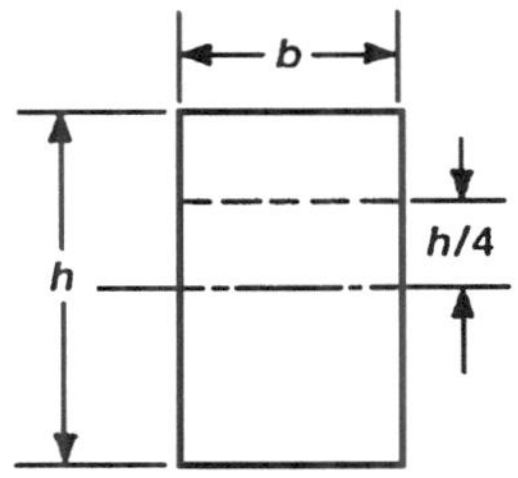

Exhibit 14

Solution

The longitudinal shear stress will be the greatest at the neutral axis. Using the above relationship for longitudinal shear,

$$Q = yA = (h/4) \times (bh/2) = bh^2/8$$
$$I = bh^3/12$$
$$t = b$$
$$S_s = VQ/It = V \times (bh^2/8)/[(bh^3/12) \times b] = 3V/2bh$$

The transverse shear stress equals V/bh, so the longitudinal shear stress at the neutral axis is 50 percent greater than the transverse shear stress for a rectangular beam.

For a circular beam, see Exhibit 15.

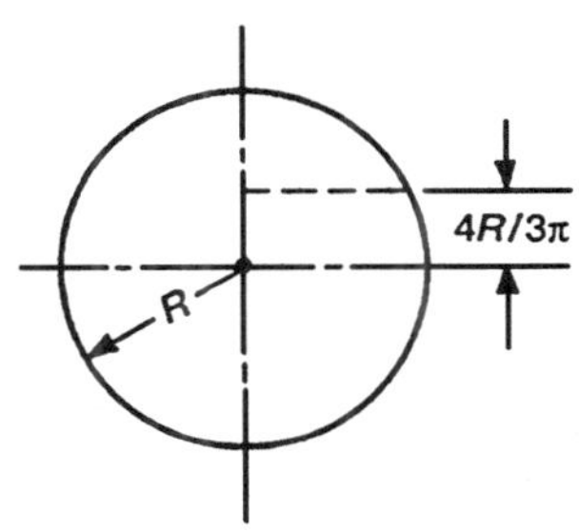

Exhibit 15

$$Q = yA = (4\pi/3\pi) \times (\pi r^2/2) = 2r^3/3$$
$$I = \pi r^4/4$$
$$t = 2r$$
$$S_s = VQ/It = V \times (2r^3/3)/[(\pi r^4/4) \times 2r] = 4V/(3\pi r^2)$$

So, the longitudinal shear at the neutral axis of a beam with a circular cross section is one-third greater than the transverse shear.

Example 7.23

Two wooden planks 5 cm by 15 cm by 2.0 meters long are layered as shown in Figure 7.5. They support a mass 28 kg at a point midway between the end supports. It is desired to stiffen the beam since the assembly sags too much in the center, so the two planks are to be fastened together with nails. If one nail is capable of withstanding a shearing force of 650 newtons, what should be the spacing of the nails?

Solution

First determine the transverse shear stress. Since it is a simply supported beam the reactions at each end will hold half the mass, or the force at each end will equal $14g = 137.4$ N. The transverse shear stress of the assembled beam will equal $137.4/(0.10)(0.15) = 9153$ Pa. The shear stress at the neutral axis of the beam will thus equal $(1.50)(9153) = 13{,}730$ Pa. The assembled beam is 0.15 meters wide, so the longitudinal shear force will equal $(0.15)(13{,}730) = 2060$ N per meter of length. This means that $2060/650 = 3.17$ nails will be required for every meter of beam length, or one nail every 31.5 cm.

COMPOSITE BEAMS

The beams so far considered have all been of one homogeneous material. There are many beams, however, that are made of two materials, principally reinforced-concrete beams and steel-reinforced wood beams. Such beams are called "composite beams"; because of the difference in the moduli of elasticity of the two materials, they require a slightly different type of analysis than that used for homogeneous beams. One type of composite beam is shown in Figure 7.6. It is made up of a 15- by 20-cm timber, 3.0 meters in length, with a 15-cm-wide strip of steel 3.2 mm thick attached to the bottom edge. To withstand the applied load, the beam will deflect, the upper part of the wooden portion will compress, and the lower part of the wood and the steel will strain in tension. The stress in any part of the beam will equal the unit deformation in that particular portion times the modulus of elasticity of the material of which that portion is made. It is assumed that the steel will strain the same amount as the assumed wood fibers on the bottom of the beam. Then the steel will be stressed in the ratio

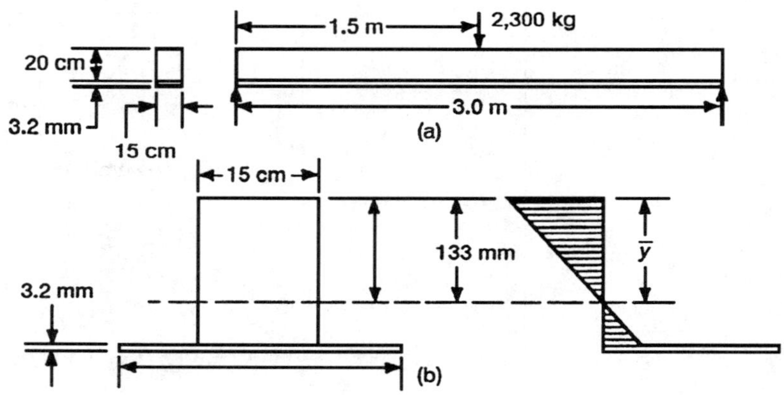

Figure 7.6

E_s/E_w times the stress in the assumed bottom-most wood fibers. Since $E_s = 2.1 \times 10^{11}$ Pa and $E_w = 7.0 \times 10^9$, the stress in the steel will be thirty times the stress in the equivalent bottom wood fibers. This means that the tensile load in the steel will be considerably greater than that in the wood, and the neutral axis will be below the geometric center of the beam.

Another way of looking at this problem is that since, Force = stress × area and Stress = $E \times \varepsilon$, the steel strip could, theoretically, be replaced by a piece of wood with an area equal to E_s/E_w × area of steel, or thirty times the area (width) of the steel, but with the same thickness. The equivalent wooden beam is shown in the figure, and the internal-force diagram showing the distribution of $S \times A$ over the cross section of the beam is also shown.

$$E_s/E_w = n$$

Example 7.24

What is the stress in the steel band shown in Figure 7.6?

Solution

To determine the stress in the, wood and in the steel, we need to know the distance to the neutral axis y. This can be determined by taking moments about a horizontal axis through the upper edge of the beam; one must remember that a slice of the equivalent beam section would balance on a fulcrum placed at the neutral axis (for an all-wood beam).

$$(15)(20)(10) + (0.32)(450)(20 + 0.16) = [300 + (450)(0.32)]y$$
$$y = 5903.04/444.0 = 13.295 \text{ cm}$$

$S = Mc/I$ where I is the moment of inertia of the equivalent bema about the neutral axis. Utilizing the parallel-axis theorem,

$$I = (0.15)(0.20)^3/12 + 0.300(0.13295 - 0.100)^2 + (4.50)(0.0032)(0.2016 - 0.1330)^2$$
$$I = 0.000100 + 0.00003257 + 0.00006777 = 0.00020034 \text{ m}^4$$

$$\text{Maximum moment, } M = (2300g/2)(3/2) = 16.917 \text{ kJ}$$
$$\text{For steel strip, } c = 0.2030 - 0.1330 = 0.0702 \text{ m}$$

The tensile stress in the bottom fibers of the equivalent wooden beam would equal

$$S = (16{,}917)(0.0702)/0.000200 = 5.938 \text{ MPa}$$

and the stress in the steel would thus be equal to

$$S = 30 \times 5.938 = 178 \text{ MPa}$$
$$n = 210/7 = 30$$

The commonest type of composite beam is the reinforced-concrete beam, which can also be handled by the method described. One difference, however, is that the concrete is assumed to withstand no tensile stress and the portion of the concrete on the tension side of the neutral axis is disregarded in the stress calculations. A reinforced-concrete beam is shown in Figure 7.7.

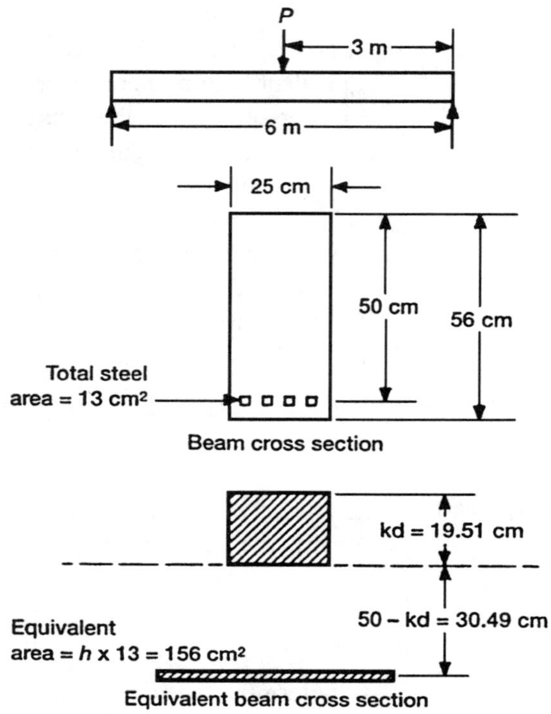

Figure 7.7

Example 7.25

Determine the allowable concentrated load, P, which the reinforced concrete beam show in Figure 7.7 will hold if $n = 12$ and the maximum allowable stresses are 6.90 MPa in compression for the concrete and 138 MPa in tension for the steel.

$$n = E_{steel}/E_{concrete}$$

Solution

The equivalent beam cross section is drawn; as the concrete will take only compressive loading, the effective concrete area will extend only to the neutral axis. The location of the neutral axis is found by taking moments about the neutral axis. The concrete section is shown shaded in the figure. It is assumed that the portion of the concrete that will be stressed in compression will occupy the top "kd" distance of the beam with an area of $kd \times 25$ cm^2. The moment of this area about the neutral axis must equal the moment of the "tensile" concrete ($n \times$ the area of the steel reinforcing bars) about the neutral axis.

$$25kd(kd/2) = 156(50 - kd)$$

where 156 equals the area of the steel reinforcement times n, or

$$12 \times 13 = 156$$

$$(kd)^2 + 12.48\ kd - 624 = 0$$

Using the general solution for a quadratic equation in the *FE Handbook* gives

$$kd = [-12.48 \pm \sqrt{(156 + 2{,}496)}]/2$$

$$kd = (-12.48 + 51.50)/2 = 19.51 \text{ cm}$$

$$I = (25 \times 19.51^3)/12 + (25 \times 19.51) \times (19.51/2)^2 + 156 \times (50 - 19.51)^2$$

$$= 206{,}909 \text{ cm}^4 = 2.07 \times 10^{-3} \text{ m}^4$$

Note that the moment of inertia of the steel portion about its own neutral axis is disregarded since it is so small compared with the other components.

$$S = MC/I$$
$$M_{concrete} = 6.90 \times 10^6 \times .00207/0.1951 = 73{,}209 \text{ Nm}$$
$$M = (P/2) \times 3 \text{ so,}$$
$$P = 2M/3$$

P, as limited by compressive stress in the concrete, is equal to 48,810 N or 4977 kg.

$$M_{steel} = 138 \times 10^6 \times 0.00207/(n \times 0.3049) = 78{,}075 \text{ Nm}$$
$$P = 52{,}050 \text{ N, or } 5307 \text{ kg}$$

Note the use of the n in the denominator for determining the allowable moment for the steel.

The concentrated load, P, is limited by the stress in the concrete and is equal to 4977 kg.

RADIUS OF GYRATION

The moment of inertia is defined as $\int r^2 dm$ or $\int r^2 dA$ and is discussed in Chapter 2. If all the mass, or all the area, were concentrated at a distance "r" from the point about which the moment of inertia was to be calculated, then "I" would become just r^2m or r^2A and "r" would equal the radius of gyration. The radius of gyration is usually indicated by "k" rather than "r".

$$I = k^2m, \text{ or}$$
$$I = k^2A$$

For a circular cross section the radius of gyration equals $D/4$.

$$I = \pi D^4/64 = AD^2/16$$

and, since $I = k^2A$ where k = radius of gyration, $k = D/4$.

It should be noted that, in general, the mass of a body cannot be considered as acting at its center of mass for the purpose of calculating the moment of inertia. This is discussed in Chapter 2.

The moment of inertia about an axis in the plane of an area is called the "plane moment of inertia"; the moment of inertia of an area about an axis perpendicular to the area is termed the "polar moment of inertia" which is discussed in Chapter 2.

COLUMNS

A column with a high slenderness ratio, L/r, length divided by the radius of gyration of the column, is prone to buckling, whereas a short column can be treated by normal methods. If the L/r ratio is greater than 30 it is safer to treat the column as a "long column" and use Euler's Formula to determine whether or not the loading is critical.

Example 7.26

A 50-mm diameter, 3.00 m long, steel rod is pinned at both ends and is unstressed at 20°C. What is the highest temperature to which the bar may be heated before it will buckle? The thermal coefficient of expansion of steel is 11.7×10^{-6}/°C.

Solution

The radius of gyration equals 0.050/4 = 0.0125 and the *L*/*R* ratio equals 3.00/0.0125 = 240 so it is a "long column" and should be treated by Euler's formula, which, from the *FE Handbook* is

$$P_{cr} = \pi^2 EI/(kL)^2, \text{ and}$$
$$P_{cr}/A = \pi^2 E/[k(L/r)]^2$$

The value for the constant *k* from the table in the *FE Handbook* is equal to 1.00.

$$L/r = 240.0$$
$$P_{cr}/A = 9.870 \times 2.1 \times 10^{11}/(1 \times 240.0)^2 = 35.98 \text{ MPa}$$

The compressive stress induced in the steel rod due to an increase in temperature will equal

$$S = \Delta T(11.7 \times 10^{-6})(2.1 \times 10^{11}) \text{ Pa, so}$$
$$\Delta T = 35.98/2.457 = 14.64\ °\text{C, and}$$
$$T = 34.64\ °\text{C}$$

RIVETED JOINTS

Riveted joints can be generally divided into two general classes—joints in which the line of action of the applied force passes through the centroid of the rivet pattern and those in which the force exerts a moment on the rivet pattern. In a simple continuous joint such as is shown in Figure 7.8 the load is assumed to be

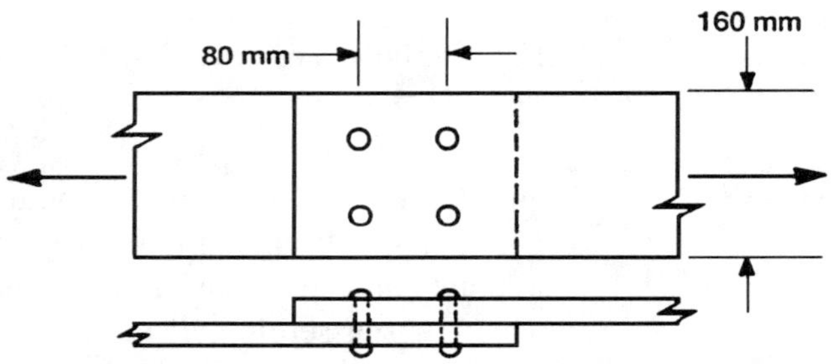

Figure 7.8

distributed equally among the rivets making up the joint. In the sample shown one-quarter of the load would be assumed to be held by each rivet. The small moment force due to the displacement of the lapping plates is ignored. However, if the load is offset from a line of action passing through the centroid of the rivet pattern as shown in Figure 7.9, the force due to the effect of the moment must also be included.

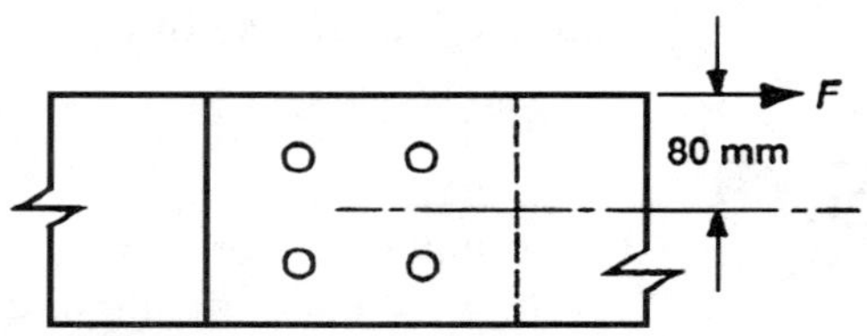

Figure 7.9

Example **7.27**

The rivets in Figure 7.9 are 12.0 mm in diameter and form a square pattern with sides of 80 mm located centrally in the 160 mm wide plates. The 80 mm is measured between the centers of the rivets at the corners of the square. The load is shown applied at the edge of the 160 mm wide plate. This load will apply a moment load to the rivet pattern equal to $F \times 0.080$ which will be resisted equally by each of the rivets. Each rivet is $0.040 \times \sqrt{2} = 0.05657$ m from the centroid of the rivet pattern. The resisting force required to counteract the moment would equal

$$0.080 \cdot F/(4)(0.05657) = 0.3535\ F \text{ each rivet}$$

this force would add geometrically to the direct force of $F/4$ applied to each rivet (see Exhibit 17). This would increase the force on two of the rivets and reduce the force on two others. The maximum force would equal (Exhibit 18)

$$([(0.250 + 0.707(0.3535)]^2 + [(0.707)(0.3535)]^2)^{1/2} = 0.559\ F$$

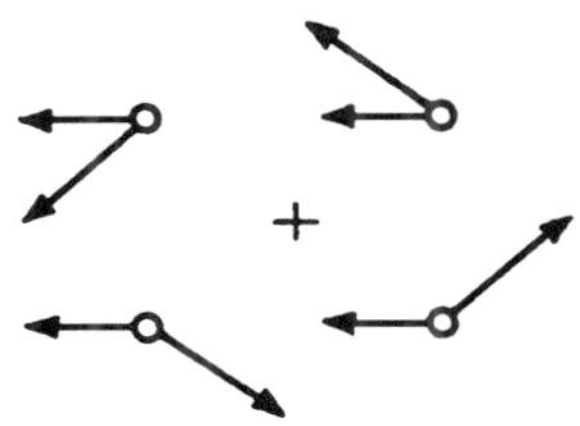

Exhibit 17

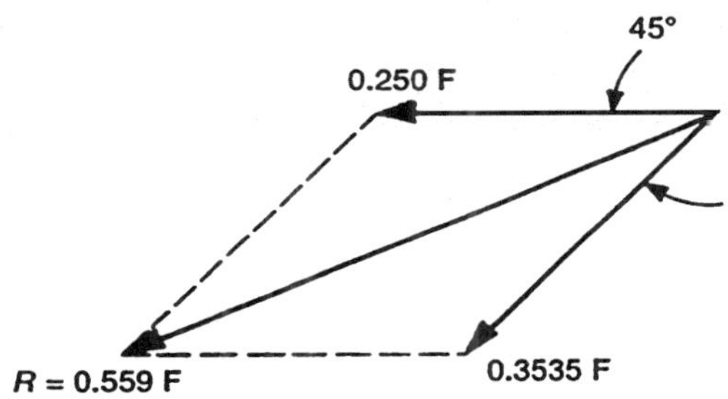

Exhibit 18

The eccentricity of the load application would more than double the maximum stress in the highest stressed rivet in this particular case.

WELDED JOINTS

The size of a weld is generally specified by the size of the fillet. For the joint shown in Figure 7.10 the fillet is given as 6.0 mm. The shear area, however, is the thickness of the throat of the fillet or, for this case

$$A = 6 \times 0.707 \times \text{length}$$

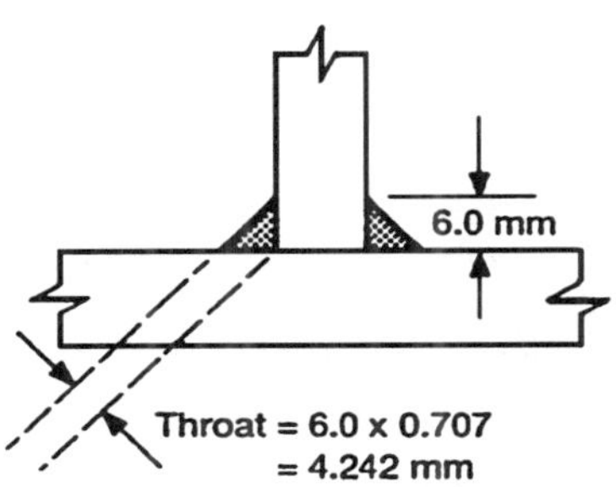

Figure 7.10

Example 7.28

For the joint shown in Figure 7.10, what is the maximum force which could be applied to the vertical member per centimeter of width if the allowable shear stress of the weld material is 100 MPa?

Solution

For each centimeter length of the joint there would be

$$2 \times 0.006 \times 0.707 \times 0.010 = 84.84 \times 10^{-6}\ \text{m}^2 \text{ of weld area}$$

$$F = 84.84 \times 100.00 = 8{,}480\ \text{N}$$

PROBLEMS

7.1 A 45,340-kg mass is supported by a 102-mm diameter steel cylinder surrounded by a copper tube of equal length with an outside diameter of 203 mm, which fits snugly around the steel cylinder. The assembly is shown in Exhibit 7.1. The weight is distributed uniformly over the surfaces of the supporting members. What is the stress in the copper tube?

$$E(\text{steel}) = 2.1 \times 10^{11} \text{ Pa}$$
$$E(\text{copper}) = 1.12 \times 10^{11} \text{ Pa}$$

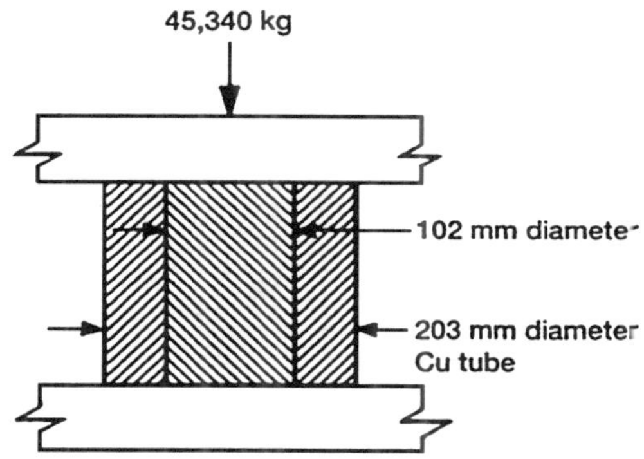

Exhibit 7.1

a. 10.58 MPa
b. 10.92 MPa
c. 11.26 MPa
d. 11.67 MPa

7.2 A 5.0-cm-diameter steel rod 3.0 meters long is pinned at both ends, and is unstressed at 15°C. See Exhibit 7.2. Which of the following most nearly equals the highest temperature to which the rod may be heated before it will buckle?

$$\alpha = 11.7 \times 10^{-6} \text{ m/m} \cdot {}^\circ\text{C}$$

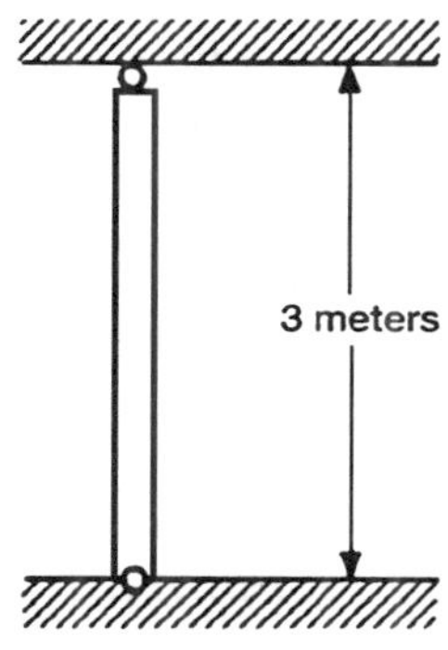

Exhibit 7.2

a. 29.6 C
b. 30.5 C
c. 31.6 C
d. 45.6 C

7.3 Three bars of different materials, as shown in Exhibit 7.3, are to be compressed equally by a plate that is to remain horizontal. Which of the following most nearly is equal to the distance x?

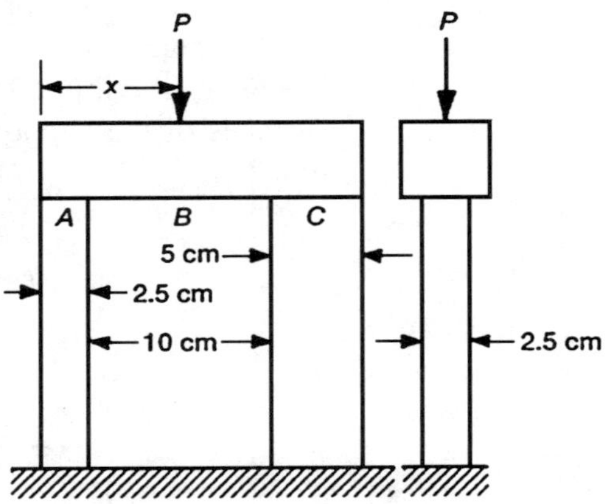

Exhibit 7.3

Bar	*E*(GPa)	Allowable Stress (MPa)
A	210	138
B	70	69
C	140	104

a. 6.0 cm
b. 6.5 cm
c. 7.3 cm
d. 7.9 cm

7.4 A 10-cm-diameter, horizontal, solid steel shaft is 2.50 meters long and is rigidly held at both ends. A torque of 16,000 N·m is applied to a pulley that is keyed to the shaft 60 cm from the left end. Which of the following most nearly equals the maximum torsional stress in the shaft?

a. 58 MPa
b. 62 MPa
c. 66 MPa
d. 69 MPa

7.5 The shaft shown in Exhibit 7.5 is made of one material and is held rigidly at the ends. Which of the following most nearly equals the maximum end torque reaction?

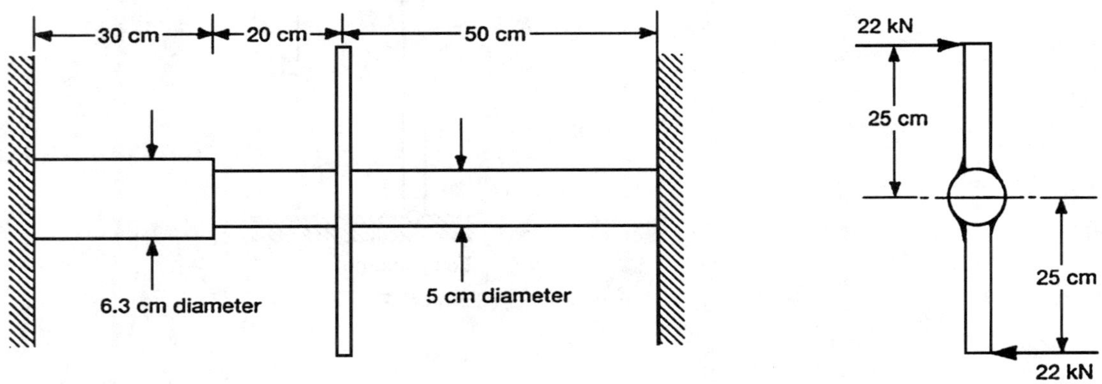

Exhibit 7.5

a. 4.3 kJ
b. 4.9 kJ
c. 6.3 kJ
d. 6.7 kJ

7.6 A massless, bimetallic, cylindrical bar is twisted by a torque of 1900 N·m at the change of section. Which of the following most nearly is equal to the maximum end reaction? (See Exhibit 7.6.)

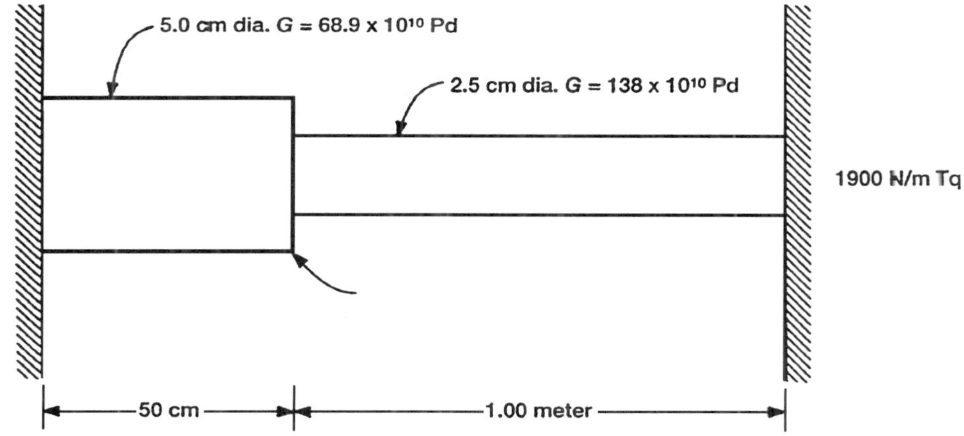

Exhibit 7.6

a. 0.9 kJ
b. 1.3 kJ
c. 1.8 kJ
d. 2.2 kJ

7.7 A mass of 450 kg was originally held up by two 2.5-cm-diameter steel tie rods as shown in Exhibit 7.7a. When it was desired to increase the mass to 900 kg a proposal was made to add two 2.5-cm-diameter cables as shown. Assume no stress before the weight was applied, and no slack in the system. Which of the following most nearly is equal to the amount of the added load that would be held by the cables? For steel cable $E = 84$ GPa

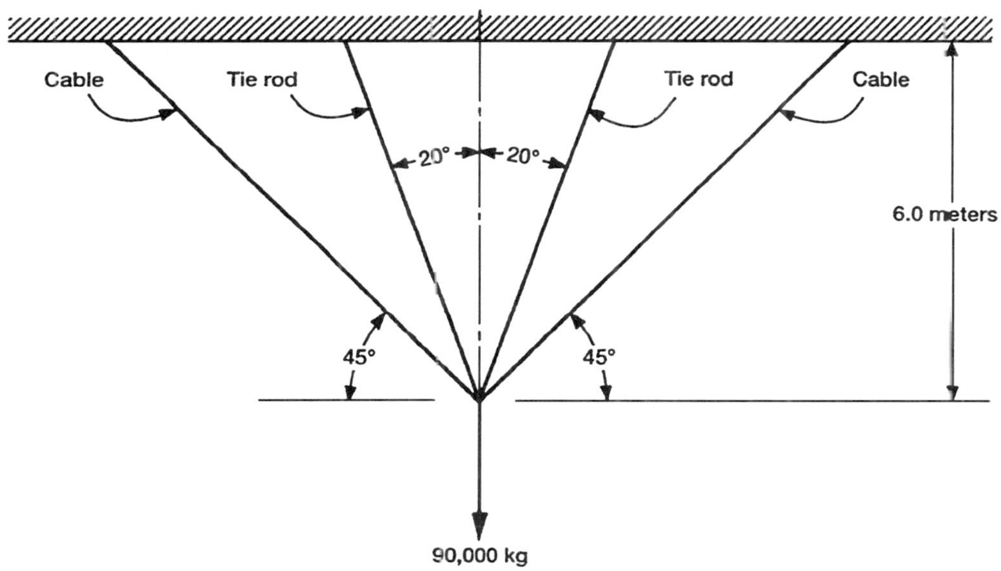

Exhibit 7.7a

a. 29%
b. 21%
c. 18%
d. 15%

7.8 A 25.0-mm-thick steel plate 0.25 m wide is loaded as shown in Exhibit 7.8. What is the maximum stress in the plate?

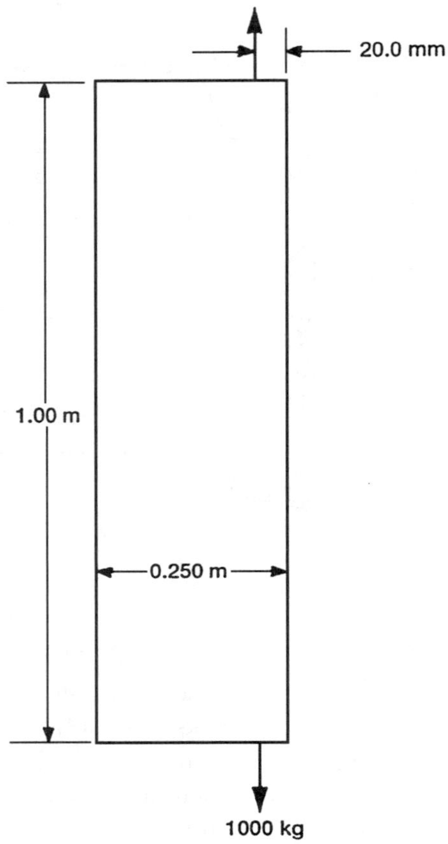

Exhibit 7.8

a. 1.60 Mpa
b. 2.91 MPa
c. 4.22 MPa
d. 5.55 MPa

7.9 A closed-end tube with a 152.4-mm *OD* and a 2.54-mm wall thickness contains a fluid at a pressure of 3.448 MPa. It is subjected to a torsional load of 678 N·m. What is the maximum principal stress in the tube?

a. 93 MPa
b. 98 Mpa
c. 101 MPa
d. 105 MPa

7.10 Mohr's circle is used to determine

a. Shear stress
b. Poisson's ratio
c. Residual stress
d. Combined stress

7.11 What would be the maximum stress in a cantilever beam holding a mass of 500 kg at its end if it is 3.00 meters long and is made of a 200-mm wide by 300-mm high wooden beam?

a. 6.8 MPa
b. 6.1 Mpa
c. 5.5 MPa
d. 4.9 MPa

7.12 A water tank is filled to a depth of ten meters. It is 35 meters in diameter and is made of steel 8 mm thick. What is the stress in the tank wall at the bottom?

a. 214 MPa c. 199 MPa
b. 206 Mpa d. 192 MPa

7.13 What is the magnitude of the force acting on member *CD* in Exhibit 7.13?

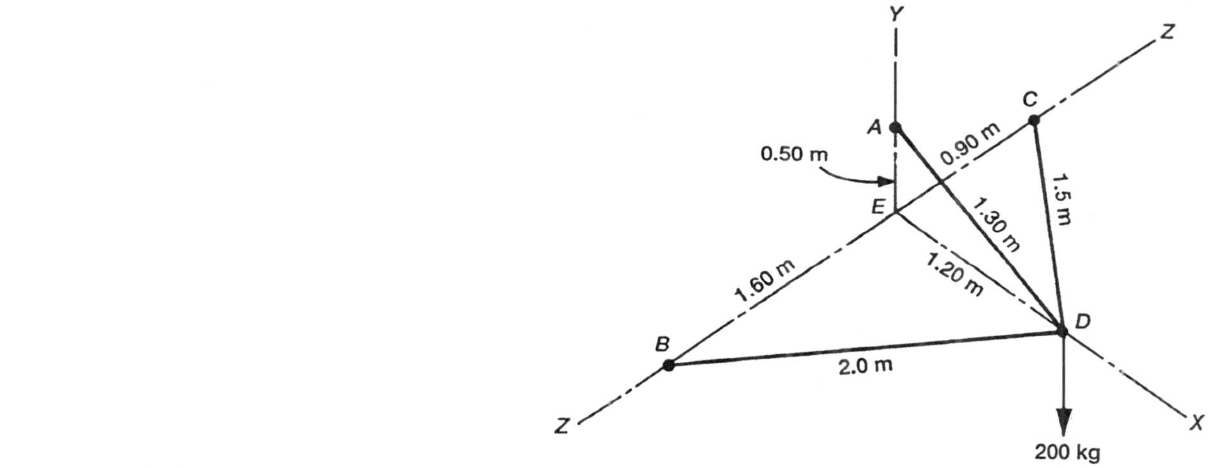

Exhibit 7.13

a. 2.9 kN c. 3.8 kN
b. 4 kN d. 4.3 kN

7.14 Given the beam shown in Exhibit 7.14, loaded as shown, which of the following most nearly is equal to the maximum moment, measured in N·m?

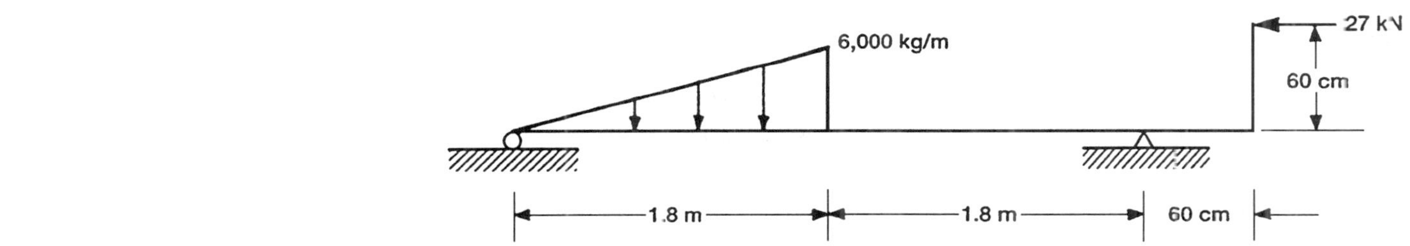

Exhibit 7.14

a. 41 kJ c. 48 kJ
b. 44 kJ d. 52 kJ

7.15 A cantilever beam supports a uniformly varying load as shown in Exhibit 7.15. Which of the following most nearly is equal to the maximum bending stress?

a. 8.7 MPa
b. 9.2 MPa
c. 10.6 MPa
d. 11.8 MPa

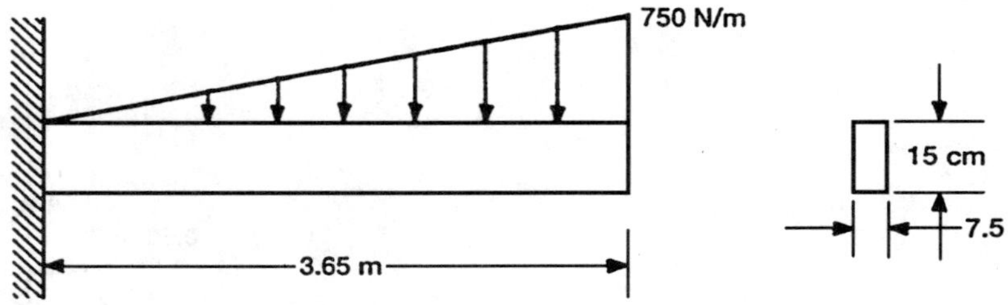

Exhibit 7.15

7.16 Which of the following most nearly is equal to the maximum moment for the beam that is loaded as shown in Exhibit 7.16?

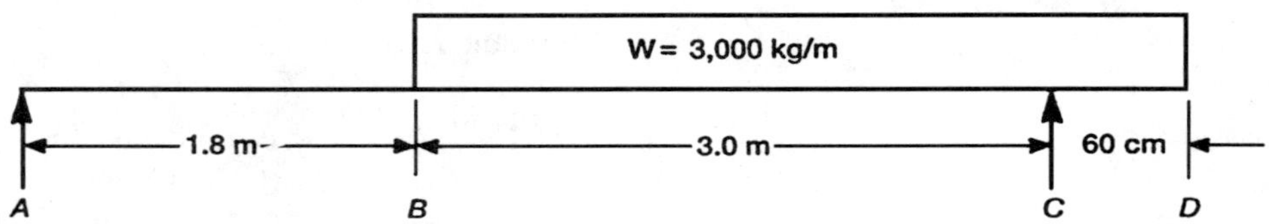

Exhibit 7.16

a. 56.5 kJ
b. 59.7 kJ
c. 62.1 kJ
d. 64.3 kJ

7.17 For the beam in Exhibit 7.17, which is loaded as shown, which of the following most nearly is equal to the maximum moment in the beam?

a. 180.8 kJ
b. 153.7 kJ
c. 132.6 kJ
d. 89.8 kJ

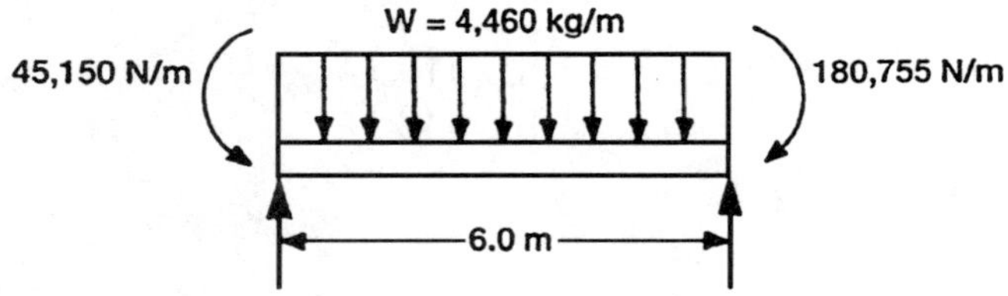

Exhibit 7.17

7.18 A section of a rectangular water tank is as shown in Exhibit 7.18. The beams are in pairs and are spaced 1.0 meter apart along the length of the tank.

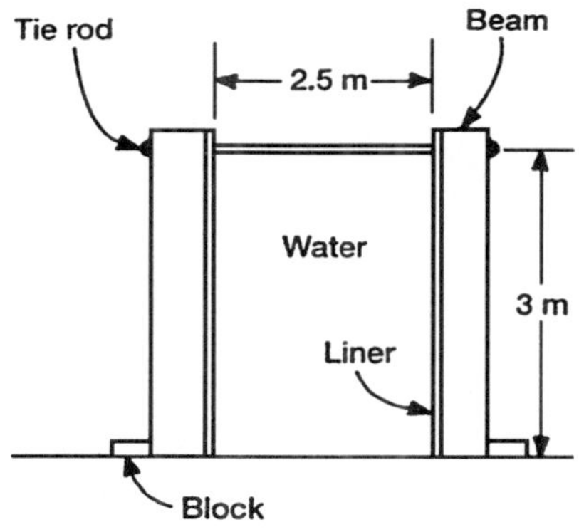

Exhibit 7.18

Assume that the section shown is near the middle portion of the tank and that no end (of the tank) effects act on the section shown. Which of the following most nearly equals the maximum moment in the beam shown in the figure?

a. 29.4 kJ
b. 22.7 kJ
c. 16.3 kJ
d. 13.1 kJ

7.19 Which of the following most nearly is equal to the maximum moment that occurs in the beam that is loaded as shown in Exhibit 7.19?

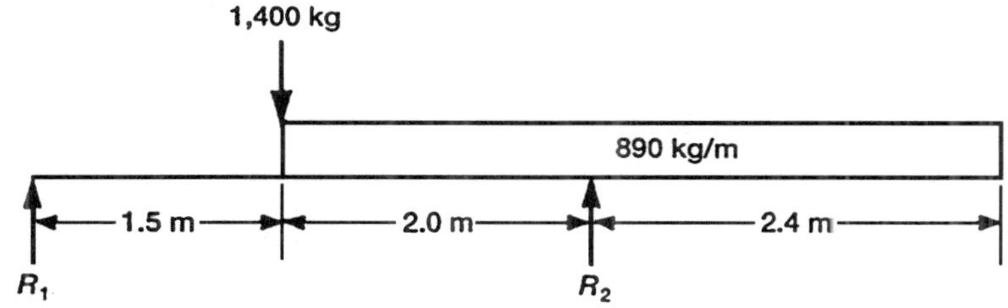

Exhibit 7.19

a. 8.5 kJ
b. 12.7 kJ
c. 18.3 kJ
d. 25.1 kJ

7.20 The composite beam shown in Exhibit 7.20 is subjected to a maximum moment of 17,000 N·m. Which of the following most nearly is equal to the maximum tensile stress in the steel? $E_{wood} = 8.4$ GPa.

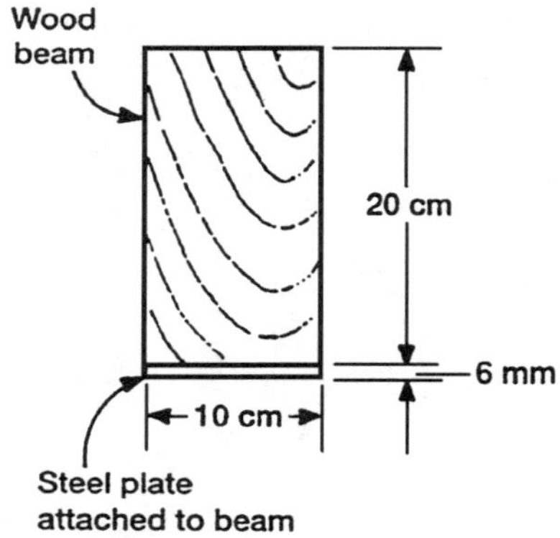

Exhibit 7.20

a. 167 MPa
b. 123 MPa
c. 92 MPa
d. 7 MPa

7.21 Which of the following most nearly is equal to the deflection of point A in Exhibit 7.21 if the parts are all of steel?

a. 4.1 mm
b. 5.2 mm
c. 6.6 mm
d. 7.3 mm

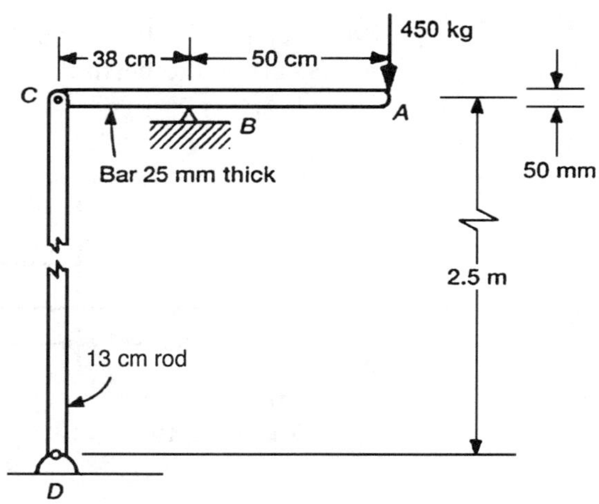

Exhibit 7.21

7.22 A short steel column with a 25-cm *ID*, and a 30-cm *OD* is filled with concrete. If the maximum allowable stress for the steel is 276 MPa and that for the concrete is 24 MPa, which of the following most nearly is equal to the maximum axial load that the composite column can safely hold? $E_{concrete} =$ 19.6 GPa.

a. 684 Mg
b. 652 Mg
c. 621 Mg
d. 598 Mg

7.23 Which of the following most nearly is equal to the maximum mass P that could be supported by the steel beam shown in Exhibit 7.23 if the compressive stress is limited to 13.8 MPa in the plastic support block?

Assume $E_{cable} = 84$ GPa and $E_{plastic} = 10.5$ GPa.

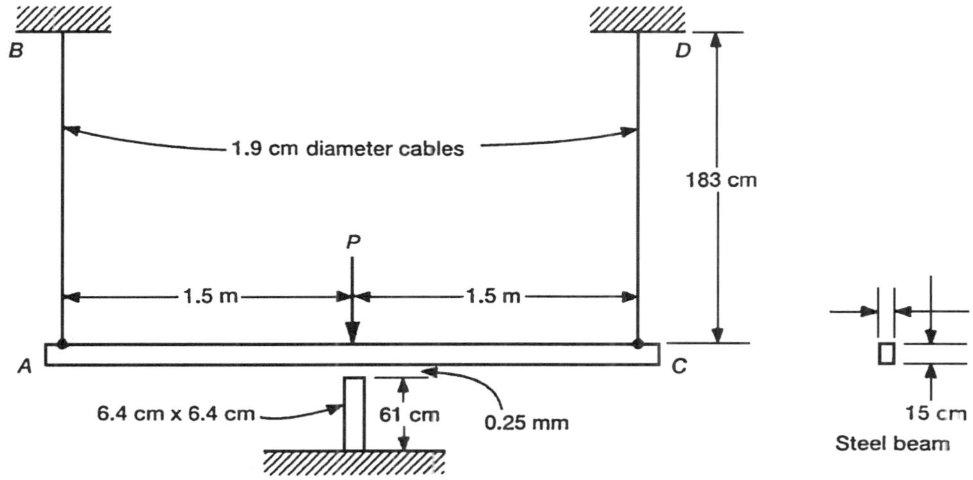

Exhibit 7.23

a. 4.9 Kg c. 5.9 Kg
b. 5.5 Kg d. 6.3 Kg

7.24 A beam is simply supported at the two ends and is uniformly loaded along its length with a load of 225 kg/m. If the beam is 4.5 meters long and is made of a 20-cm by 20-cm wooden timber, which of the following most nearly is equal to the deflection at a point 122 cm from one end? $E_{wood} = 5.0$ GPa.

a. 5.9 mm c. 6.9 mm
b. 6.4 mm d. 7.2 mm

7.25 A reinforced concrete beam, 30 cm wide by 76 cm deep and 6 meters long, is simply supported at the ends. The beam is reinforced with four bars, each with a cross-sectional area of 3.9 cm^2 and placed 8 cm from the bottom of the beam. The beam supports its own mass plus an applied load of 3000 kg per linear meter. Which of the following most nearly is equal to the stress in the steel rebar?

$$n = 10 \text{ and}$$
$$\rho_{concrete} = 2370 \text{ kg/m}^3.$$

a. 141 MPa c. 151 MPa
b. 149 MPa d. 158 MPa

SOLUTIONS

7.1 **c.** The area of the steel core cylinder is equal to $(0.102)^2 \times \pi/4 = 0.008171\ \text{m}^2$ The outer area of the copper sleeve equals is $0.032366\ \text{m}^2$, so the area of the copper put into compression is equalto $0.02419\ \text{m}^2$.

The load to be held is equal to 45,340 kg = 444,650 N. Then,

$$\varepsilon[(0.008171)(2.1) + (0.02419)(1.12 \times 10^{11}) = 444{,}650\ \text{N, and}$$
$$\varepsilon = 4.4465/(0.044251 \times 10^6) = 100.5 \times 10^{-6}\ \text{m/m}$$

The stress in the copper tube is

$$\sigma = (100.5 \times 10^{-6})(1.12 \times 10^{11}) = 11.26\ \text{MPa}$$

7.2 **a.** Euler's formula for long columns is, from the *FE Handbook*

$$P_{cr} = \pi^2 EI/(kl)^2$$

For this case, from the tables given in the *FE Handbook*

$$k = 1, \quad \text{and}$$
$$I = \pi r^4/4$$

for a solid circular bar, or, as given in the *FE Handbook* in the section on columns

$$I = r^2 \text{A}$$

where r = radius of gyration, which is equal to the radius of a circular bar divided by 2, which again gives

$$I = \pi r^4/4$$

where r is the radius of the solid round bar.

$$I = 0.30680 \times 10^{-6}\ \text{m}^4$$

The critical load is then

$$P_{cr} = (9.870)(2.1)(0.03068 \times 10^6)/(3^2) = 70.205\ \text{kN}$$

which indicates a stress of

$$\sigma = 70{,}205/0.001963 = 35.764\ \text{MPa}$$

which corresponds to a strain of

$$\varepsilon = 35{,}764{,}000/(2.1 \times 10^{11}) = 170.304 \times 10^{-6}\ \text{m/m}$$

The increase in temperature to produce this amount of strain is

$$\Delta T = \varepsilon/\alpha = (170.304 \times 10^{-6})/(11.7 \times 10^{-6}) = 14.55\,°\text{C}$$
$$T = 15 + 14.6 = 29.6\,°\text{C}$$

7.3 **d.** The strains in the three blocks will be equal, so the forces exerted by each block are as follows

$$P_A = \varepsilon(210 \times 10^9)(625 \times 10^{-6}) = (131.25 \times 10^6)\varepsilon$$
$$P_B = \varepsilon(70 \times 10^9)(2500 \times 10^{-6}) = (175.00 \times 10^6)\varepsilon$$
$$P_C = \varepsilon(105 \times 10^9)(1250 \times 10^{-6}) = (131.25 \times 10^6)\varepsilon$$

The total force is then

$$P_{total} = (437.5 \times 10^6)\varepsilon$$

Take moments about the left end

$$M = P_A(0.0125) + P_B(0.075) + P_C(0.15) = P_{total}x$$
$$x = [(1.641 + 13.125 + 19.6875) \times 10^6]/(437.5 \times 10^6)$$
$$= 0.0788 \text{ m}, \quad \text{or} \quad 7.88 \text{ cm}$$

7.4 **b.** The angle of twist in the 60-cm length will be the same as the angle of twist in the 190-cm length.

$\phi_1 = \phi_2$, and
$\phi = TL/ZGJ$ (See *FE Handbook*)
$J = \pi a^4/2 = 9.817 \times 10^{-6}$ m^4
$G = 8.3 \times 10^{10}$ Pa, as given in the *FE Handbook*
$T_1 + T_2 = 16{,}000$ J
$T_1(0.60) = T_2(1.90)$, so
$T_1 = 3.167T_2$
$T_1 + T_2 = 4.167\,T_2$
$T_2 = 16{,}000/4.167 = 3{,}840$ J, and
$T_1 = 12{,}160$ J

Torsional stress is equal to Tc/J, so the maximum stress occurs in the 60-cm length of the shaft and is

$$\sigma = (12{,}160)(0.05)/(9.817 \times 10^{-6}) = 61.933 \text{ MPa}$$

7.5 **d.**

$$\text{Torque} = 2(22{,}000)(0.25) = 11{,}000 \text{ N}\cdot\text{m}$$

The angle of twist of the section to the left of the applied torque will be equal to the angle of twist of the section to the right of the applied torque, or $\phi_L = \phi_R$.

$$J = \pi r^4/2 \text{ for a solid round bar.}$$

For a 6.3-cm-diameter bar

$$J = \pi(0.0315)^4/2 = 1.546 \times 10^{-6} \text{ m}^4$$

For the 5-cm-diameter section

$$J = 0.6136 \times 10^{-6} \text{ m}^4$$

From the *FE Handbook*

$$\phi = TL/GJ, \text{ and}$$
$$G = 83 \text{ GPa}$$
$$T_L + T_R = \text{applied torque}$$
$$T_L(0.30)/(1.546 \times 10^{-6})G + T_L(0.20)/(0.6136 \times 10^{-6})G$$
$$= T_R(0.50)/(0.6136 \times 10^{-6})G$$
$$T_L(0.5200)/83{,}000 = T_R(0.8149)/83{,}000$$
$$T_L = 1.567T_R$$
$$2.567T_R = 11.00 \text{ kJ}$$
$$T_R = 4{,}285 \text{ J, and}$$
$$T_L = 6{,}715 \text{ J}$$

7.6 **c.** The angles of twist for the left and right sections will be equal, or $\phi_L = \phi_R$. $\phi = TL/JG$, as given in the *FE Handbook.*

$$J = \pi r^4/2$$

For the 5.0-cm-diameter section

$$J = 0.6136 \times 10^{-6} \text{ m}^4$$

and for the 2.5-cm diameter section

$$J = 0.0383 \times 10^{-6} \text{ m}^4$$
$$T_L + T_R = \text{applied torque} = 1900 \text{ J}$$
$$T_L(0.500)/(68.9 \times 10^{10})(0.6136 \times 10^{-6})$$
$$= T_R(1.00)/(138 \times 10^{10})(0.0383 \times 10^{6})$$
$$T_L = (0.1892/0.01184)T_R = 15.98T_R$$
$$16.98T_R = 1900 \text{ J}$$
$$T_R = 112 \text{ J}$$
$$T_L = 1790 \text{ J}$$

7.7 **a.** First, sketch a figure like that shown in Exhibit 7.7b. When the 90,000-kg load is applied the distance H will increase, and the tie rod L and the cable M will strain. Distances D and E will remain constant. The stresses in L and M will be proportional to the strains in these two members. The relative strains can be calculated as follows

$$L^2 = D^2 + H^2$$
$$M^2 = E^2 + H^2$$

Differentiating both equations gives

$$2LdL = 2HdH, \quad \text{and}$$
$$2MdM = 2Hdh$$

Remembering that D and E remain constant,

$$\delta_L = dL/L = (H/L^2)dH, \quad \text{and}$$
$$\delta_M = dM/M = (H/M^2)dH$$

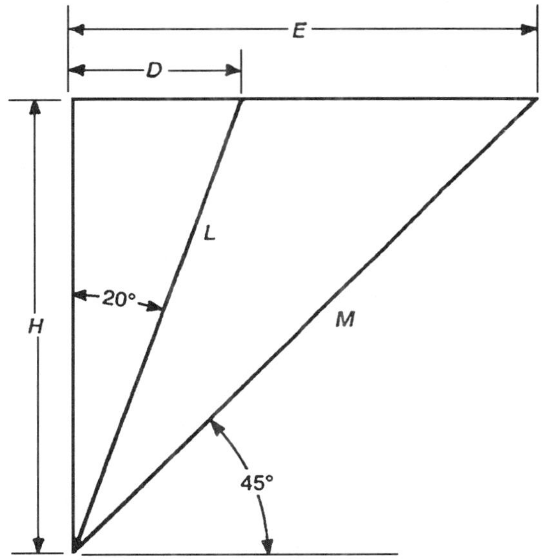

Exhibit 7.7b

then

$$dH = (L^2/H)(dL/L) = (M^2/H)(dM/M)$$
$$\delta_L = (M^2/L^2)\delta_M$$
$$M = H/(\cos 45°) = 1.4142H$$
$$L = H/(\cos 20°) = 1.0642H$$
$$\delta_L = 1.766\delta_M$$

The system is symmetrical about the centerline, so L and M, together, will support half of the load.

$$F_L \cos 20° + F_M \cos 45° = 45{,}000g$$
$$F_L = \delta_L E_L A_L$$

where $A_L = 490.9 \times 10^{-6}$ m^2

Similarly,

$$F_M = \delta_M E_M A_M$$

where A_M is also equal to 490.9×10^{-6} m^2.

$$1.766\delta_M(210 \times 10^9)(490.9 \times 10^{-6})\cos 20°$$
$$+ \delta_M(84 \times 10^9)(490.9 \times 10^{-6})\cos 45° = 441{,}300 \text{ N}$$
$$(171{,}076 + 29{,}150)\delta_M(1000) = 441{,}300$$
$$\delta_M = 0.002204 \text{ m/m}$$
$$\delta_L = 1.766\delta_M = 0.003892 \text{ m/m}$$
$$F_M = \delta_L E_M A_M = 0.002204(84 \times 10^9)(490.9 \times 10^{-6})$$
$$= 90{,}880 \text{ N force in each cable}$$

The vertical component is equal to $90{,}880 \cos 45° = 64{,}250$ N.

$$F_L = \delta_M E_L A_L = 0.003892(210 \times 10^9)(490.9 \times 10^{-6})$$
$$= 401{,}220 \text{ N force in each rod}$$

The vertical component is equal to 401,220 cos 20° = 377,020 N

The cables support 64,250/441,300 = 0.1456, or the cables support 14.6% of the total load and, thus, 14.6% of the added load.

7.8 **d.** The stress will consist of two components: direct stress and moment stress, since the plate is loaded eccentrically. The direct stress is

$$\sigma = (1000)(9.807)/(0.250)(0.025) = 1.569 \text{ MPa}$$

The load is located 0.125 – 0.020 = 0.105 meters away from the centroid of the plate, so the load will produce a moment equal to 0.105(9807) = 1030 N·m in the plate. The equation for stress due to an applied moment is $\sigma = Mc/I$ where $I = bh^3/12$. For this case

$$I = (0.025)(0.250)^3/12 = 32.55 \times 10^{-6} \text{ m}^4$$

The maximum bending stress,

$$\sigma = Mc/I = (1030)(0.125)/(32.55 \times 10^{-6}) = 3.955 \text{ MPa}$$

This would add to the direct stress, giving a maximum stress in the outer fibers of 5.55 MPa.

7.9 **c.** The maximum principal stress will be calculated using the method of Mohr's circle, as given in the *FE Reference Handbook* supplied by the NCEES. Assume a thin-walled tube, then $\sigma = PD/2t$.

The tube ID = 152.4 – 5.08 = 147.32 mm or 0.14732 m

Hoop stress, $\sigma_y = (3.448 \times 10^6)(0.14732)/0.00508 = 99.992$ MPa

$$\text{Longitudinal stress, } \sigma_x = PA_{\text{ID}}/\text{Area of tube wall}$$
$$A_{\text{ID}} = (\pi/4)(0.14732)^2 = 0.017046 \text{ m}^2$$
$$\text{Area of tube wall} = (\pi/4)(OD^2 - ID^2) = 0.001195 \text{ m}^2$$
$$\sigma_x = (3.448 \times 10^6)(0.017046)/0.001195 = 49.184 \text{ MPa}$$
$$\text{Shear stress } S_s = Tc/J$$
$$\text{where } J = (\pi/32)(OD^4 - ID^4) = 6.7152 \times 10^{-6} \text{ m}^4$$
$$S_s = (678)(0.0762)/(6.7152 \times 10^{-6}) = 7.6935 \text{ MPa}$$

From the figure of Mohr's circle in the *FE Handbook*

$$\tau_{\text{max}} = \sqrt{\left[S_s^2 + \{(\sigma_x - \sigma_y)/2\}^2\right]} = \sqrt{(7.6935^2 + 25.404^2)} = 26.54 \text{ MPa}$$
$$\sigma_1 = (\sigma_x + \sigma_y)/2 + \tau_{\text{max}} = 74.588 + 26.45 = 101 \text{ MPa}$$

7.10 **d.** Mohr's circle is used to determine the stress resulting from the stresses on two other surfaces that are perpendicular to each other.

7.11 **d.** Stress, $\sigma = Mc/I$
where
$M = (500)(9.8066)(3.00) = 14{,}710\ \text{N}\cdot\text{m}$, or J
$I = bh^3/12 = (0.200)(0.300)^3/12 = 4.500 \times 10^{-4}\ \text{m}^4$
$c = 0.150$ in

$$\sigma = (14{,}710)(0.150)/(4.500 \times 10^{-4}) = 4.903\ \text{MPa}$$

7.12 **a.**

$$\text{Hoop stress} = PD/2t$$

$$\text{Pressure} = \rho \times g \times \text{depth} = (1000)(9.8066)(10)$$
$$= 98.07\ \text{kPa at the bottom of the tank}$$
$$\text{Stress} = (98{,}070)(35.0)/0.16 = 214.5\ \text{MPa}$$

7.13 **c.** The force exerted by a 200-kg mass = (200)(9.8066) = 1961 N

$$\text{Force in member } AD = (13/5)(1961) = 5099\ \text{N}$$
$$ED = X \text{ component of force in } AD = (1.2/1.3)(5099) = 4707\ \text{N}$$

The Z components of forces in member CD and BD in the X-Z plane are equal—

$$(0.9/1.5)F_{CD} = (1.6/2.0)F_{BD}, \text{ so}$$
$$F_{BD} = 0.75(F_{CD})$$

The sum of the X components of F_{CD} and F_{BD} is equal to the X component of the force in F_{AD}

$$(1.2/1.5)F_{CD} + (1.2/2.0)F_{BD} = (1.2/1.3)F_{AD}$$
$$0.800\ F_{CD} + (0.600)(0.750)F_{CD} = 4707\ \text{N}$$
$$F_{CD} = 4707/1.250 = 3765.6\ \text{N}$$

7.14 **a.** Determine the reaction at the support points. Take moments about the right support

$M = (0.600)(27{,}000) + (\tfrac{1}{2})(6000)g(1.80)(1.8 + 0.600) = 3.600R_L$
$R_L = 39{,}408$ N

The area of the triangular load distribution at the left end of the beam is equal to $(\tfrac{1}{2})(6000)g(1.8)$ m = 52,958 N, and its center of force is 1.8 + 1.8/3 = 2.4 m from the right support.

From $\Sigma F_y = 0$

$$R_R = 52{,}958 - 39{,}408 = 13{,}550\ \text{N}$$

The maximum moment occurs at the point where the shear is zero.

The shear, starting at the left support, would equal 39,408 minus the applied load. The total applied load to the left of any point on the beam would be equal to $(X/2)X(6000)g/1.8) = 16{,}345X^2$ where X is the distance from the left end of the beam, since the magnitude of the load in the first 1.8 meters of the beam to the right of the left support is equal to $X(6000)g/1.8$ for the first 1.8 meters of the beam. The total load acting down on the beam to the left of a point X is equal to the area of the load triangle from the left end of the beam to that point. The point at which the applied load is equal to 39,408 N, the magnitude of the left support, is $39{,}408 = 16{,}345X^2$, giving $X = 1.553$ meters. At this point the magnitude of the shear is equal to zero, so the moment is a maximum. Summing the moments to the left of the point gives

$$M = (-39{,}408)(1.553/3) + (39{,}408)(1.553) = 40.800\ \text{N} \cdot \text{m}, \text{ or } 40{,}800\ \text{J}$$

7.15 **d.** The maximum bending stress is equal to Mc/I where M is the maximum moment, c is equal to the distance from the neutral axis to the outermost fibers of the beam, and I is equal to the moment of inertia of the beam for a beam of constant cross-section. For this case

$$I = bh^3/12 = (0.075)(0.150)^3/12 = 21.094 \times 10^{-6}\ \text{m}^4$$
$$c = 0.075$$
$$M = (3.65/2)(750)(\tfrac{2}{3})(3.65) = 3331\ \text{N}\cdot\text{m}$$
$$\text{Stress} = (3331)(0.075)/(21.094 \times 10^{-6}) = 11.84\ \text{MPa}$$

7.16 **b.** First, determine the reactions at the two support points, A and C. Take moments about the left support

$$M = 4.8R_C = (3.6)(3.6)(3000)g$$
$$R_c = 79{,}440\ \text{N}$$

From $\Sigma y = 0$

$$R_A = (3.6)(3000)g - 79{,}440 = 26{,}520\ \text{N}$$

Determine the point of zero shear. Starting from the left end of the applied load

$$26{,}520 = 3000gX$$
$$X = 0.901\ \text{meters}$$

So the point of zero shear occurs at a distance of 1.8 + 0.9016 = 2.701 m from the left support. The moment is a maximum at the point where the shear is zero. Summing the moments to the left of the point of zero shear gives

$$M = -(0.901)(3000)g(0.901/2) + (26{,}520)(2.701)$$
$$= 59{,}689\ \text{N}\cdot\text{m}, \quad \text{or} \quad 59{,}689\ \text{J}$$

7.17 **d.** First, determine the reactions at the two supports. Calculate the effects of the end moments. Treating the left end moment alone

$$R_R = 45{,}150/6.0 = -7527\ \text{N, giving}$$
$$R_L = 7527\ \text{N from } \Sigma y = 0$$

Similarly, treating the right end moment alone

$$R_L = -180{,}755/6.0 = -30{,}126 \text{ N, and}$$
$$R_R = 30{,}126$$

Treating the applied load alone

$$R_L = (6.0)(4460)g/2 = 131{,}218 \text{ N, and}$$
$$R_R = 131{,}218 \text{ N}$$

Summing the three partial loads gives

$$R_L = 131{,}218 + 7527 - 30{,}126 = 108{,}619 \text{ N}$$

Similarly

$$R_R = 131{,}218 + 30{,}126 - 7527 = 153{,}817 \text{ N}$$

To find the point of zero shear, sum the applied load from the left support

$$108{,}619 - 4460gX = 0$$

Which gives $X = 2.483$ meters. Sum the moments to the left of this point

$$M = (108{,}619)(2.483) - (4460)g(2.483)(2.483/2)$$
$$= 134{,}869 \text{ N·m due to the load.}$$

Subtracting the moment applied at the left support gives

$$M = 134{,}869 - 45{,}150 = 89{,}719 \text{ N·m}$$

Check by summing the moments about the right end

$$M = (3.517)(153{,}817) - (4.460)g(3.517)(3.517/2)$$
$$= 270{,}462 \text{ N·m due to the load.}$$

Subtracting the moment applied at the right support gives

$$M = 270{,}460 - 180{,}755 = 89{,}707 \text{ N·m,}$$

which is the maximum moment at the mid-portion of the beam; however, the applied moment at the right support of 180,755 N·m is greater.

7.18 **c.** The loading on the vertical support beam would increase linearly with depth, from zero at the top. The pressure of the water increases from zero at the top to $(3.0)(1000)g = 29{,}421$ Pa at the bottom of the tank. Each vertical beam would withstand the force exerted by a 1.0-m length of the tank wall. The beam would thus be loaded with a varying load, ranging from zero at the top to 29,421 N/m at the lower end due to the pressure of the water. The beam is hinged at the top and bottom, and is thus simply supported at the ends. Calculate the end reactions.

The total load withstood by the beam due to the triangular load pattern is equal to one-half the length times the height = (½)(3.0)(29,421) = 44,132 N. The centroid of a triangle is one-third the distance from the base to the apex, so the reaction at the top of the beam, taking moments about the bottom end, would be

$$R_T = (44{,}132)(3)(1/3)/3 = 14{,}711 \text{ N}$$

Similarly, taking moments about the top support, gives

$$R_B = (44{,}132)(3)(2/3)/3 = 29{,}422 \text{ N}$$

Taking the top support as a reference point, the loading on the beam at a depth D would be

$$P_D = (44{,}132/3)D$$

The point of zero shear can be determined

$$29{,}422 - (44{,}132/3)D(\tfrac{1}{2})D = 0$$
$$D^2 = 29{,}422/7{,}355$$
$$D = 2.00 \text{ m}$$

Sum the moments acting on the beam about a point 2.00 m below the surface of the water.

Loading on the beam at a depth of 2.00 m is

$$w = (2.00)(1000)g(1.00) = 19{,}614 \text{ N/m}$$

This will decrease linearly to zero at the surface, giving a triangular load. Total force over the 2.00 top meters of depth is

$$P = (19{,}614)(2.00)(\tfrac{1}{2}) = 19{,}614 \text{ N.}$$

The moment arm to the center of force triangle = 2.00/3 = 0.667 m. Summing the moments acting on the beam at 2.00 meters depth yields

$$M = (2.00)(14{,}711) - (0.667)(19{,}614) = 29{,}422 - 13{,}083 = 16{,}340$$

7.19 **d.** Determine reactions at the two support points. Sum moments about R_1

$$M = R_2(3.5) - (1400)(1.50)g - (4.40)(890)g(3.70) = 0$$
$$R_2 = 162{,}690/3.50 = 46{,}483 \text{ N}$$
$$R_1 = [1400 + (4.40)(890)]g - 46{,}483 = 52{,}134 - 46{,}483 = 5651 \text{ N}$$

Find the point of zero shear. The shear load to the right of R_1 would be constant at 5651 N to the point of the applied load of 1400 kg and would then change abruptly to 5651 – 1400g = –8079 N, passing through zero. The moment at this point would be equal to (1.50)(5651) = 8477 N·m.

There would also be an abrupt change in the shear loading at support R_2, where the shear loading would again pass through zero. At this point the moment would be

$$M = (2.40)(890)g(1.20) = 25{,}137 \text{ N}\cdot\text{m,}$$

which is larger than the moment at the other point of zero shear, so this moment would control, 25,137 N·m = 25,137 J.

7.20 **a.** First, locate the neutral axis and calculate n

$$n = 210/8.4 = 25$$

Take moments about the upper edge.

$$y = [(10)(20)(10) + (20.3)(0.60)(10)(25)]/[(200 + (0.60)(250)]$$
$$= 5045/350 = 14.414 \text{ cm} = 0.14414 \text{ m}$$
$$I = bh^3/12 + A_w s_1^{\,2} + (\text{Equivalent Area}) s_2^{\,2}$$
$$\text{Equivalent Area} = (10.0)(0.60)(25) = 150 \text{ cm}^2$$

Using the parallel axis theorem (see *FE Handbook*) and ignoring the moment of inertia of the wood-equivalent area because it is so small

$$I = (10)(20)^3/12 + [200(14.414 - 10)^2] + [150(20.3 - 14.414)^2]$$
$$= 6667 + 3897 + 5197 = 15,761 \text{ cm}^4$$
$$= 1.5761 \times 10^{-4} \text{ m}^4$$
$$\text{Stress} = Mc/I$$

where c = distance from neutral axis to the outermost fibers.

c = 0.1441 m to the top fibers in the wood, or
c = 20.6 – 14.414 = 6.186 cm = 0.06186 m to the outermost steel fibers

If the strain were in the wood the stress would be

$$S = (17,000)(0.06186)/(1.5761 \times 10^{-4}) = 6.672 \text{ MPa}$$

However, the strain is in steel, so the stress is

Stress = n × wood stress = (25)(6.672) = 166.8 MPa = stress in steel.

7.21 **c.** The beam will bend, and the rod will stretch. The force acting on the rod is

$$F_R(38) = 450g(50)$$
$$F_R = 450g(50/38) = 5807 \text{ N}$$
$$\text{Strain in the rod} = L \times S/E$$
$$\delta_L = (5807)/(132.7)/(210 \text{ GPa}) = (250)(43.760 \text{ MPa}/210,000 \text{ MPa})$$
$$= 0.0521 \text{ cm}$$

Treat the beam as two cantilever beams with the ends at the fulcrum.

For a cantilever beam with load concentrated at the free end, the deflection is

$$\delta = PL^3/(3EI) \text{ (see } FE\ Handbook)$$
$$I = bh^3/12 = (0.025)(0.050)^3/12 = 0.2604 \times 10^{-6} \text{ m}^4$$

Deflection at the left end is

$$\delta_L = (5807)(0.38)^3(3)(210\text{ GPa})(0.2604 \times 10^{-6})$$
$$= (318.64/164{,}000) = 1.943\text{ mm}$$

Deflection at the right end is

$$\delta_R = 450g(0.50)^3/164{,}000 = 3.364\text{ mm}$$

Total equivalent deflection at the left end is

$$\delta_{L,\text{equiv.}} = 0.521 + 1.943 = 2.464\text{ mm}$$

This would give a deflection of the right end in inverse proportion to the lengths of the two sections of the beam as for similar triangles

$$\delta_{R,\text{equiv.}}(50/38)(2.464) = 3.242\text{ mm}$$

The right end of beam would deflect 3.242 mm due to bending of the left end and the extension of the rod.

Deflection at Point A is

$$\delta_A = 3.242 + 3.364 = 6.606\text{ mm}$$

7.22 **a.**

$$\text{Area of steel} = (30^2 - 25^2)(\pi/4) = 216\text{ cm}^2 = 0.0216\text{ m}^2$$
$$\text{Area of concrete} = 0.0491\text{ m}^2$$

Allowable stress in the steel is equal to 275 MPa, and strain = 275/210,000 = 0.00131 m/m.

Allowable stress in the concrete = 24 MPa, and strain = 24/19,600 = 0.00122 m/m.

Therefore, the stress in the concrete controls.

Load held by steel = (0.00122)(210 GPa)(0.0216) = 5,533,920 N
Load held by concrete = (0.00122)(19.6 GPa)(0.0491) = 1,174,080 N

The total permissible load is equal to 6,708,000 N, or 684,000 kg.

7.23 **d.** First, calculate the permissible deflection and load ability of the plastic block

$$A = (6.4)(6.4) = 40.96\text{ cm}^2 = 0.004096\text{ m}^2$$
$$P = (0.004096)(13{,}800 \times 10^6) = 56{,}525\text{ N}$$
$$\delta = (0.61)(13.8 \times 10^6)/(10{,}500 \times 10^6) = 8.017 \times 10^{-4}\text{ m}$$

The beam and cables can thus deflect 0.00025 + 0.0008017 = 0.001052 m. See the *FE Handbook* for the equation for beam deflection at the center with an applied load, P

$$\delta = PL^3/(48EI)$$
$$I = bh^3/12 = (0.06)(0.15)^3/12 = 16.875 \times 10^{-6}\text{ m}^4$$

Beam deflection

$$\delta_{beam} = P(27)/(48)(210{,}000)(16.875)$$
$$= P(0.1587 \times 10^{-6}) \text{ m where } P \text{ is in N.}$$
$$\delta_{cable} = LP/AE$$
$$A = 2(0.019)^2(\pi/4) = 568.2 \times 10^{-6} \text{ m}^2 \text{ for the two cables}$$
$$\delta_{cable} = 1.83P/(568.2)(84{,}000) = P(0.03834 \times 10^{-6}) \text{ m}$$

The beam and cables constitute two springs in series, where

$$I/k = 1/k_1 + 1/k_2 = (0.1587 + 0.0383) \times 10^{-6} = 0.1970 \times 10^{-6} \text{ m/N}$$

The load to deflect the beam-cable assembly 0.01052 m is

$$P = 0.001052/(0.1970 \times 10^{-6}) = 5340 \text{ N}$$

Add to this the load withstood by the plastic block to get a total load of 5340 + 56,525 = 61,865 N, or 6308 kg

7.24 **b.** The beam deflection formula given in the *FE Handbook* is

$$\delta = (w_0x/24EI)/(L^3 - 2Lx^2 + x^3)$$

where

w_0 is the load per unit length = 225g = 2207 N/m

x = 1.22 meters

L = 4.5 m

$I = (0.20)(0.20)^3/12 = 133.3 \times 10^{-6} \text{ m}^4$

$\delta = [(2{,}207)(1.22)/(24)(10{,}500)(133.3)](91.125 - 13.396 + 1.816)$

$\delta = (80.155 \times 10^{-6})(79.545) = 0.00638$ m, or 6.38 mm

7.25 **d.** Concrete is assumed to take no load in tension. The equivalent "tensile concrete" area (area of steel rebar times n) is equal to 4(3.90)(10) = 165 cm^2 or 0.0165 m^2 with the center of the section taken as 68 cm, or 0.68 m, from the top of the beam. Determine the location of the neutral axis.

Taking moments about the top of the beam gives

$$y = [(0.30y)(y/2) + (0.68)(0.0165)]/(0.30y + 0.0165)$$

which reduced to

$$y^2 + 0.115y - 0.0747 = 0$$

Solving for y, using the general solution for a quadratic equation in the *FE Handbook*, gives

$$y = 0.2180 \text{ m}$$
$$I = (bh^3/12) + A_1S_1^2 + A_2S_2^2$$

where S is equal to the distance from the CG of the area to the neutral axis.

$$I = (0.30)(0.218)^3/12) + (0.30)(0.218)(0.218/2)^2$$
$$+ 0.0165(0.680 - 0.218)^2 = 0.004558 \text{ m}^4$$

The maximum moment is

$$M_{max} = wL^2/8$$
$$w = (0.30)(0.76)(2370 + 3000)g = 34{,}720 \text{ N/m}$$
$$M_{max} = (34{,}720)(6)^2/8 = 156{,}240 \text{ J}$$

Stress in equivalent "tensile concrete" = (156,240)(0.462)/0.004558 = 15.837 MPa

Tensile stress in rebar = 10(15.837) = 158.37 MPa, where $n = 10$

CHAPTER 8

Dynamic Systems, Vibration, Kinematics

OUTLINE

A good starting point for reviewing these concepts is the "Dynamics" chapter in the *Fundamentals of Engineering (FE) Supplied-Reference Handbook*. Throughout this chapter we will reference specific portions of the "Dynamics" chapter that would be good to review in conjunction with this book.

Perhaps the best way to start a review of dynamic systems is to review Newton's laws, since the principles expounded by Newton govern most non-nuclear mechanical actions. The three laws can be expressed as follows:

1. When a body is at rest or moving with a constant speed in a straight line, the resulof all the forces acting on the body is zero. (ΣForces = 0)
2. The rate of change of the momentum of a body is proportional to the force acting upon it and it is in the direction of the force. ($F = ma$)
3. Whenever one body exerts a force on another the second always exerts on the first a force which is equal in magnitude, but oppositely directed, and collinear. (Action and reaction are equal and opposite.)

LINEAR MOTION

The basic relationships of motion should also be reviewed: Velocity, acceleration, and distance as discussed in the section titled KINEMATICS in the *Handbook*.

Velocity equals rate of change of distance with time, or $v = ds/dt$, which is a vector quantity, for example, a velocity has both magnitude and direction. If the rate of change of position has only magnitude but no direction, it is a scalar quantity—speed.

For straight-line motion and constant velocity, distance, s, equals $s_o + vt$, where s_o is the position at time zero.

Acceleraton, a, is the rate of change of velocity, or $a = dv/dt = d^2s/dt^2$. If a constant acceleration acts on a particle then $ds/dt = v_o +$ at giving $ds = v_o dt + atdt$

Distance: Integration gives $s = s_o + v_o t + \frac{1}{2} at^2$ for straight-line motion.

Another important relationship is $v^2 = v_o^2 + 2a\Delta s$ where $\Delta s = (s - s_o)$, the distance traveled from time $t = 0$ to the instant when the new velocity is to be determined.

The four basic relations of linear (straight line) motion with constant acceleration can thus be stated as follows

$$v = v_o + at \qquad v^2 = v_o^2 + 2v_o at + a^2t^2$$
$$v^2 = v_o^2 + 2a\Delta s \qquad 2as = 2a(v_o t + \tfrac{1}{2}at^2)$$
$$2as = 2v_o at + a^2t^2 \qquad v_{\text{avg}} = (v + v_o)/2$$
$$s = s_o + v_o t + \tfrac{1}{2}at^2 \quad \text{or} \quad s = v_{\text{avg}}t \text{ (For } S_o = 0)$$

Example 8.1

The speed of an automobile starting from rest increased in 18 seconds to 22 m/s. What was its final speed in km/hr? What was its average speed in km/hr during the 18 seconds its speed was increasing?

Solution

Assume a uniform rate of acceleration. Find also the rate of acceleration and the distance traveled.

From: $v = v_o + at$, $v = 22$ m/s, $v_o = 0$, $t = 18$s $a = 22/18 = 1.222$ m/s^2
the final speed in kilometers per hour equals—22 m/s × (3,600 s/hr)/1,000 = 79.2 km/hr

the average speed = (79.2 + 0)/2 = 39.6 km/hr
the distance traveled, $s = v_{avg} \times t$

$$s = 39.6 \times 18/3{,}600 = 0.198 \text{ km} \quad \text{or} \quad 18 \times 22/2 = 198 \text{ m}$$

or,

$$s = \tfrac{1}{2}at^2 = \tfrac{1}{2} \times 1.222 \times 18^2 = 198 \text{ m}$$

Relative Velocity

Velocity is defined as the rate of change of distance (from some arbitrarily selected point) with time, or $v = ds/dt$. Thus a velocity possessed by an object must be in reference to some other object or reference system. That is, *all velocities are relative*. Usually we think of a velocity (i.e., any noncelestial velocity) in reference to the surface of the earth and tend to think of this as *the* velocity of an object. That can become confusing in some cases. Take, for example, the following problem.

Example 8.2

A runaway railroad car is moving along a (frictionless) horizontal stretch of track at a constant velocity of 50 km/h. A locomotive starts out in pursuit and reaches a speed of 100 km/hr when it is exactly 1.60 km behind the runaway car. The engineer (locomotive engineer, that is) starts to decelerate the engine at a constant rate so that when it touches the car to couple with it, the locomotive will be going at the same speed as the car. What should be the rate of deceleration and what additional distance would the runaway car travel before being caught?

Solution

This problem can be worked by relating all velocities to the surface of the earth. If this approach is used the distance the car will travel can be denoted by D_c and the distance the locomotive will travel will thus equal D_l where $D_l = D_c + 1.6$ Then $100t + \tfrac{1}{2}at^2 = 50t + 1.6$ which gives: $50t + \tfrac{1}{2}at^2 = 1.6$. With constant deceleration (negative acceleration) the average velocity of the locomotive would equal $v_{avg} = (100 + 50)/2 = 75$ km/h.

The velocity of the car will remain constant. $D_l = D_c + 1.6$ gives $75 \cdot t = 50 \cdot t + 1.6$ or $t = 0.064$ hr $s = 50 \times 0.064 + 1.6 = 100 \times 0.064 + \tfrac{1}{2}a \times 0.064^2$, which gives $a = -781.25$ km/hr^2.

$$D_c = 50 \times 0.064 = 3.2 \text{ km}$$

It is more straightforward and simpler to use relative velocities. The initial relative velocity of the locomotive in reference to the run-away car equals

$$v_r = 100 - 50 = 50 \text{ km/hr} = v_o$$

The final relative velocity equals zero.

$$v^2 = v_o^2 + 2as \quad \text{to obtain the final velocity}$$
$$0 = 50^2 + 2a(1.6) \quad \text{giving } a = -781.25\text{km/hr}^2$$
$$s = v_o t + \tfrac{1}{2}at^2 \quad \text{or} \quad s = v_{avg} \times t$$
$$v_{avg} = 50/2 = 25 \text{ km/hr} \quad s = 1.6 \text{ km} \quad t = 0.064 \text{ hr}$$

As before $D_c = 0.064 \times 50 = 3.2$ km

FLIGHT OF A PROJECTILE

(See Section PROJECTILE MOTION in the *Handbook*.)

Another example of the use of equations of linear motion is afforded by a type of past examination problem which asked for details regarding the flight of a projectile.

Example 8.3

A bullet leaves the muzzle of a gun at a velocity of 825 m/s at an angle of 30° with the horizontal, angle θ in the figure for projectile motion in the *Reference Handbook*. What is the maximum height to which the bullet will travel and how far will the bullet travel horizontally measured along the same elevation as the gun muzzle? Disregard air resistance.

Solution

First determine the vertical and horizontal components of the velocity

$$v_v = 825 \sin 30° = 412.5 \text{ m/s} \qquad v_h = 825 \cos 30° = 714.5 \text{ m/s}$$

The vertical component of the velocity of the bullet will be subjected to the acceleration of gravity, and if air resistance is disregarded, the time required for the bullet to reach its apogee can be determined from the fact that at the top of the bullet's flight the vertical velocity will equal zero.

$$v = v_o + at \quad \text{or} \quad 0 = v_v - gt \quad t = v_v/g$$
$$\text{time to rise to apogee} = 412.5/9.807 = 42.06 \text{ s}$$

The height to which the bullet will rise will equal

$$\text{height} = v_{\text{avg}} \times \text{t} = (412.5/2) \times 42.06 = 8675 \text{ m}$$

or, from

$$v^2 = v_o^2 + 2ah, \text{ which gives } h = v_v^2/2g$$
$$h = 412.5^2/(2 \cdot 9.807) = 8675 \text{ m } (a = -g)$$

The distance the bullet will travel horizontally, the range, equals the time of flight times the horizontal velocity. The total time of flight will equal the time to rise to the apogee plus the time to fall back to earth, or twice the time to rise.

$$R = 2 \times 42.06 \text{ s} \times 714.5 \text{ m/s} = 60{,}103 \text{ m}$$

ROTARY MOTION

(See Section KINETICS in the *Handbook*.)

For plane circular motion the relationships corresponding to linear motion are

θ, radians, corresponds to linear distance, s
$\omega = d\theta/dt$, radians per second, rate of rotation, corresponds to v, linear velocity
$\alpha = d\omega/dt$ or $d^2\theta/dt^2$, radians per second squared, corresponds to α, linear acceleration
$\omega = \omega_o + \alpha t$

In addition, the term for mass in linear motion is replaced by the moment of inertia, I, in rotational motion. This feature is discussed in detail later on in this chapter.

$$\omega^2 = \omega_o^2 + 2\alpha\Delta\theta$$
$$\omega_{avg} = (\omega + \omega_o)/2$$
$$\theta = \theta_o + \omega_o t + \tfrac{1}{2}\alpha t^2$$

FORCE

(See Sections CONCEPT OF WEIGHT and CENTRIFUGAL FORCE in the *Handbook*.)

Newton's second law states $F = ma$ and the acceleration of gravity acts on a mass at the earth's surface in accordance with that relationship where $a = g = 9.8066\ m/s^2$. The force of gravity exerted by a mass $m = m \times 9.8066$ in newtons, thus the units of a newton

$$N = kg \cdot m/s^2.\ \text{Rearranging terms gives } kg = N \cdot s^2/m$$

There are also forces that are exerted due to rotation—the centripetal force (force acting inward toward the center of rotation) and the centrifugal force (force acting outward—centrifugal means fleeing a center). These forces are equal and opposite and the force is usually referred to as a centrifugal force. An object moving in a circular path must have an acceleration toward the center of rotation or it would fly away from the circular path at a tangent as did the stone from David's sling as he pelted Goliath. Analysis shows that the acceleration of a particle in a circular path toward the center of revolution equals the square of the angular velocity times the radial distance or since force equals mass times acceleration, the centrifugal force, $F_c = \omega^2 r \cdot m$ or

$$F_c = (v^2/r) \cdot m$$

D'ALEMBERT PRINCIPLE

A sometimes useful relationship is termed the d'Alembert principle. This states that there is an "inertial force" which acts on any mass being accelerated, and equals ma. This derives from Newton's second law. Since $F = ma$, then $F - ma = 0$. The use of this term is illustrated in Example 8.4 on the following page.

FRICTION

(See Section FRICTION in the *Handbook*.)

Friction is a passive, resisting force. It is equal to the coefficient of friction times the force *normal* to the surface and is independent of the area of contact. If the applied force parallel to the surface is greater than the force of friction, then the net force acting to cause motion will equal the applied parallel force minus the force of friction. If the friction force is greater than the component of the applied force parallel to the surface, no motion will occur and the system will be static. A frictional force is always a static resisting force; it is never an active force. The application of friction is illustrated in Example 8.4.

KINETIC ENERGY

(See Section KINETIC ENERGY in the *Handbook*.)

A mass in motion has energy, termed kinetic energy. This energy equals $\frac{1}{2}\,mv^2$ and is expressed in joules: $\text{kg} \times \text{m}^2/\text{s}^2 = \text{N}\cdot\text{s}^2/\text{m} \times \text{m}^2/\text{s}^2 = \text{N}\cdot\text{m}$ = joules. For example, a 10 kg mass moving at 8 m/s possesses

$$\tfrac{1}{2} \times 10 \times 8^2 = 320 \text{ J energy}$$
$$(\text{N}\cdot\text{s}^2/\text{m} \times \text{m}^2/\text{s}^2 = 320 \text{ N}\cdot\text{m})$$

Similarly a rotating mass contains rotational kinetic energy in the amount of $KE = \frac{1}{2}I\omega^2$, where I equals the moment of inertia of the mass about the center of rotation, and ω is the angular velocity in radians per second.

Example 8.4

An example illustrating many of these principles is illustrated with the aid of Exhibit 1. Assume a frictionless pulley and massless connecting line. For the system shown calculate (a) velocity of the 25-kg mass at impact, and (b) the final distance of the 100-kg mass from the edge.

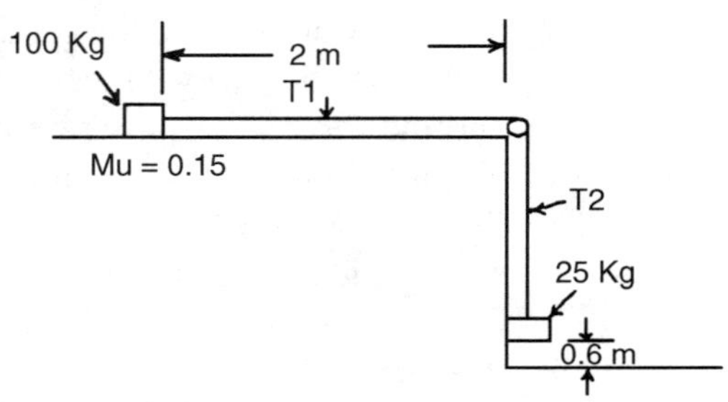

Exhibit 1

Solution

The tension in the line equals $m_1 \times a + f$ where f equals the frictional force = $100 \times 9.8066 \times 0.15 = 147.10$ N and $m_1 a$ is the d'Alembert, or resisting, force. so

$$T_1 = 100\,a + 147.10, \qquad T_2 = mg - ma = 25 \times 9.8066 - 25\,a,$$

where mg is the force downward due to the force of gravity acting on the 25 kg mass, and ma is the d'Alembert, or resisting, force.

$$T_1 = T_2 \quad 125\,a = 98.06 \text{ N} \quad \text{and} \quad a = 0.7845 \text{ m/s}^2$$

Velocity at impact

$$v^2 = v_o^2 + 2as = 0 + 2 \times 0.7845 \times 0.60 \quad v = /\ 0.9414 = 0.97303 \text{ m/s (a)}$$

At the instant of impact the 100 kg mass would possess the same velocity, and the kinetic energy would equal: $KE = \frac{1}{2} \times 100 \times 0.9703^2 = 47.07$ J This energy would be expended by work done against friction, $s = 47.07/147.10 = 0.320$ m

DIST from the edge = 2.00 – 0.320 – 0.6 = 1.080 m (b)

Example 8.5

What is the minimum force F to move the 50-kg mass over the level surface in Exhibit 2 if the coefficient of friction is 0.4?

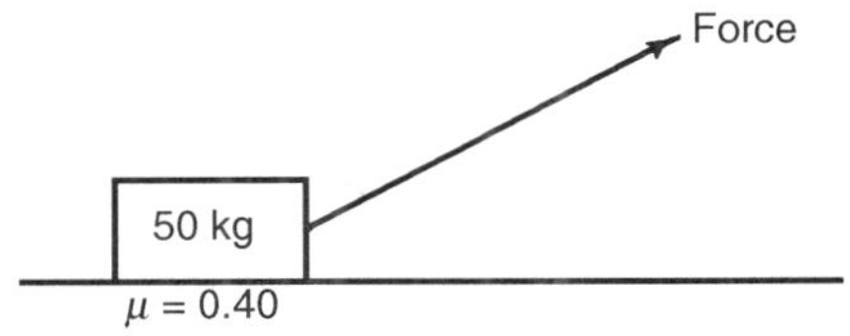

Exhibit 2

μ = coefficient of friction

Solution

Many engineers will merely glance at the problem and assume that the force required would equal $0.4 \times 50g$, but that is not correct because the force F has two components—a horizontal component that will cause the mass to slide, and a vertical component that will reduce the vertical force on the surface. Let the angle the force F makes with the horizontal equal α.

Then $(mg - \mathrm{F} \sin\alpha)\mu = F \cos\alpha$

The force will be a minimum when $dF/d\alpha = 0$ expand and differentiate, giving

$$-F\mu \cos\alpha \, d\alpha - \mu \sin\alpha \, dF = -F \sin\alpha d\alpha + \cos\alpha \, dF$$

combining terms gives

$$dF/d\alpha = [F(\sin\alpha - \mu \cos\alpha)]/(\mu \sin\alpha + \cos\alpha) = 0$$

$$F(\sin \alpha - \mu \cos\alpha) = 0 \quad \text{and} \quad \sin\alpha = 0.4 \cdot \cos\alpha$$

$$\tan\alpha = 0.4 \quad \alpha = 21.80° \quad (50g - 0.371F) \times 0.4 = 0.928\ F$$

$$F = 0.4 \times 50g/1.076 = 182.3\ \mathrm{N}$$

POTENTIAL ENERGY

(See Section POTENTIAL ENERGY in the *Handbook.*)

Potential energy is energy contained by a body (mass) due to its position, or by a system due to its condition in which it possesses contained energy such as a spring, a fluid under pressure, an explosive, or similar arrangement. Two examples are a mass at an altitude and a compressed spring. If a mass is lowered or dropped a distance s the change in energy equals $IFds$ or $Imgds$ and is measured in joules. Another form of potential energy is contained by a compressed spring. The force exerted by a spring = ks, where k = the spring constant, N/m. The potential energy equals $IFds$ or $Iksds$ which equals, for $s_o = 0$, $\frac{1}{2} ks^2$, newton meters, or joules.

Example 8.6

Verify the velocity calculated in Example 8.4 by comparing the work done with the change in potential energy.

Solution

$$\Delta PE = \Delta KE + \text{work done} = 25 \times 9.8066 \times 0.600 = 147.10\ \mathrm{J} = \Delta KE - W_f$$

$$W_f = \text{work done against friction} = 0.600 \times 147.10 = 88.26\ \mathrm{J}$$

$$\text{so } \Delta KE = 147.10 - 88.26 = 58.84 = \tfrac{1}{2} \times 125 \times v^2 \quad m = 100 + 25 = 125\ \mathrm{kg}$$

$$v = /(58.84 \times 2/125) = 0.9703\ \mathrm{m/s} \quad \text{as before.}$$

Example 8.7

Determine how much a spring with a spring constant of 40 N/cm will compress if a mass of 50 kg is dropped onto it from a height of 2.00 m above the top of the spring.

Solution

The loss of potential energy of the mass will equal: $50 \times 9.8066 \times (2.00 + s)$, where s equals the amount the spring compresses. This will equal the increase in potential energy in the spring

$$\Delta PE = \tfrac{1}{2} \times 4{,}000 \times s^2 \text{ where } s \text{ is in meters. } 490.33 \times (2.00 + s) = 2{,}000\, s^2$$

which reduces to

$$s^2 - 0.2452s - 0.4903 = 0$$

Using the general solution for a quadratic equation as given in the *Handbook* gives

$$s = [0.2452 \pm (0.0601 + 1.9612)]/2 \quad s = 0.8335 \text{ m}$$

IMPULSE AND MOMENTUM

(See Section IMPULSE AND MOMENTUM in the *Handbook*.)

Momentum, the momentum of a body equals the product of its mass times its velocity, or $M = mv$, assuming a constant mass. Impulse = $IFdt$ or, since $F = ma$, Impulse = $Im(dv/dt)\ dt = Imdv$, or impulse = change in momentum. If no external force acts on a system of bodies, the linear momentum is constant, this is expressed as the conservation of momentum. It should be noted that linear momentum is a vector quantity, since it is a vector quantity, v, times a scalar quantity, m. In addition, the angular momentum of a system of bodies about a fixed axis remains constant if no external moment acts about the axis. It is also true that the center of mass of a system remains constant if no external force acts on the system. These concepts can be applied to a number of different situations.

The ballistic pendulum used to measure the velocity of a bullet is one application.

Example 8.8

If a bullet with a mass of one gram is fired at and becomes embedded in a block of mass two kg which is suspended by a cord 2.00 m long and causes the supporting cord to form an angle of 3.50° with the vertical. What is the velocity of the bullet at impact?

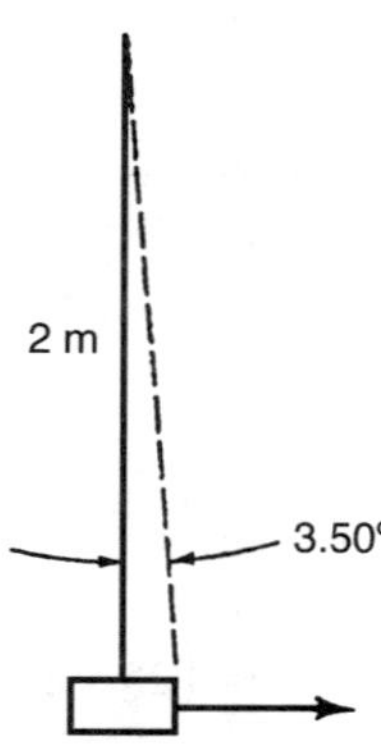

Exhibit 3

Solution

The law of conservation of momentum states that the linear momentum of a system of bodies before impact equals the final linear momentum after impact if there is no resultant external force acting on the system, or $m_1v_1 = (m_1 + m_2)\cdot v_2$ the pendulum block will rise, $2 \times (1 - \cos 3.5°) = 0.003730$ m. The change in potential energy equals the change in kinetic energy.

Therefore $v = /2gh$, or $v = /(19.6132 \times 0.003730) = 0.2705$ m/s

$$v_1 = 2.001 \times 0.2705/0.001 = 541.2 \text{ m/s}$$

COEFFICIENT OF RESTITUTION

Many impacts are not plastic and in many cases one of the impacting bodies will "bounce." If a ball rebounds upward after being dropped onto a fixed plate, the mass of the plate being much greater than that of the ball, then the velocity after impact will be less than the velocity at impact, and the coefficient of restitution, e, will equal the negative of the ratio of the two velocities.

$$e = -v_2/v_1 = /(h_2/v_1) \quad (v_2 \text{ is opposite in sense to } v_1)$$

The coefficient of restitution for two bodies whose masses do not differ widely is equal to the ratio of the velocity of separation, $v_1 - v_2$, divided by the velocity of approach, $u_2 - u_1$, or

$$e = (v_1 - v_2)/(u_2 - u_1), \quad \text{as noted in the } \textit{Handbook}.$$

Example 8.9

A ball with a mass of 4.0 kg traveling with a velocity of 6 m/s overtakes and strikes a 6.0-kg ball traveling in the same direction with a velocity of 2 m/s. If the coefficient of restitution equals 0.75, what would be the velocities of the two balls after the impact?

Solution

$$e = 0.75 = -(v_2 - u_2)/(6 - 2) \text{ which gives } u_2 - v_2 = 3 \text{ but } m_1v_1 + m_2u_1$$
$$= m_1v_2 + m_2u_2 \text{ so } 4.6 + 6.2 = 4\cdot v_2 + 6\cdot u_2 \quad \text{and} \quad 9 = v_2 + 1.50\cdot u_2$$

Combining the two equations gives

$$v_2 = 1.80 \text{ m/s} \qquad u_2 = 4.80 \text{ m/s}$$

Another example of the principles of impulse and momentum is afforded by the following problem.

Example 8.10

Two barges, one of total mass 10,000 kg and another of 20,000 kg are connected by a cable in still water. The two barges are initially connected by a cable 30 m long and are at rest. If the cable is drawn in by a winch on the larger barge until the distance separating the two barges is 15 m, what would be the distance moved by the 10,000 kg barge? Assume friction is negligible.

Solution

The key point in this problem is that the center of gravity of a system cannot change if no external force is applied to the system.

The impulse will be the same on each of the two barges and the center of gravity of the system will remain the same since no external force acts on the system. Initially the CG will be a distance of 30/3 = 10 m from the larger barge and 20 m from the smaller barge.

$$10{,}000g \times s = 20{,}000g \times (30 - s) \quad s = 20 \text{ m},$$

so initially the CG will lie 2/3 the distance between the small and the large barge. The location of the CG will not change so the small barge will move 10 m and the larger barge will move 5 m.

Another illustration of the principles of impulse and momentum is afforded by a golf ball that is struck by a club.

Example 8.11

High-speed photography shows that the club is in contact with the ball for approximately one-half-thousandth of a second and the velocity of the ball was 75 m/s immediately after being struck. What force is exerted by the club if the mass of the ball is 47 grams?

Solution

Assume the force is constant during the period of impact, so the impulse will equal $0.0005F$. The change in momentum of the golf ball will equal Δmv, which gives

$$\Delta mv = 47 \times 75 = 3{,}525 \text{ gm}\cdot\text{m/s} \quad \text{or} \quad 3.525 \text{ kg}\cdot\text{m/s}$$

Since the change in momentum is equal to the impulse

$$0.0005F = 3.525 \quad F = 7{,}050 \text{ N} \quad \text{kg}\cdot\text{m/s} = F\cdot s \quad \text{or} \quad F = \text{kg}\cdot\text{m/s}^2 = \text{newtons}$$

It should be noted that the concept of conservation of momentum does *not* mean conservation of energy. This is demonstrated by the following problem.

Example 8.12

A plastic body with a mass of 45 kg is moving with a velocity of 6 m/s and overtakes another plastic body with a mass of 70 kg, moving in the same direction with a velocity of 4 m/s. What is the final velocity of the combined masses and what is the loss in kinetic energy?

Solution

From the principle of conservation of momentum we have $m_1v_1 + m_2v_2 = (m_1 + m_2)v_r$

The resulting velocity equals—

$$v_r = [(45)(6) + 70\,(4)]/(45 + 70) = 4.783 \text{ m/s}$$
$$KE \text{ at start} = \tfrac{1}{2}(45)(6)^2 + \tfrac{1}{2}(70)(4)^2 = 1370 \text{ J}$$
$$\text{Final } KE = \tfrac{1}{2}[(115)(4)783]^2 = 1315 \text{ J}$$

The loss of kinetic energy equals 55 J or 4.0%, which amount of lost mechanical energy is converted to thermal energy that results in heating up the two bodies.

MOMENT OF INERTIA

(See Section MOMENT OF INERTIA in the "Statics" chapter of the *Handbook*; also review table listing *Area & Centroid, Moment of Inertia, (Radius of Gyration)*^{2}, and *Product of Inertia* for a number of different figures in that chapter).

Note that $J = Ir^2dA$ (here J is the polar moment of inertia), and $J = k^2A$ where k = radius of gyration. So in the table k^2 is given (the symbol r is used in the *Handbook* rather than k). The values given in the table are for plane figures. To obtain the values of I for a solid it is necessary to introduce the mass of the solid. For example the mass moment of inertia for a cylinder that is rotating about an axis through its center of gravity, or the center of the circle shown, equals $\tfrac{1}{2}Mr^2$. The value of J for the plane polar moment of inertia equals Ir^2dA, whereas the mass polar moment of inertia J_M equals Ir^2dM, but Mass = density × volume, or $\rho \times L \times A$.

Then $dM = \rho L dA$ and $J_M = \rho L \pi r^4/2 = \rho V r^2/2$ or $J_M = \frac{1}{2}Mr^2$

It might also be noted that the polar moment of inertia equals the sum of the moments of inertia about any two axes at right angles to each other in the plane of the area and intersecting at the pole. It is interesting to note how this concept simplifies the calculation of the transverse moment of inertia of a circular section about a transverse axis through its center in the plane of the surface. The solution of the relationship Ix^2dA about an axis in the plane of the disk is quite complex. It is relatively easy to calculate the polar moment of inertia of the disk, and half of the polar moment of inertia equals the transverse moment of inertia since the moments of inertia of a circular disk at right angles are the same. Then the transverse moment of inertia equals

$$\tfrac{1}{2}J = \tfrac{1}{2}Ir^2dA = \tfrac{1}{2}Ir^2 \cdot 2\pi r dr$$

since $dA = 2\pi r dr$, which integrated from 0 to R gives

$$J = \pi R^4/2 \quad \text{or} \quad AR^2/2 \quad \text{and} \quad I\ (\text{transverse}) = \pi AR^2$$

The corresponding relationships for linear and rotary motion are described above, however there are other relationships for rotary motion that are also important. The quantity of mass in a rotational problem is replaced by the polar moment of inertia.
where
$I = Ir^2dm$ $(J = \frac{1}{2}MR^2)$

which gives the relationship of $T = I\alpha$ corresponding to $F = ma$,
where
T is the torque in Nm.

In addition, $KE = \frac{1}{2}I\omega^2$, for rotary motion which corresponds to the linear $KE = \frac{1}{2}mv^2$. Note that it is customary to refer to the moment of inertia in rotary motion as I, rather than J, as, for example in $T = I\alpha$.

Another important relationship in rotary motion is the concept of centrifugal force, F_c, or force outward on an object due to its rotation about some point, for example, a satellite rotating about the earth.

$$F_c = \omega^2 rm \quad \text{or} \quad F_c = (v^2/r)m$$

where
ω is the rotational velocity in radians per second
v is the linear velocity in m/s, since $v = \omega r$

Note that r, the radius, is the distance from the center of mass of the object to the center of rotation of the system. In the case of an earth's satellite, the center of rotation would be the center of the earth, and the distance r would be the distance from the center of the earth to the center of mass of the satellite.

As noted previously, the mass rotational or polar moment of inertia,

$$I = Ir^2dm \quad \text{and} \quad I = k^2m$$

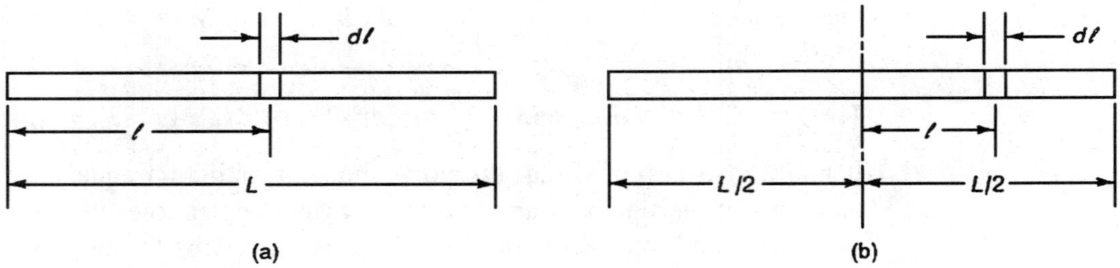

Figure 8.1

where

k is the radius of gyration, or the distance from the axis of interest at which all of the mass can be considered concentrated as, for example, a hoop with negligible thickness

It should be noted that, in general, the mass of a body cannot be considered as acting at its center of mass for the purpose of calculating the moment of inertia. This can easily be shown by calculating the moment of inertia of a uniform slender rod about an axis through one end and perpendicular to the rod, see Figure 8.1.

The mass of the rod will equal ρAL where ρ is the density of the rod material in kg per unit volume and A is the cross-sectional area and L is the length. The dm of the relationship for $I = Ir^2dm$ will then equal $\rho Ad\ell$ giving

$$I = I\ell^2\rho Ad\ell \text{ for } \ell = 0 \text{ to } \ell = L$$

$$\text{I then equals } \rho AL^3$$

and since the mass, M, of the rod $M = \rho AL$ $\quad I = ML^2$ rather than $ML^2/4$ $[M(L/2)^2 = ML^2/4]$

The distance from the end of the rod to the center of mass equals $L/2$, but the length of the radius of gyration from the end of the rod equals $L/3$.

Similarly, the moment of inertia of a slender, uniform rod about an axis through its center and perpendicular to the rod will equal $I\ell^2dm$ from $-L/2$ to $+L/2$ giving

$$I = (1/12)ML^2$$

and the moment of inertia of a flat, rectangular surface about an axis through its center lying in the plane of the surface will equal (see Figure 8.2)

$$I = Ix^2b\ dx \text{ from } x = h/2 \text{ to } x = -h/2$$

which gives $I = bh^3/12$

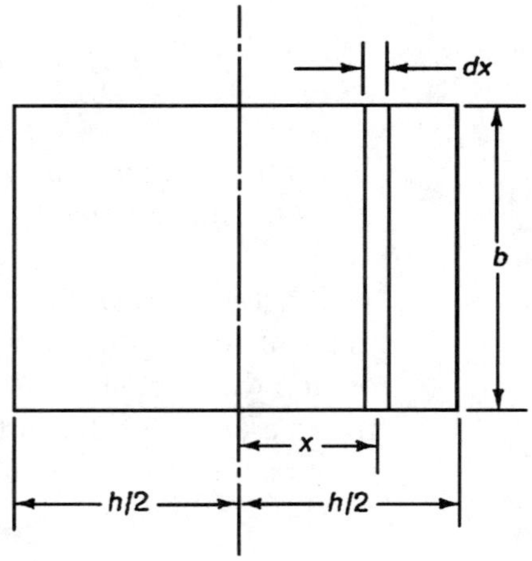

Figure 8.2

Example 8.13

An armature is keyed to a shaft 10 cm in diameter. It rests transversely on two parallel steel rails inclined to the horizontal with a slope of 1:12. See Exhibit 4. The shaft rolls, without slipping and without rolling friction, a distance of two meters along the rails from rest in one minute. What is the radius of gyration of the armature and its shaft about the assembly's longitudinal axis?

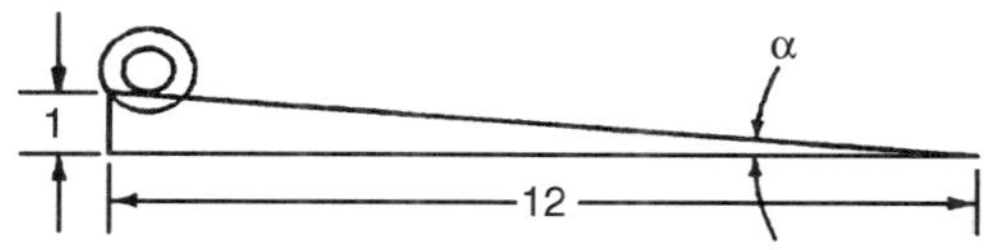

Exhibit 4

Solution

The kinetic energy possessed by the assembly at the end of one minute equals—

$$KE = \tfrac{1}{2}mv^2 + \tfrac{1}{2}I\omega^2 \quad \text{and} \quad v_{avg} = 2/60 = 0.0333 \text{ m/s}$$

Since there is a constant acceleration, the final velocity will equal twice the average velocity or

$$v = 0.0667 \text{ m/s} \quad \omega = v/r$$

so

$$\omega = 0.0667/0.050 = 1.3333 \text{ rad/s}$$
$$KE = \tfrac{1}{2}m(0.0667^2) + \tfrac{1}{2}I(1.3333^2) \quad \text{but} \quad I = k^2m$$

where
k equals the radius of gyration.

So

$$KE = \tfrac{1}{2}m(0.004445) + \tfrac{1}{2}(k^2m)1.7779 = m(0.002222 + 0.8888k^2)$$

The increase in kinetic energy must equal the decrease in potential energy since no work is done against friction. The angle of the slope, α, equals $\tan^{-1}(1/12) = 4.76°$

$$\Delta PE = \text{mg} \sin \alpha \times 2.0 = 9.807 \times 0.08305 \times 2 \times \text{m} = 1.6289 \times \text{m}$$

ΔPE equals ΔKE so

$$0.002222 + 0.8888\, k^2 = 1.6289 \text{ giving } k = /1.8304 = 1.353 \text{ meters}$$

Example 8.14

The same result can be obtained by determining the acceleration of the armature assembly with the relationships that $Tq = I\alpha$ and $I = mk^2$. See Exhibit 5.

$$s = v_o t + \tfrac{1}{2}at^2$$

where $v_o = 0$ and $t = 60$ s

$$2 = \tfrac{1}{2}a \cdot 60^2 \qquad a = 0.00111 \text{ m/s}^2$$

Solution

The force acting on the assembly to cause it to accelerate down the incline will equal its mass times the gravitational constant times the sine of the slope angle,

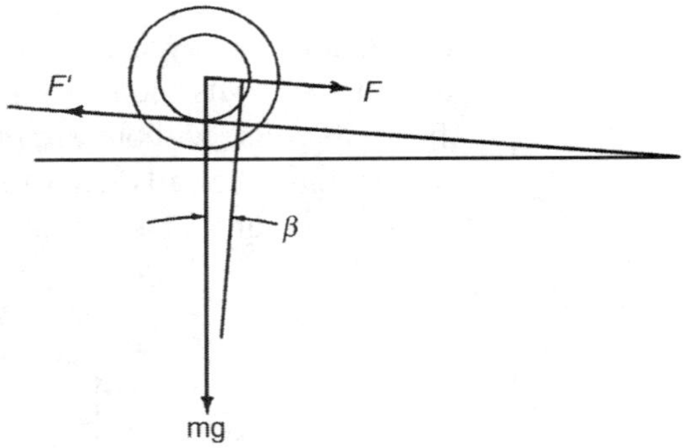

Exhibit 5

and since there is no slippage, the resisting force applied by the rails will be equal and opposite to the applied force, F, giving $F = F'$

$$F = \text{mg} \sin \alpha$$
$$= m \cdot 9.807 \cdot 0.08305 = 0.8145 \cdot m \qquad Tq = I\alpha = I(a/r)$$
$$Tq = F' \times r = F \times r$$
$$\text{so } 0.8145 \cdot m \cdot r = k^2 \cdot m \cdot (a/r) \text{ which gives } k^2 = 0.8145 \cdot r^2/a$$
$$= 0.8145 \cdot 0.050^2/0.00111 = 1.8345$$

which gives $k = 1.354$ meters as before.

Example 8.14 brings up an important point about the summation of the forces acting on the armature, or any rotating body subjected to rotational movement along a surface. In the illustrated case the frictional force, F', was just equal to the applied force, F, and was balanced by it. There was no excess force to cause a sliding, linear acceleration and, α the rotational acceleration equaled the linear acceleration divided by the rolling radius. If, however, F had exceeded F', the armature would have slipped as well as rolled and there would have been a linear acceleration in addition to the angular acceleration. The effect of the greater force can be illustrated by a past exam problem as follows:

Example 8.15

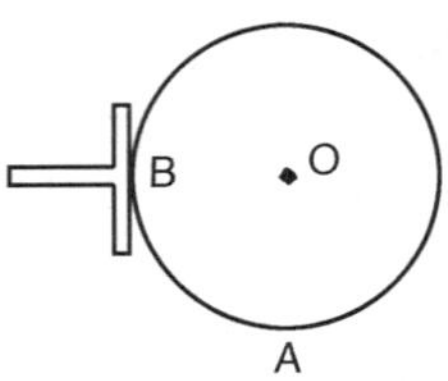

Exhibit 6

A cylinder with a mass of 30 kg and a radius of 1.25 m is pushed by a moving bulldozer; see Exhibit 6. The coefficient of friction between the blade of the bulldozer and the cylinder and between the cylinder and the surface it rests upon is 0.20. Determine the minimum acceleration of the bulldozer for the cylinder to slide along the surface without rotating, and the maximum acceleration of the bulldozer if the cylinder is to roll without slipping.

Solution

The maximum frictional force between the cylinder and the surface it rests upon will equal $F' = 0.20 \cdot \text{mg}$. The net torque acting on the cylinder will equal (see Exhibit 7)

$$F' \times R - F_B \times R,$$

where F_B is the frictional force between the blade of the bulldozer and the cylinder at the point of contact. The blade is assumed to be flat and it contacts the cylinder at a point at the same elevation as the center of the cylinder. From the equation for the torque acting on the cylinder it can be seen that if F_B is greater than F'

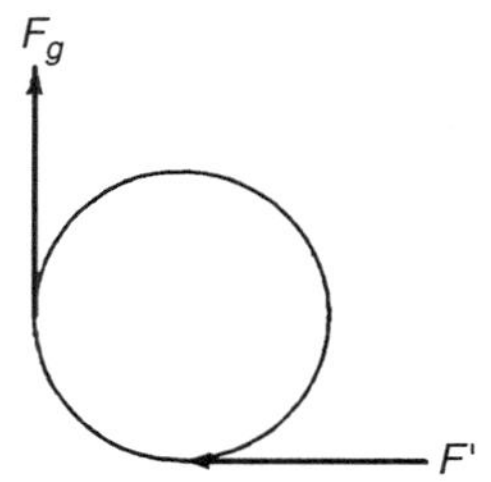

Exhibit 7

the cylinder will not roll. Similarly if F' is greater than F_B the cylinder will roll rather than slide. The critical acceleration, where the cylinder will either slide or roll with a change in the acceleration of the bulldozer, will occur when $F' = F_B$. At this point the force exerted by the blade of the bulldozer on the cylinder must equal the same as the gravitational force exerted by the cylinder on the surface it rests on. The inertial, or d'Alembert, force must then equal mg or the critical rate of acceleration of the bulldozer equals the acceleration of gravity. At any lower rate of acceleration the cylinder will roll, and at any greater rate of acceleration the cylinder will slide.

TRANSFER OF MOMENT OF INERTIA

As discussed in the *Handbook*, the moment of inertia of a body or an area about another point or axis parallel to the axis through the area's centroidal axis can be determined by means of the parallel axis theorem which states the I_p, the moment of inertia about a point p equals the moment of inertia of that body about its centroidal axis plus it mass (or area) times the square of the distance from the centroidal axis. This principle is of more importance in the determination of stresses in the study of mechanics of materials than it is in this chapter, and it will be treated in more detail there.

Example 8.16

How high must a satellite be located to rotate around the earth once each day?

Solution

The force out would equal the centrifugal force, this would be balanced against the gravitational force exerted by the earth. The force of gravity existing between two objects equals a constant times the product of the two masses divided by the square of the distance between the two centers of mass: $F_g = K(m_1 m_2/s^2)$. The force of gravity on a mass at the earth's surface equals 9.8066 times the mass. The radius of the earth is approximately 6437 km.

$$F = mg \quad \text{or} \quad F = (m_1 \times 9.8066) \text{ newtons}$$
$$F = K(m_1 m_e/6{,}437{,}000^2) \quad \text{so} \quad K = 4.063 \times 10^{14}/m_e$$

where

m_e = mass of the earth in kg

$\omega = 2\pi/(24 \times 3600) = 0.00007272$ rad/s

The centrifugal force, $f_c = 5.288 \times 10^{-9}\ s \cdot m$, which equals the gravitational force exerted by the earth.

$$5.288 \times 10^{-9} \times s \cdot m = [4.063 \times 10^{14}/m_e] \times m m_e/s^2$$

s = distance between the CG of the earth and the satellite.

$$s^3 = 0.7683 \times 10^{23} \quad \text{or} \quad s = 42{,}506{,}000 \text{ meters or } 42{,}506 \text{ km},$$

so the height of the satellite must be 42,506 – 6,437 = 36,089 km above the surface of the earth (6437 km = approx. radius of earth).

Example 8.17

A marble of radius r is allowed to roll down an incline at the bottom of which it enters a loop-the-loop of radius R. How high above the bottom of the loop-the-loop must the top of the incline be, distance h, for the marble to remain in contact with the loop at the top? I for a sphere equals $2/5mr^2$. Assume the marble rolls without slipping and no energy is lost to friction.

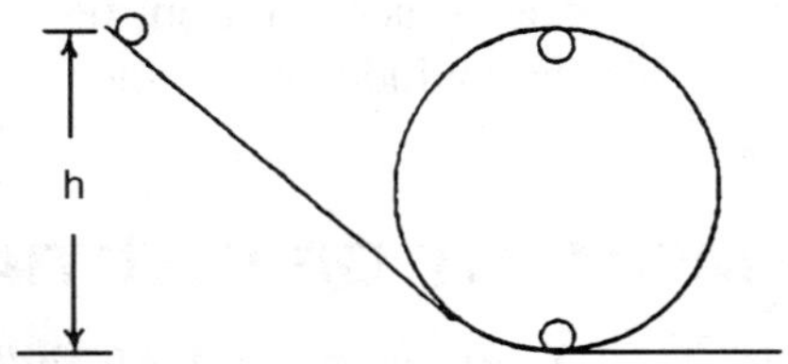

Exhibit 8

Solution

At the top of the loop, the energy contained by the rolling marble will equal its translational energy due to its velocity plus it rotational energy. The centrifugal force must equal the gravitational force

$$\omega^2 r_e m = mg = (v^2/r_e) \times m$$

where

$r_e = R - r$ (R = radius of loop and
r = radius of marble) then

$$v_T = /[g \times (R - r)]$$

to determine the velocity at the bottom, the energy will increase an amount equal to the change in potential energy

$$\Delta KE = \Delta PE$$
$$\Delta PE = mg\Delta h = mg \times 2(R - r) = \tfrac{1}{2} mv^2 + \tfrac{1}{2} I\omega^2$$
$$\text{at top } KE = \tfrac{1}{2} mv_T^2 + \tfrac{1}{2} \times (2/5 \times mr^2)v_T^2/r^2$$

where

$\omega = v/r$

Giving an amount of energy at the top of loop

$$KE = 0.700\ m \cdot v_T^2 = 0.700m \times g(R - r)$$

at bottom of loop, energy equals

$$KE \text{ at top} + mg \times 2(R - r) = 2.70 \times mg \times (R - r)$$
$$mgh = 2.70\ mg \times (R - r) \quad \text{or} \quad h = 2.70 \times (R - r)$$

Example 8.18

A homogeneous cylinder of radius **r** rolls down a plane inclined at an angle of 30° with the horizontal. (a) What is the rate of acceleration of the center, and (b) what is the minimum coefficient of friction, μ, to prevent the cylinder from slipping? See Exhibit 9.

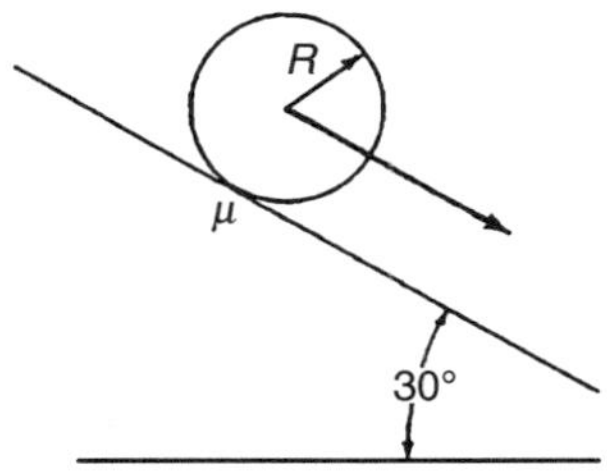

Exhibit 9

Solution

The force acting at the CG of the cylinder parallel to the plane equals mg sin 30° = 0.500 mg. Torque = $I\alpha$. The torque equals 0.500 $mg \times r$. The moment of inertia about the point of contact, I_c, can be calculated by means of the parallel axis theorem.

$$I_c = \tfrac{1}{2}mr^2 + mr^2 = 1.5\ mr^2$$
$$\text{torque} = 0.500\ mg \times r = 1.5\ mr^2\ \alpha \qquad \alpha = a/r \qquad a = g/3$$
$$\text{the answer to part (a) } a = 3.27\ \text{m/s}^2$$

For part (b) ΣF on cylinder $0.500\ mg - \mu mg \times 0.866 = ma = mg/3$
giving $\mu = 0.1667/0.866 = 0.1925$ (b)

BANKING OF CURVES

Another type of problem utilizing the principles of centrifugal (or centripetal) force is the calculation of the superelevation (slope of roadway) required in the construction of a highway for an assumed particular speed of automobile.

Example 8.19

What superelevation, β, is required on a highway where the automobile design speed is taken as 100 km/hr and the curve of the radius is to be 765 m, if a driver is to exert no sideward force on the car seat? If the mass of the driver equals 90 kg, how much force will he exert perpendicular to the seat?

Solution

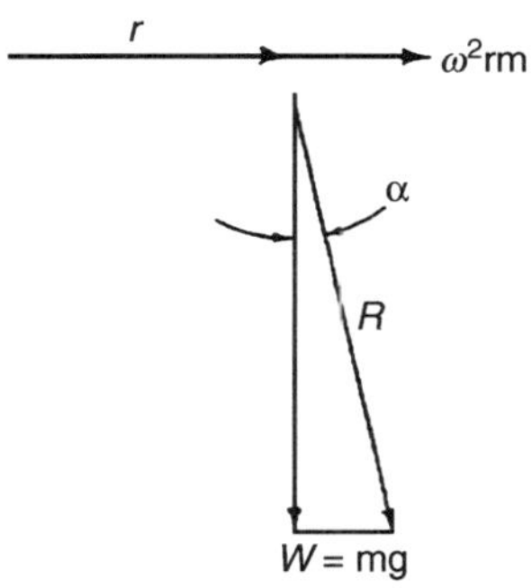

Exhibit 10

The forces acting on the car will equal the force of gravity down and the centrifugal force acting outward. The two forces will form the legs of a right triangle whose angle will equal the required angle of supereleveation. See Exhibit 10.

The acceleration acting down would equal

$$F = m \cdot g = 9.807 \cdot m$$

The centrifugal force acting at a right angle to the force of gravity would equal

$$\omega^2 rm \quad \text{or} \quad (v^2/r) \cdot m$$
$$(v^2/r) \cdot m = (100{,}000/3600)^2/765$$
$$= (1.008\ \text{m/s}^2) \cdot m$$

The required angle of superelevation would equal

$$\beta = \tan^{-1}\ 1.008/9.807 = 5.87°$$

Combining terms and reducing the relationship to its simplest form gives $\beta = \tan^{-1} v^2/(gr)$ which is the relationship given in the *Handbook*. $\tan^{-1}(1.008/9.807) = 5.87°$ as before.

Example 8.20

The force the driver would exert on the seat would equal the square root of the sum of the squares of the two components of the acceleration times his mass.

Solution

$$\text{Force on seat} = /(9.807^2 + 1.008^2) \times m$$
$$= /(96.177 + 1.016) \times 90 = 9.859 \times 90 = 887.28 \text{ N}$$

or an increase of 4.7 N, which equals 0.5%

VIBRATION

Free vibration is the vibration that takes place when an elastic system vibrates under the influence of forces within the system itself. Forced vibration is the vibration that takes place due to the influence of external forces.

Linear Motion

For a simple single degree of freedom such as is shown in Figure 8.3, if the mass is displaced a distance x from its point of equilibrium it will be acted upon by a restoring force equal to $-kx$. The net force acting on the mass equals

$$F = mg - k\,(x + \Delta x) = ma \quad \text{but} \quad mg = k\Delta \quad \text{so} \quad F = -\,kx$$
$$\text{or } -kx = ma = md^2x/dt^2$$

which is a second order differential equation, the general solution to which is

$$x = A \sin \omega t + B \cos \omega t$$

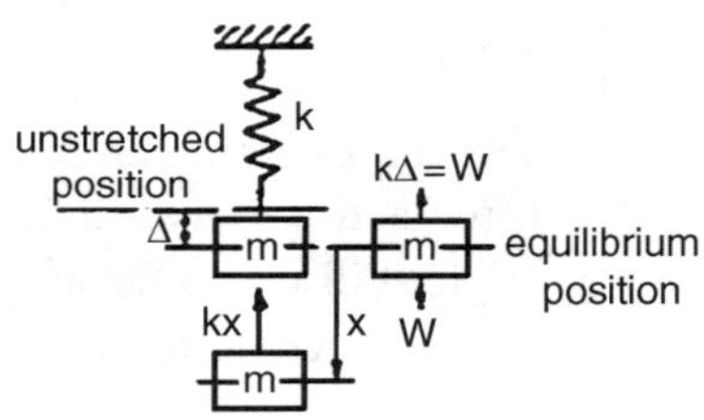

Spring mass system and free-body diagrams.

Figure 8.3

SIMPLE HARMONIC MOTION

Simple harmonic motion is defined as the motion of a point in a straight line such that the acceleration of the point is proportional to the distance of the point from its equilibrium position, or

$$a = -kx = -d^2x/dt^2$$

An example of simple harmonic motion is shown by the suspended mass shown in Figure 8.3 and described previously.

Substitution of the proper values in the basic equation gives the undamped natural frequency, f, of simple harmonic motion equals the inverse of the period, T, or $f = 1/T$

Then $f = 1/2\pi \times /[\text{g/static deflection}] = 1/T$

It can be shown that if a point moves with constant speed in a circular path the motion of the projection of the point on a diameter of the reference circle is that

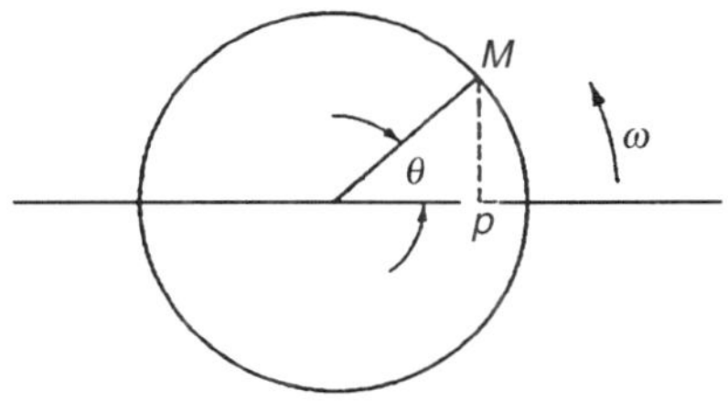

Figure 8.4

of simple harmonic motion. If ω is the constant speed of rotation, and θ is the angle the radius, r, makes with the diameter, then

$$\theta = \omega \cdot t \quad \text{and} \quad x = r \cdot \cos\theta = r \cdot \cos\omega t$$

See Figure 8.4.

The velocity of the point on the reference diameter, that is, of the object subjected to simple harmonic motion, $v_p = dx/dt = -\omega r \sin\omega t = -\omega y$ and the acceleration of the point (or body) $a_p = dv_p/dt = -\omega^2 r \cos\omega t = -\omega^2 x$, which is the equation for simple harmonic motion. The time for the point P to make one complete oscillation is the same as the time for the point M to make one complete revolution, or $T = 2\pi/\omega$, and the frequency equals $1/T = \omega/2\pi$.

Example 8.21

A 50-g mass is resting on a spring with a spring constant of $k = 100$ N/m, what is the natural frequency of the system? See Exhibit 11

Solution

The force, F, exerted on spring equals

$$F = 0.050 \times 9.8066 = 0.4903 \text{ N}$$
$$\text{Deflection} = 0.4903/100 = 0.004903 \text{ m}$$

and frequency

$$f = 1/2\pi \times /(9.8066/0.004903) = 7.12 \text{ c.p.s.}$$

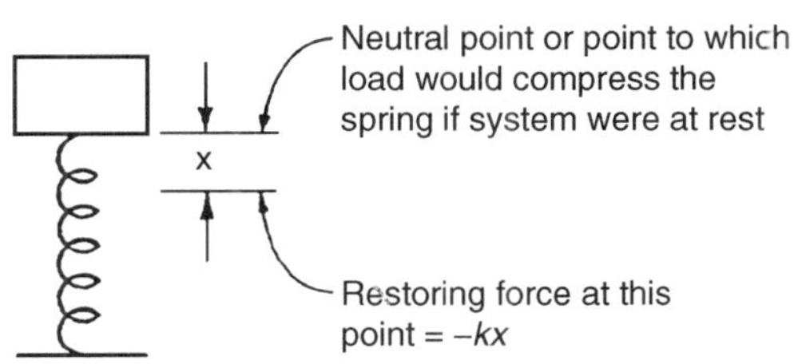

Exhibit 11

Example 8.22

The base of a one-kg instrument is set on four rubber mounts, each of which is rated at 5.0 mm deflection per kg of load. What is the natural frequency of vibration?

Solution

The deflection of the four rubber mounts together is

$$0.005/4 = 0.00125 \text{ m}$$
$$f = (1/2\pi) \times /(9.8066/0.00125) = 14.10 \text{ c.p.s.}$$

Example 8.23

A truck body lowers 15 cm when a load of 12 metric tons is placed on it. What is the frequency of vibration of the loaded truck if the total spring-borne mass is 17 metric tons?

Solution

Frequency = $1/T = \omega/2\pi$, and the acceleration of the mass $a = \omega^2 x$ but $F = ma$, and the force acting on the mass equals $-kx$, so $-kx = -\omega^2 mx$ and $\omega = /(k/m)$

$$f = \omega/2\pi = (1/2\pi)/(k/m) \quad \text{or} \quad \text{static deflection} = (17/12) \times 0.15 = 0.2125 \text{ m}$$

The spring constant equals $12{,}000 \cdot 9.807/0.15 = 784{,}560$ N/m

Combining terms gives

$$f = (1/2\pi)/[(9.807/0.2125)] = 1.081/\text{s}$$

which gives

$$f = \text{constant} \times /[(\text{N/m})(\text{m/s}^2)/\text{N}] = \text{oscillations/s}$$

Note that the restoring force acting on the truck body equals $-kx$, where x = the distance the spring is compressed past the neutral point.

Example 8.24

If the lever is assumed to be mass-less in the system shown in Exhibit 12, at what frequency would the mass of 12 kg move up and down after it had been displaced a small amount vertically? The spring constant, k, equals 75 N/m.

Solution

This is an example of undamped free vibration with a single degree of freedom. For small displacements the system will vibrate in conformance with the relationship for simple harmonic motion in the same way as the suspended mass shown in Figure 8.3. Thus, the motion will satisfy the differential equation— $d^2x/dt^2 + (k/m) \cdot x = 0$ as given in the *Handbook*, or in this case $d^2y/dt^2 + (k/m) \cdot y$ since the displacement is vertical.

The general solution of this equation gives $\omega^2 = k/m$, where ω is measured in radians per second. A circular function repeats itself in 2π radians, so one cycle of vibratory motion is completed when

$$\omega T = 2\pi$$

where T is the time for one complete cycle or T = the period of motion.

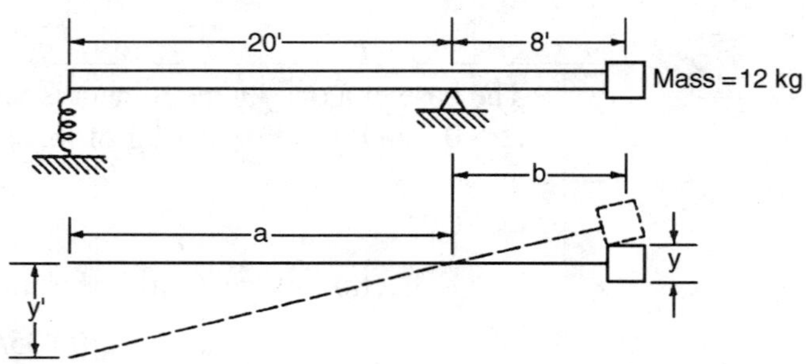

Exhibit 12

The frequency, f, in cycles per second would then equal $1/T$, or $f = \omega/2\pi$.
In addition, acceleration = d^2y/dt_2, and from Newton's second law

$$F = ma \quad \text{and} \quad (\text{acceleration}) = F/m$$

Combining these gives

$$F/m + (k/m)y = 0 \quad \text{or} \quad F = -ky$$

where the minus sign is explained by the fact that the force is opposite in sense to the displacement, y, of the mass.
The force acting on the mass in the figure would equal

$$-ak_{sp}y' = bF$$

Where y' is the displacement (extension or compression) of the spring and k_{sp} is the spring constant of the spring.

$$b/y = a/y' \quad \text{so} \quad y' = (a/b)\cdot y$$

and the force acting on the mass,

$$F = -\left(\frac{a}{b}K_{sp}\right)y'$$

But force, F, also equals $-ky$
where k is the equivalent spring constant acting on the mass, so

$$k = (a/b)^2 \cdot k_{sp}, \quad \omega^2 = k/m,$$
$$\text{and} \quad f = \omega/2\pi \quad f = (1/2\pi)\cdot\surd(k/m) = (1/2\pi)\cdot\surd\,[(a/b)^2\cdot k_{sp}/m]$$
$$(a/b)^2 = (20/8)^2 = 6.25 \quad f = (1/2\pi)\cdot\surd(6.25\cdot 75/12)$$
$$= 0.995 \text{ cycles per second}$$

ROTARY VIBRATION

Angular oscillation considers torsional stiffness of the supporting member and the moment of inertia of the vibrating member, rather than mass and linear spring constant. The undamped natural frequency of torsion of the system shown in Exhibit 13 is given by the relationship

$$f = (1/2\pi) \times \surd(K/J),$$

where
K = joules/radian = stiffness of supporting shaft and
J = polar mass moment of inertia of supported member.

Example 8.25

A 1.00 kg circular disk 17.0 cm in diameter is rigidly attached to a 9.00 mm diameter steel rod 20.00 cm long. The rod is fixed at its other end. What is the natural frequency of the system?

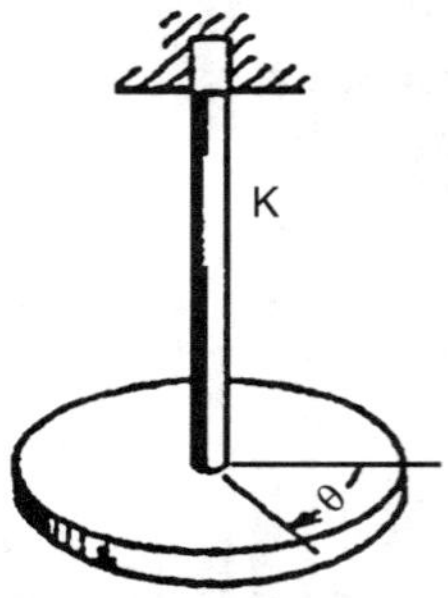

Exhibit 13

Solution

The torsional spring constant of the rod = GJ_{rod}/ℓ where

G = shear modulus, steel = 8.3×10^{10} Pa (*FE Handbook*)
$J_R = \pi d^4/32 = (\pi \times 0.009^4)/32 = 6.441 \times 10^{-10}$
$K = GJ/\ell = 6.441 \times 8.3/0.200 = 267.3$ Nm/rad
$I_{Disk} = \frac{1}{2}mr^2 = \frac{1}{2} \times 1.00 \times 0.085^2 = 0.00361$ Nm/^{2}
$f = (1/2\pi) \times /(267.3/0.00361) = 272.11/6.283 = 43.3$
Frequency = 43.3 cycles per second

PROBLEMS

8.1 The d'Alembert force is
a. Force of gravity in France.
b. Force due to inertia.
c. Resisting force due to static friction.
d. Atomic force discovered by d'Alembert.

8.2 The base of a 1 kg instrument is set on four rubber mounts, each of which is rated at 5.0 mm deflection per kg of load. What is the natural frequency of vibration?
a. 14 Hz
b. 16 Hz
c. 18 Hz
d. 20 Hz

8.3 A car traveling at 90 km/hr goes around a curve in the road that has a radius of 500 m. What is the lateral acceleration acting on the car?
a. 0.04 g
b. 0.07 g
c. 0.10 g
d. 0.13 g

8.4 A pile driver uses a mass of 500 kg to drive a pile into the ground. If the average resisting force of friction equals 135 kN, the mass of the pile equals 400 kg, and the 500 kg driver drops 6.0 m, which of the following most neatly equals the depth of penetration of the pile? Assume a perfectly inelastic impact between driver and pile.
a. 9 cm
b. 13 cm
c. 17 cm
d. 21 cm

8.5 A horizontal force is applied to a block moving in a horizontal guide. When the block is at a distance x from the origin, the force is given by the equation $F = x^3 - x$. Which if the following most nearly equals the work that is done by moving the block from one meter to the left of the origin to one meter to the right of the origin?
a. 0 J
b. 0.25 J
c. 0.50 J
d. 0.75 J

8.6 In the system shown in Exhibit 8.6 the mass, M, equals 10 kg. It is observed to vibrate 10 times in 15 seconds when displaced a small amount from its neutral position. If the mass of the lever arm is disregarded, which of the following most nearly equals the stiffness coefficient of the spring?

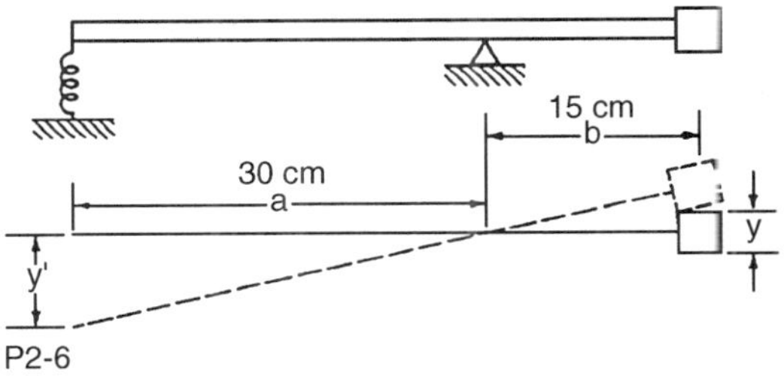

P2-6
Exhibit 8.6

a. 57.0 N/m
b. 52.3 N/m
c. 48.2 N/m
d. 43.1 N/m

8.7 An unbalanced force in newtons varies with time in seconds in accordance with the relationship $F = 3500 - 500\,t$ acts on a body with a mass of 1500 kg for 5.0 seconds. If the body has an initial velocity in the direction of the applied force of 3.0 m/s, which of the following most nearly equals its final velocity?

a. 9.8 m/s c. 11.2 m/s
b. 10.5 m/s d. 11.9 m/s

8.8 A wheel 4.0 m in diameter has a mass of 30.0 kg and a radius of gyration equal to 25.0 cm. If it starts from rest and rolls without slipping down a 30° plane, which of the following most nearly equals its speed when it has rolled 10.0 m measured along the surface of the plane?

a. 11.0 m/s c. 10.2 m/s
b. 10.6 m/s d. 9.8 m/s

8.9 A rectangular door one meter wide swings horizontally. It has a mass of 30 kg uniformly distributed. The door is supported by frictionless hinges installed along one of its vertical sides. It is controlled by a spring that exerts a torque on it proportional to the angle through which the door is turned from its closed position. If the door is opened 90° and released, the torque due to the spring is then 4.0 J. Which of the following most nearly equals the angular velocity at the time the door has swung back and is passing through the 45° position?

a. 0.61 rad/s c. 0.79 rad/s
b. 0.69 rad/s d. 0.85 rad/s

8.10 A body is moving in a straight line according to the relationship $S = \frac{1}{4}t^4 - 2t^3 + 4t^2$ where S equals the distance traveled in time t. Which of the following most nearly equals the time interval during which the body is moving backward?

a. 0 to 2 s c. 4 to 6 s
b. 2 to 4 s d. 1 to 3 s

8.11 A belt 20 cm wide and 6.0 mm thick drives a pulley 30 cm in diameter at a speed of 1650 rev/min. The maximum tension the belt can tolerate is 1.40 MPa. If the slack side has three-fourths the tension of the tight side, which of the following most nearly equals the power that can be transmitted?

a. 7.6 kW c. 9.8 kW
b. 8.7 kW d. 10.9 kW

8.12 A kite is 40 meters high with 50 meters of cord out. If the kite is moving horizontally away from the man flying it at a rate of 6.50 km/h, which of the following most nearly equals the rate at which the cord is being paid out in meters per second?

a. 0.8 m/s c. 1.4 m/s
b. 1.1 m/s d. 1.7 m/s

8.13 A bullet leaves the muzzle of a gun with a velocity of 900 meters per second and an angle of 45° with the horizontal. Which of the following most nearly equals the maximum distance the bullet will travel horizontally, measured along the same elevation as the gun muzzle? Disregard air resistance.

a. 68 km
b. 73 km
c. 78 km
d. 83 km

8.14 An airplane has a true air speed of 380 km/hr and is heading true north. A 65-km/hr head wind is blowing 45° from true north from the northeast. Which of the following most nearly equals the true ground speed?

a. 337 km/h
b. 351 km/h
c. 366 km/h
d. 380 km/h

8.15 The rifling in an eight mm caliber rifle causes the bullet to turn one revolution for each 250 mm of barrel length. If the muzzle velocity of the bullet is 885 m/s, which of the following most nearly equals the rate at which the bullet is rotating at the instant at which it leaves the muzzle?

a. 200,000 RPM
b. 204,000 RPM
c. 208,000 RPM
d. 212,000 RPM

8.16 A faster moving body is directly approaching a slower moving body traveling in the same direction. The speed of the faster body is 95 km/hr and the speed of the slower body is 55 km/h. At the instant when the separation of the two bodies is 30 meters, the faster body is given a constant deceleration. Which of the following most nearly equals the deceleration so that the two bodies just touch at the instant of impact with no shock?

a. 0.21 g
b. 0.25 g
c. 0.29 g
d. 0.31 g

8.17 At the beginning of the drive a golf ball has a velocity of 270 km/h. If the club remains in contact with the ball for 1/25 s, which of the following most nearly equals the average force on the ball if the ball has a mass of 45 g?

a. 94 N
b. 89 N
c. 84 N
d. 79 N

8.18 An artificial satellite is circling the earth at a radius of 15,000 km. Which of the following most nearly equals the number of revolutions it makes per day? Assume the radius of the earth is 6437 km, and assume no air resistance to the travel of the satellite. (The elevation above the earth would equal 8563 km.)

a. 2.5/day
b. 3.6/day
c. 4.8/day
d. 5.8/day

8.19 A balloon is ascending vertically at a uniform rate for one minute. A stone falls from it and reaches the ground in 5.0 s. Which of the following most nearly equals the height from which the stone fell?

a. 114 m
b. 124 m
c. 134 m
d. 144 m

8.20 A body takes twice as long to slide down a plane at 30° to the horizontal as it would if the plane were smooth. Which of the following most nearly equals the coefficient of friction?

a. $\mu = 0.43$
b. $\mu = 0.50$
c. $\mu = 0.25$
d. $\mu = 0.37$

8.21 Which of the following most nearly equals the distance from the center of a phonograph record turning at 78 rpm that a pickle can lie without being thrown off if the coefficient of friction is 0.30?

a. 5 cm
b. 2 cm
c. 3 cm
d. 4 cm

8.22 A cast aluminum drum 60 cm in diameter and 30 cm in length is supported by an axial shaft on smooth bearings. A light cord wrapped around the drum is pulled with a constant force of 90 N until five meters of cord is unwrapped and is then released. The density of aluminum is 2.70 g/cm^3 (as given in the *Handbook*). Which of the following most nearly equals the speed of rotation when the pull on the cord has ceased?

a. 78 RPM
b. 89 RPM
c. 98 RPM
d. 67 RPM

8.23 A body with a mass of 90 kg starts from rest at the top of a plane, which is at an angle of 60° with the horizontal. After sliding 2.5 meters down the plane, the body strikes a spring with a constant of 17.5 kN/m. The coefficient of friction is 0.25. The body remains in contact with the plane throughout. Which of the following most nearly equals the compression of the spring?

a. 35 cm
b. 40 cm
c. 54 cm
d. 47 cm

8.24 The friction surface on a friction disk type of clutch has an outside diameter of 25.0 cm and an inside diameter of 7.0 cm. The coefficient of friction between the two surfaces equals 0.30. Which of the following most nearly equals the force that must be applied to the plate if the clutch is to transmit 75.0 kW at 3300 rev/min?

a. 8.2 kN
b. 7.5 kN
c. 6.9 kN
d. 6.4 kN

8.25 A motorcar has a maximum speed of 110 km/hr on the level, at which instant the motor is transmitting 45 kW to the wheels. If the tractive effort of the wheels remains constant while the resistance to motion varies as the square of the speed, which of the following most nearly equals the angle of slope up which the car can maintain a speed of 80 km/hr if the mass of the car is 1800 kg?

a. 1.90°
b. 2.25°
c. 2.75°
d. 3.10°

8.26 A ball, A, has caromed off the cushion at the end of a pool table and is moving at a velocity of 8 m/s in the X direction (see Exhibit 8.26). The cue ball hits it at an angle of 20° with a velocity of 15 m/s as shown in the figure. Both balls have the same mass. The coefficient of restitution is 0.60. Which of the following most nearly equals the velocity of ball A after the impact?

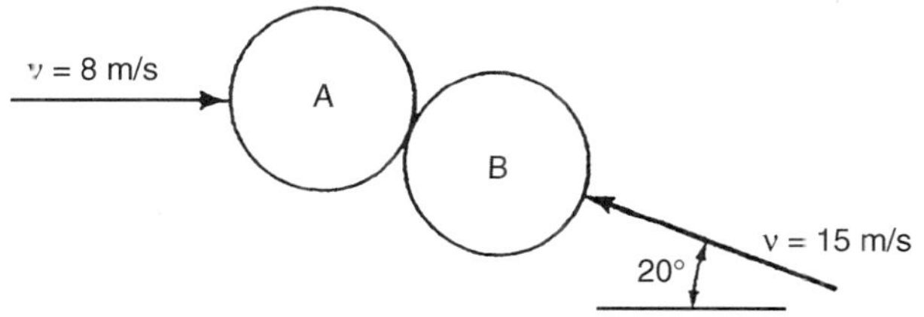

Exhibit 8.26

a. –9.7 m/s
b. –8.5 m/s
c. –7.3 m/s
d. –6.9 m/s

SOLUTIONS

8.1 **b.** Newton's second law states that force equals mass times acceleration. Thus, when a mass is accelerated there is a resisting, or inertial, force equal to *ma*. This is termed the d'Alembert force.

8.2 **a.** The deflection of the four rubber mounts together is

$$0.005/4 = 0.00125 \text{ m}$$
$$f = 1/2\pi \times /(9.8066/0.00125) = 14.10 \text{ c.p.s.}$$

8.3 **d.** Centrifugal force equals

$$m \times \omega^2 \times \mathrm{R}$$
$$\text{or } F = m \cdot v^2/R = m \times (90 \times 10^3/3{,}600)^2/500 = 1.25 \times m$$
$$\text{or } F = 0.1275\ g$$

8.4 **b.** $M_1v_1 = (M_1 + M_2)v_r$ velocity at impact, v_1, equals $/(2gh) = 10.85$ m/s
$500 \times 10.85 = 900 \times v_r$
The resulting velocity of the 900 grams equals 6.03 m/s.
The net force acting on the assembly equals

$$900g - 135\,\text{kN} = -126.17\,\text{kN}$$

the rate of acceleration would equal

$$a = F/m = -126,\ 170/900 = -140.2 \text{ m/s}^2$$

from

$$v^2 = v_o^{\ 2} + 2$$

as we get

$$0 = 36.36 - 2 \times 140.2 \times s$$

which gives

$s = 0.130$ m or $KE = \tfrac{1}{2} Mv^2 = \tfrac{1}{2} \times 900 \times 36.36 = 16.36$ kJ Work
= (net force) × distance or
s = *KE*/Force 16.36/126.17 = 0.130 meters

8.5 **a.**

$$\text{Work} = IF \cdot ds = I(x^3 - x)\,dx \text{ from } (x = 1) \text{ to } (x = 1)$$
$$\text{Work} = (x^4/4 - x^2/2) \text{ giving Work} = (\tfrac{1}{4} - \tfrac{1}{2}) - (\tfrac{1}{4} - \tfrac{1}{2}) = 0$$

8.6 **d.** Let $a = 30$ cm, $b = 15$ cm, and $m = 10$ kg.

$$d^2y/dt^2 + (k/m)y = 0 \quad \omega^2 = k/m \quad F = ma$$

See Exhibit 8.6

$$d^2y/dt^2 = a = F/m \quad F = -ky = -(a/b)\,k_{sp}y'$$
$$\text{Force on mass} = -(a/b)^2 k_{sp} \cdot y \quad k = (a/b)^2 \cdot k_{sp}$$

$\tau = 15$ seconds/10 vibrations = 1.5 s

$$\tau = 2\pi/[(m/k_{sp})(b/a)^2] \quad k_{sp} = 43.1 \text{ N/m}$$

8.7 **b.**

$$v = v_o + at \qquad a = F/m \qquad a = (2.333 - 0.333t)\ \text{m/s}^2$$
$$Idv = I(a\ dt) \qquad \Delta v = (2.333 - 0.333\text{t})\ dt \text{ from } t = 0 \text{ to } t = 5$$
$$\Delta v = 2.333t - 0.167t^2 = 7.49\ \text{m/s} \qquad v = v_o + \Delta v = 10.49\ \text{m/s}$$

8.8 **d.**

$$I = mk^2 = 30 \times 0.25^2 = 1.875\ \text{kg}\cdot\text{m}^2 \qquad \Delta PE = \Delta KE$$
$$\Delta KE = \tfrac{1}{2}mv^2 + \tfrac{1}{2}I\omega^2 = \tfrac{1}{2}mv^2 + \tfrac{1}{2}I(v/r)^2 = v^2(15 + 0.234)\ \Delta PE$$
$$= 30 \times 0.5 \times 10 \times g = 1471\ \text{N}\cdot\text{m} \qquad v = /(1471/15.234) = 9.83\ \text{m/s}$$

8.9 **b.**

$$\text{mass} = 30\ \text{kg} \qquad Tq = I\alpha = k\theta$$

where k = spring constant

$$I = Ir^2dm \text{ from } r = 0 \text{ to } r = 1 \quad I = Ir^2dm = I(r^2 \cdot 30 \cdot dr) = 10.0$$
$$k = 4.0/1.571 = 2.546\ \text{J/rad from } \alpha = (d\omega/dt \text{ and } \omega = (d\theta/dt)$$

we get $\alpha = \omega(d\omega/d\theta)$

$$Tq = -k\theta = -2.546\theta = 10 \cdot \omega(d\omega/d\theta)$$
$$3.928 \cdot I\omega d\omega = -I\theta d\theta \quad 3.928 \cdot \omega^2 = -\theta^2 + C$$
$$\omega = 0 \text{ at } \theta = \pi/2 \quad \text{So } C = 2.467$$
$$3.928 \cdot \omega^2 = -(\pi/4)^2 + 2.467 \qquad \omega = 0.686\ \text{rad/s}$$

8.10 **b.**

$$S = {}^1\!/_4 t^4 - 2t^3 + 4t^2 \quad v = ds/dt = t^3 - 6t^2 + 8t$$
$$v = t(t-4)(t-2) \qquad v = 0 \text{ at } t = 0,\ t = 4, \text{ and } t = 2$$
$$\text{at } t = 3 \qquad v = -3\ \text{m/s}$$

Thus the body moves backward between $t = 2$ s and $t = 4$ s.

8.11 **d.** Maximum belt tension = $0.20 \times 0.006 \times 1.40 \times 10^6 = 1{,}680$ N

$$Tq = 0.25 \times 1{,}680 \times 0.30/2 = 63.0\ \text{N}\cdot\text{m}$$
$$\text{Power} = 2\pi \cdot Tq \cdot \text{rev/s}$$
$$\text{Power} = 2\pi \times 63.0 \times 1{,}650/60 = 10{,}886\ \text{W or } 10.9\ \text{kW}$$

8.12 **b.** Take L = length of cord and S = horizontal distance along the ground. Then dL/dt = rate at which cord is being paid out. $S_o = (50^2 - 40^2) =$ 30 m 6.50 km/hr = 1.806 m/s

$$L = /[40^2 + (30 + 1.806t)^2] \qquad d/x = \tfrac{1}{2}x^{-1/2}dx$$
$$dL/dt = (108.36 + 6.524t)/[2 \times /(2{,}500 + 108.36t + 3.262t^2)]$$
$$\text{at } t = 0 \quad dL/dt = 108.36/(2 \times /2{,}500) = 1.084\ \text{m/s}$$

8.13 **d.** Height of apogee, h, equals $v_y^2/2g = (900 \cdot \sin 45°)^2/2g \quad h = 20{,}648.5$ m time to reach apogee equals h/(avg. vert velocity) = $20{,}648.5/[(900 \cdot \sin 45°)/2] = 64.892$ s

$$\text{horizontal travel} = 2 \times 64.892 \times 900 \times \cos 45° = 82{,}594\ \text{m}$$

8.14 **a.** Backward component of speed $65 \cdot \sin 45° = 45.96$ km/hr
component of speed westerly $65 \cdot \cos 45° = 45.96$ km/hr
Ground speed = $[(380 - 45.96)^2 + 45.96^2] = 337.2$ km/hr

8.15 **d.** Velocity equals 885,000 mm/s bullet makes one rev in 250 mm

$$885{,}000/250 = 3{,}540\ \text{rev/s} \quad \text{or} \quad 212{,}400\ \text{rpm}$$

8.16 **a.** Use relative velocities. $U^2 = U_o^2 + 2as$

$$95 - 55 = 40 \text{ km/hr} = 11.111 \text{ m/s}$$
$$0 = 11.111^2 + 2 \cdot a \cdot 30$$
$$a = -2.058 \text{ m/s}^2 \quad \text{or} \quad 0.210\, g$$

8.17 **c.** Impulse = momentum

$$F \cdot t = m \cdot v = 270{,}000/3{,}600 = 75.0 \text{ m/s}$$
$$m = 45/1{,}000 = 0.045 \text{ kg}$$
$$F = 0.045 \cdot 75.0/(1/25) = 84.375 \text{ N}$$

8.18 **c.** The force outward equals $\omega^2 R m_s$. This is balanced by the gravitational force inward which equals $k \cdot m_s \cdot m_e / R^2$ where R equals the distance from the center of mass of the satellite to the center of mass of the earth, k, equals the gravitational constant, m_s is the mass of the satellite, and m_e is the mass of the earth. The value of k can be determined from the acceleration of gravity at the surface of the earth and the radius of the earth, $r \cdot m_o \cdot g = k \cdot m_o \cdot m_e / r^2$. The radius of the earth is approximately 6,437 km or 6,437,000 meters.

So k equals $r^2 \cdot g/m_e = 4.063 \times 10^{14}/m_e$.

This gives

$$\omega^2 = k/R^3 = (4.063/3.375) \times 10^{-7}$$

which gives

$$\omega = 0.0003497 \text{ rad/s } (0.0003497/2\pi) \times 24 \text{ h} \times (3{,}600 \text{ s/hr})$$
$$= 4.808 \text{ or } 4.81 \text{ revolutions about the earth per day.}$$

8.19 **c.** Take upward velocity, v, positive. Height, h, of balloon at one minute $h = 60 \cdot v$. The initial velocity of the stone equals the velocity of the balloon upward. When it is dropped it will be subjected to a negative acceleration of gravity.

$$60 \cdot v = 5 \cdot v - \tfrac{1}{2} g t^2 \quad 55 \cdot v = 122.59 \quad v = 2.229 \text{ m/s}$$
$$h = 60 \times 2.229 = 133.74 \text{ meters (N.B. } g = -9.807 \text{ m/s}^2)$$

8.20 **a.** Force perpendicular to plane = $mg \cdot \cos 30° = 0.866 \cdot mg$

so the frictional force resisting motion equals $\mu \cdot 0.866 \cdot mg$. The net force parallel to plane equals $mg \cdot \sin 30° - \mu \cdot 0.866 \cdot mg = mg(0.500 - 0.866\mu)$ for friction.

$$\text{Net force for no friction} = 0.500\, mg \quad S = v_o t + \tfrac{1}{2} a t^2$$
$$v_o = 0 \quad t_f = 2 \cdot t_{\text{smooth}} \quad \tfrac{1}{2} a_s t^2 = \tfrac{1}{2} a_f \cdot 4t^2$$

so

$$a_f = \pi\, a_s \quad a = \text{F/m and}$$
$$F_s = 4F_f \text{ parallel to surface of plane } 0.500 \cdot mg$$
$$= 4 \cdot (0.500 - 0.866\mu) mg \quad \mu = 1.50/3.464 \quad \mu = 0.433$$

8.21 **d.** Force on pickle = $\omega^2 \cdot Rm$

$$\omega = (78/60) \times 2\pi = 8.168 \text{ rad/s Resisting force} = 0.30 \cdot mg$$
$$8.168^2 \cdot Rm = 0.30 \cdot m \cdot 9.807$$
$$R = 0.0441 \text{ m} \quad \text{or} \quad 4.41 \text{ cm}$$

8.22 **b.**

$$\text{Volume} = (\pi/4) \cdot 60^2 \cdot 30 = 84{,}823 \text{ cm}^3$$
$$\text{Mass} = 229.0 \text{ kg}$$
$$I = \tfrac{1}{2} MR^2 = 0.5 \cdot 229.0 \cdot 0.30^2 = 10.305$$
$$KE = \tfrac{1}{2} I\omega^2 = 90 \cdot 5 \text{ Nm or } 450.0 \text{ joules}$$
$$\omega = (900/10.305) = 9.345 \text{ rad/s}$$
$$9.345 \cdot 60/2\pi = 89.24 \text{ RPM}$$

8.23 **d.** Net force, F, acting down the plane would equal

$$F = mg \cdot \sin 60° - mg \cdot \cos 60° \cdot \mu = 654.03 \text{ N}$$

so energy = 654.03 × 2.5 = 1635 J

The same can be calculated by calculating the *KE*

$$v^2 = 2 \text{ as} \quad F = mg \sin 60° - \mu \times \cos 60° \, mg$$
$$= 0.741 \times mg \quad \text{or} \quad a = 0.741 \, g$$
$$v^2 = 2 \times 2.5 \times 0.741 \times g = 36.334$$
$$KE = \tfrac{1}{2} mv^2 = \tfrac{1}{2} \times 90 \times 36.334 = 1635 \text{ J}$$

The total energy to be absorbed by the spring would equal 1,635 + S × 654.03
where S = distance spring is compressed.

$$1635 + 654 \times S = \tfrac{1}{2} \times 17{,}500 \times S^2$$

which reduces to

$$S^2 - 0.0747 \times S - 0.1868 = 0$$

Solve for S (see the *Handbook* for solution of quadratic equation)

$$S = 0.4712 \text{ meter or } 47.12 \text{ cm}$$

8.24 **a.** Watt = joule/s joule = newton·meter = torque, Tq.

power = $2\pi \cdot$ (rev/s) Tq = watts Tq = Force × radius
Force = $I(\mu p \times 2\pi r \, dr)$ so $Tq = I[(\mu p \times 2\pi r \, dr) \times r]$
75,000 = $Tq \cdot (3{,}300/60) \cdot 2\pi$ so Tq = 217.03 N · m for friction clutch
$Tq = I(2\pi r dr)\mu p r$

where
r = radius, μ = friction coefficient, and p = pressure

$Tq = 2\pi\mu p \, Ir^2 dr = 2\pi\mu p r^3/3$ between $r = 0.125$ and $r = 0.035$
$Tq = 0.001200 \cdot p = 217.03 \; N \cdot m \; p = 180{,}858$ Pa

Clutch face area = $\pi(0.125^2 - 0.035^2) = 0.04524 \text{ m}^2$

$$\text{Force} = 0.04524 \times 180{,}858 = 8182 \text{ N}$$

8.25 **d.** Since resistance to motion is proportional to the square of the speed the resistance to motion at 80 km/hr will equal $(80/110)^2$ times resistance to motion at 110 km/hr or $0.5289 \times 45 = 23.800$ kW This leaves 21.200 kW to overcome the increase in elevation going up the hill. The power required to go up the hill, ignoring the resistance to motion would equal

$$1800 \times g \times \sin \alpha \times \text{speed}$$
$$80 \text{ km/hr} = 80{,}000/3{,}600 \text{ m/s} = 22.222 \text{ m/s}$$

power requirement = $(392{,}276 \times \sin \alpha)$ watts
power available = 21,200 watts so $\sin \alpha = 21{,}200/392{,}276 = 0.0540$

$$\alpha = 3.098°$$

8.26 **a.** Momentum is conserved after an impact. Momentum is a vector quantity, so this means that momentum in the X and Y directions are both conserved. Masses are equal.

$$\text{Momentum in } Y \text{ direction} = M \cdot 15 \cdot \sin 15° = 3.882\text{M}$$

so the velocity of B in the Y direction will remain the same.

$$X \text{ direction momentum} = Mu_1 + Mv_1 = 8 - 15 \cdot \cos 20° = -6.095$$

The coefficient of restitution is 0.6. Let: $u = A$ velocity, $v = B$ velocity and initial and final velocities be denoted by subscripts 1 and 2. This gives

$$(u_2 - v_{2x})/(-14.095 - 8) = 0.6 \quad \text{and} \quad u_2 = v_{2x} - 13.257.$$

But from the conservation of momentum in the X direction

$$u_2 + v_{2x} = -6.095 \quad \text{so} \quad v_{2x} = -u_2 - 6.095$$

and

$$v_{2x} = -v_{2x} + 13.257 - 6.095,$$

which gives

$$v_{2x} = 3.581 \text{ m/s} \quad \text{and} \quad u_2 = -9.67 \text{ m/s}$$

CHAPTER 9

Automatic Controls

OUTLINE

This chapter reviews classical control systems equations in addition to an introduction to modern control systems (i.e., state variable analysis). In the past, the examination did not include questions on state variables; now, however, several questions may well be on this subject. While in the last few years many undergraduate control systems college courses have included an introduction to z-transforms, it is unlikely the examination will include any questions on this subject. One of the more popular texts widely used throughout the United States that does include both classical and modern control is *Automatic Control Systems* by Kuo (Prentice-Hall, 1991), which is written for the undergraduate level. There are many other books that cover both areas, but it's best to stick with your particular college text that you are familiar with if it includes state variables. (Caution: no need to go overboard on all of the details concerning state variables; know the notation and concepts and you should be ok.)

CLASSICAL CONTROL SYSTEMS

More than likely, the bulk of the questions will involve classical control theory, which includes: block diagram formulation and reduction, linear system stability, system error, second order system specifications (with higher ordered systems being approximated by second order ones), frequency response methods, and system compensation. The background for this material requires a good understanding of Laplace transforms, pole-zero maps, and transient response analysis.

BLOCK DIAGRAMS

The general form of the block diagrams will usually be some variation of Figure 9.1 where $R(s)$ is the referenced input,

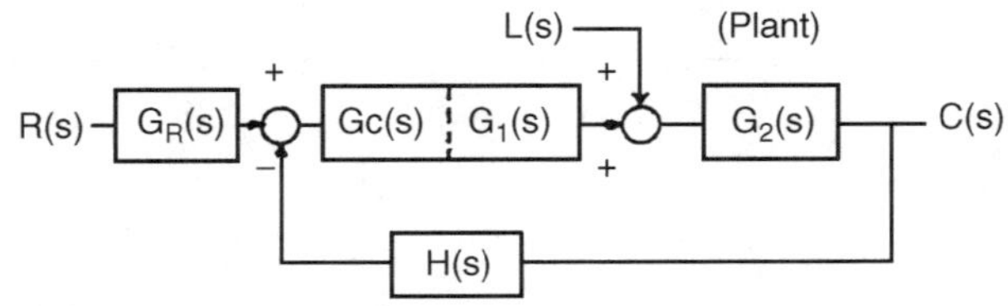

Figure 9.1 A system block diagram with negative feedback

$C(s)$ is the controlled output, and $L(s)$ may be a disturbance (or loading) signal. For a simple single loop system, the closed loop system equation is easily found by realizing that the numerator(s) are nothing more than the products of the transfer functions in the forward path(s) of the signals from source to output. The denominator(s) are merely one plus the loop gains (the product of the transfer functions in the loop). If more than one loop exists, then somewhat more formal definitions and techniques are involved; the equations are derived from Mason's rules and flow graph methods. The equation relating the input(s) to the output, using Laplace transforms (rather than time domain relationships—where multiplying now becomes convolving in the time domain) are clearly shown to be (and here, since the complex frequency domain is implied, the "of(s)" notation will be dropped) as follows

$$C = \frac{G_c G_1 G_2}{1 + G_c G_1 G_2 H} G_R R + \frac{G_2}{1 + G_c G_1 G_2 H} L \tag{9.1}$$

The denominator, $1 + G_c G_1 G_2 H$, if equal to zero defines the poles of the system (i.e., for values of s that make the denominator zero—or C equal to infinity) and, of course, is the characteristic equation of the system. It is clear that if these poles could be easily found, then system stability is known and there would be no need for the Routh-Hurwitz test or for going through the tests for root-locus paths. Also note that this characteristic equation is not a function of either R or L, which means that stability for linear systems is independent of the inputs.

ROUTH-HURWITZ CRITERION FOR STABILITY

The Routh-Hurwitz Criterion gives information as to whether any poles lie in the RHP but does not give their actual location. Another technique, the root locus method, will give the root location in both planes but is far lengthier.

However, for the Routh-Hurwitz method (frequency referred to as just the Routh Test), one needs only to set up an array from the characteristic polynomial and examine the first column of this array to determine stability. This test is used for any equation; but here it refers to the characteristic equation polynomial, which is simplified to

$$a_0 s^n + a_1 s^{n-1} + a_2 s^{n-2} + \cdots a_n = 0 \tag{9.2a}$$

The test then involves making an array out of the coefficients, a's, with additionally created new coefficients, b's. The first two rows, starting with the highest coefficient, a_0, are formed by making the first row of all even values of the a's, the second row is made up of the odd coefficients. The next row of b's are created as shown in equation 9.2b. If all of the resulting coefficients in the first column of a's and b's are the same sign and nonzero, then all of the roots will be in the left-hand portion in the s-plane (i.e., and will be positive),

s^n	a_0	a_2	a_4	a_6	…	…
s^{n-1}	a_1	a_3	a_5	a_7	…	
s^{n-2}	b_1	b_2	…			
…	…					
s^0	b_n					

where Eq. 9.2b

$$b_1 = \frac{a_1a_2 - a_0a_3}{a_1}$$

$$b_2 = \frac{a_1a_4 - a_0a_5}{a_1}$$

Consider the following simple example (refer to Figure 9.1 with $G_P = 1$, $G = G_cG_1G_2$)

$$G = 10(s + 2)/[s(s + 1)(s + 3)], \qquad H = (s + 4)$$

Then,

$$G_{\text{system}} = 10(s + 2)/[s^3 + 14s^2 + 63s + 80].$$

The characteristic polynomial, $s^3 + 14s^2 + 63s + 80$, may be arranged in a Routh array as follows

s^3	1	63
s^2	14	80
s^1	b_1	
s^0	b_2	

$b_1 = [(14)(63) - (1)(80)]/14 = 57.3$

$b_2 = [(b_1)(80) - (14)(0)]/b_1 = 80$

Since the first column is all positive, the system will be stable in closed loop. (Try this problem for H being unity and G being 3 times larger—you will get an unstable system!) Of course stability may be found by frequency response techniques (i.e., Nyquist criterion, etc.); but, if only the question of stability is to be answered, then use the easy Routh method.

STANDARDIZED SECOND-ORDER SYSTEMS

Although actual systems are usually higher than second-order, the standardized response of this kind of a system to either step or variable frequency input is customarily used as a standard. Their equations are

$\omega_D = \omega_{res} = \omega_n\sqrt{1-2\zeta^2}$, the "driven" resonant frequency. (9.3a)

$\omega_r = \omega_n\sqrt{1-\zeta^2}$, the "ringing" or damped frequency. (9.3b)

$t_p = \pi/\omega_r$, the time-to-first-peak, step response. (9.3c)

$C_p = M_p = 1 + e^{-\pi\zeta/\sqrt{1-\zeta^2}}$, the peak magnitude of the step response (9.3d)

The quantities that describe this response are usually used to describe and approximate higher order ones. From Figure 9.1, where $H = G_R = G_C = G_1 = 1$ and, the plant, $G_2 = K/[s(s + a)]$, results in Figure 9.2. Standardized curves are well known and are published for normalized second-order system; these relationships and notation will simplify problem solutions.

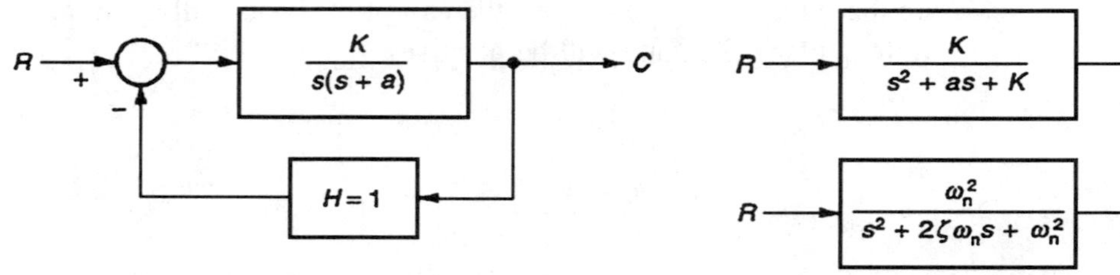

(a) System block diagram

(b) System transfer function

Figure 9.2 An ideal second-order system

As previously noted, the parameter zeta is related to whether the system is under-damped ($\zeta < 1$), critically damped ($\zeta = 1$), or over-damped ($\zeta > 1$). Furthermore, by defining the undamped natural frequency as ω_n (the frequency of a pure sinusoid), one may normalize a time axis with this value and generate the standardized curves for the ideal second-order system. For this simple system and its system transfer function (see Figure 9.2b), the solutions for step responses are as shown in Figure 9.3 with standardized notation. This notation normally becomes part of the system specifications.

Furthermore, the curves are given for the time solution (eliminating the need for finding any inverse Laplace transforms). As zeta approaches zero, the undamped curve becomes sinusoidal and the time-to-the-first peak is obviously π on the normalized axis. Clearly, for damped sinusoids, the peak will always be a number somewhat larger than 3.14; if one remembers this relationship, it will frequently give a check for an approximate answer on the examination! As a short example, assume for Figure 9.2a, $G = 25/[s(s + 2)]$, then the system transfer function is,

$$G_{sys} = 25/(s^2 + 2s + 25) = \omega_n^2/\left(s^2 + 2\zeta\omega_n s + \omega_n^2\right)$$

Here $\omega_n = 5$, $\zeta = 0.2$, from the normalized curves, one immediately determines that the percent overshoot is approximately 52% and the peak value occurs at about 3.2 on the normalized time scale; the 3.2 value is equal to $\omega_n t$, therefore $t_p = 3.2/\omega_n = 3.2/5 = 0.64$ seconds.

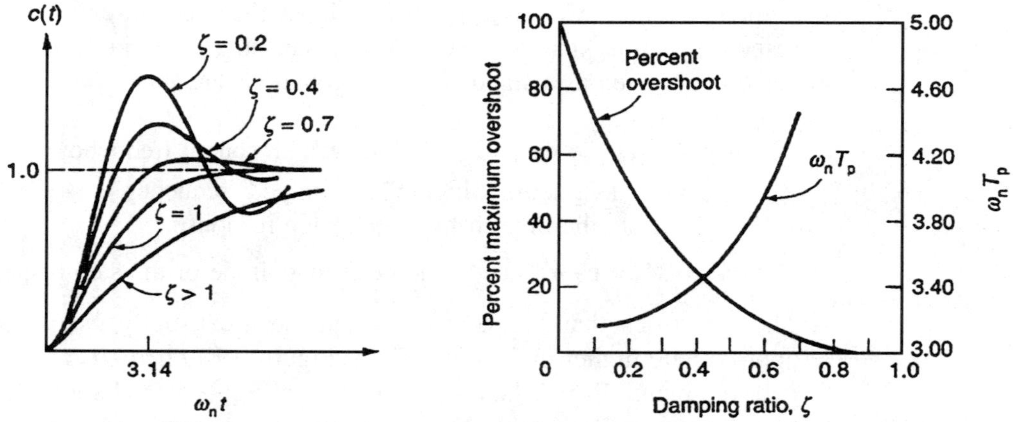

(a) Step response

(b) Percent overshoot and peak time response vs ζ

Figure 9.3 Standardized second-order curves

STEADY STATE GAIN AND SYSTEM ERROR

The steady state gain and the error of a system is found from the final value theorem (FVT); making use of the FVT requires that all poles of G are located in the LHP (i.e., negative real parts). In Figure 9.1 assume that $H = G_R = G_C = G_2 = 1$, then gain is given by equation 9.4a and for system error $(R - C)$ by 9.4b,

$$\text{dc Gain Constant, } K_p = \lim_{s\to 0} G(s) \tag{9.4a}$$

$$\text{System Error } (R - C) = \lim_{s\to 0} s\left(\frac{R(s)}{1+G(s)}\right) \tag{9.4b}$$

PROBLEMS

9.1 For a typical SISO system (refer to Figure 9.1), assume the transfer functions, $G_C G_1$, have a dc gain greater than unity (and $G_R = G_2 = 1$) and also assume a load disturbance, L. Which has the most effect on the output, C, the input, R, or the input, L?

9.2 If, for system stability, a Routh array is made for the system characteristic equation of 4th order is made, the first column indicates that all coefficients are positive except the 3rd entry of this first column. How many roots lie in the right-hand plane of the s-plane?

9.3 For a second-order control systems with $\zeta < 1$, will the damped natural resonant frequency (i.e., ringing frequency) to a step input be the same the driven resonant frequency?

9.4 A typical simple second-order system may be described by the following linear differential equation

$$\ddot{\theta}(t) + 2\dot{\theta}(t) + 25\theta(t) = 25u(t)$$

Find the values of the system matrix, A, the input distribution matrix, B, and the output matrix, C, for the following state block diagram (Exhibit 9.4). Hint: Define

$$x_1 = \theta(t) \quad \text{and} \quad x_2 = d\theta(t)/dt.$$

Also find the percentage of the maximum displacement of θ for a unit step input to the system (this is the point located on the $\theta(t)$ axis at t_2 in Figure 9.3b).

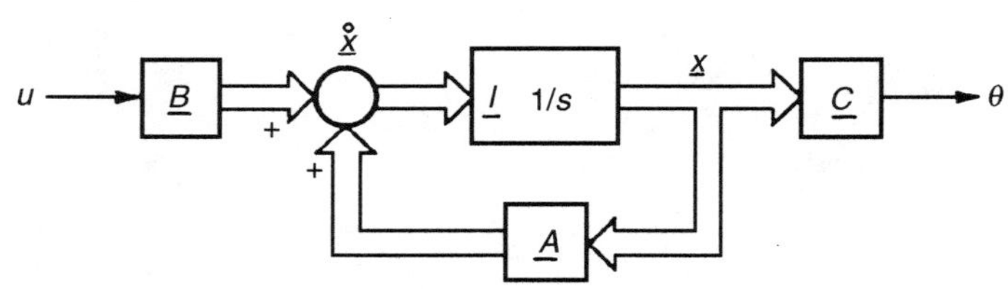

Exhibit 9.4 State block diagram

SOLUTIONS

9.1 The output C for an input at L is

$$C = [G_2/(1 + G_C G_1 G_2 H)]L,$$

whereas that for R is

$$C = [(G_C G_1 G_2)/(1 + G_C G_1 G_2 H)]R.$$

Because the denominators are the same and the numerator term of $G_C G_1$ of the second equation is >1, obviously the answer is R.

9.2 As two sign changes appear (one from the second to third term, and another from the third to fourth term), it is clear that two roots appear in the RHP.

9.3 Refer to equations 9.3a,b. The term under the radical sign will be smaller for the driven resonant frequency. Therefore $\omega_p < \omega_r$.

9.4

$$dx_1/dt = 0x_1 + 1x_2$$
$$dx_2/dt = -25x_1 - 2x_2 + 25u(t)$$

Then

$$\mathbf{A} = \begin{bmatrix} 0 & 1 \\ -25 & -2 \end{bmatrix}, \quad \mathbf{B} = \begin{bmatrix} 0 \\ 25 \end{bmatrix}, \quad \mathbf{C} = [1 \quad 0]$$

(Note: Mixed notation of time and frequency is common in the state variable world.)

Because this is a second-order system it is obvious that $2\zeta\omega_n = 2.0$, and $\omega_n^2 = 25$, $\omega_n = 5$, therefore $\zeta = 0.2$; and, from the standardized second-order curves (see Figure 9.3), the percent maximum overshoot is found to be approximately 52%.

This chapter was written by Lincoln D. Jones and also appears in *Electrical Engineering: FE Exam Preparation*.

CHAPTER 10

Instrumentation and Measurement

OUTLINE

Most engineering design requires data for use in the analysis phase of the design process. This data must be obtained through measurement and these measurements must be obtained by means of instrumentation. The generic instrumentation system consists of the primary sensing device (which is the transducer), the lead wires, and the secondary instrumentation.

TRANSDUCERS

A transducer is a device that changes one form of energy into another form. The final form of energy is usually an electric signal that is read and interpreted by secondary instrumentation (i.e., a readout device). There are many types of transducers, such as thermocouples, strain gages, piezoelectric crystals, linear variable displacement transformers (LVDT), and a host of others.

THERMOCOUPLES

A thermocouple is a device that operates on the principle of basic material property variations caused by temperature differences. There are three thermoelectric effects caused by these temperature differences. Fortunately, the most significant one is the Seebeck effect, which is the junction potential produced when two dissimilar are joined together. The emf that is produced is directly proportional to the temperature difference between the junction of the dissimilar metals and the junctions of the readout device or the reference temperature.

Example 10.1

A simple thermocouple is used to measure the temperature of boiling water at standard conditions of the environment. Determine the emf output of the thermocouple in terms of the thermoelectric coefficient α, and the temperatures.

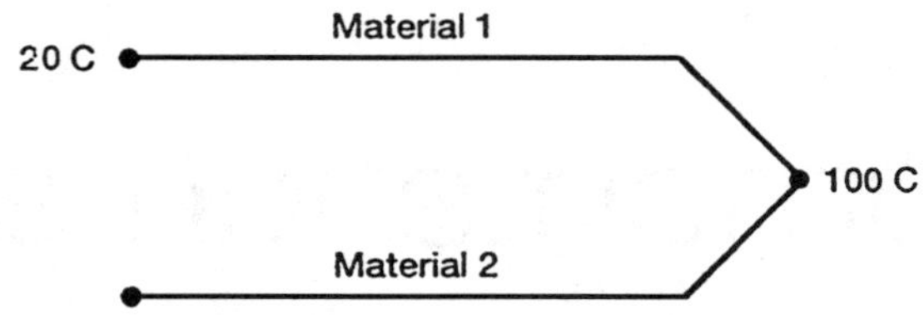

Exhibit 1

Solution

Because the electromotive force produced by the thermocouple is directly dependent upon the thermoelectric effect of the material times the temperature difference, or

$$\text{emf} = \alpha_{\text{mat }1}(100 - 20) + \alpha_{\text{mat }2}(20 - 100)$$

and because the thermoelectric effect of material one is opposite in sign to the thermoelectric effect of material two, the equation may be written as

$$\text{emf} = (\alpha_{\text{mat }1} - \alpha_{\text{mat }2})(100 - 20)$$

The above equation indicates, then, that the emf of any combination of materials is equal to the summation of the emf s produced by each of the constituents.

Example 10.2

A multijunction thermocouple (as shown hereafter) is used to determine the temperature of some unknown fluid. The measuring device is at a temperature T_1, the bath temperature is at T_2, and the unknown temperature is at T_3.

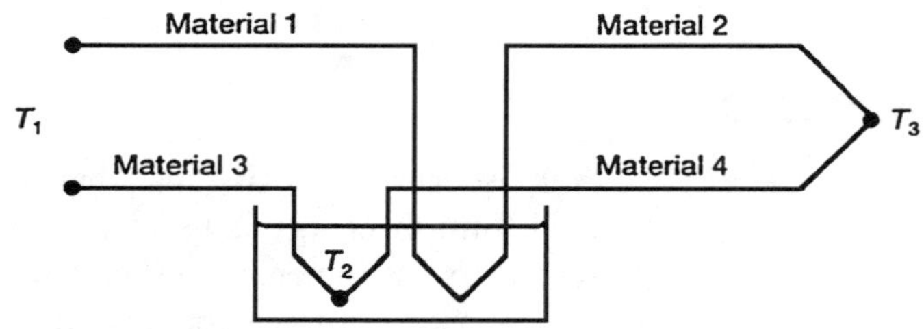

Exhibit 2

Determine whether the temperature of the measuring device's binding posts makes any difference in the output emf.

Solution

Write the equation for the summation of the emf's produced by the several combined materials.

$$\text{emf}_{\text{total}} = \alpha_{\text{mat }1}(T_2 - T_1) + \alpha_{\text{mat }2}(T_3 - T_2) + \alpha_{\text{mat }3}(T_2 - T_3) + \alpha_{\text{mat }1}(T_1 - T_2)$$

Notice that the first and fourth terms are equal and opposite each other. As a consequence the total emf obtained by the thermocouple is proportional to the difference in temperature of T_2 and T_3. We may therefore say that the binding post temperature has no effect upon the total emf.

STRAIN GAGES

A strain gage is a device that falls in the non–self-generating class of transducers, which fits the following model.

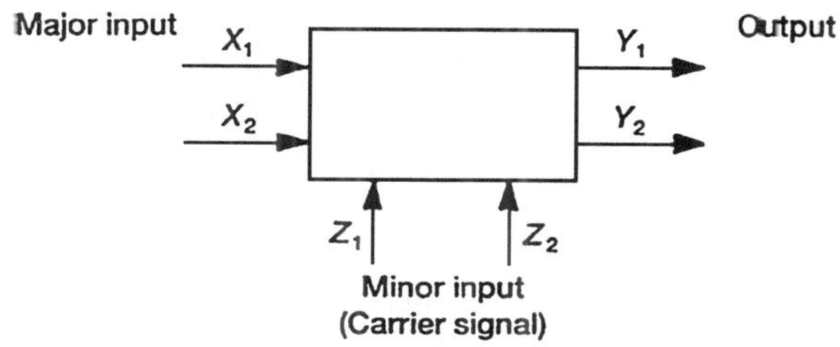

Figure 10.1

The major input is the effect caused by the environmental change that is being measured. This effect then is carried by the excitation/carrier signal and comes out as the output signal, which must then be translated into useful information. Typical strain gages operate on the principle of resistance change due to an applied force. The effect of the resistance change is measures by a bridge circuit, such as the Wheatstone bridge circuit shown in Figure 10.2.

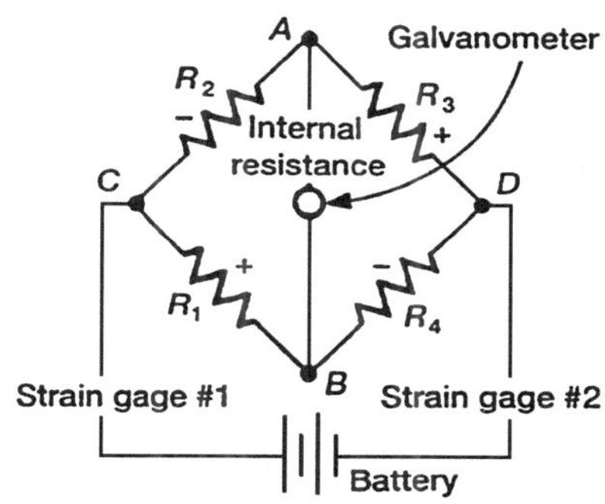

Figure 10.2

The signs of the values of each of the resistances are positive for R_1 and R_3 and negative for R_2 and R_4. These characteristics may be used to obtain various outputs of the bridge circuit.

Example 10.3

Determine the output of the cantilever beam in Exhibit 3, wired up to a simple two-arm bridge.

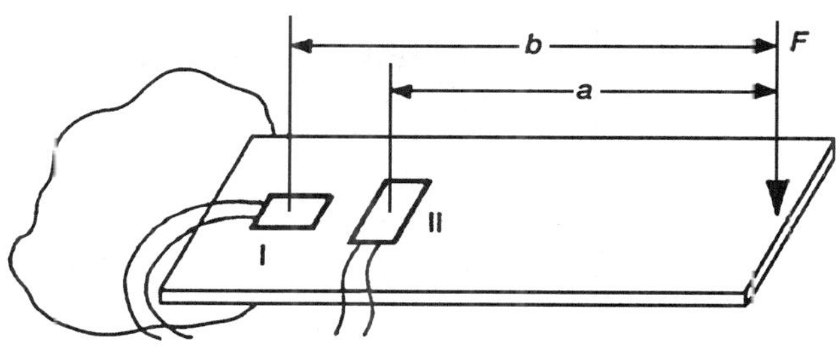

Exhibit 3

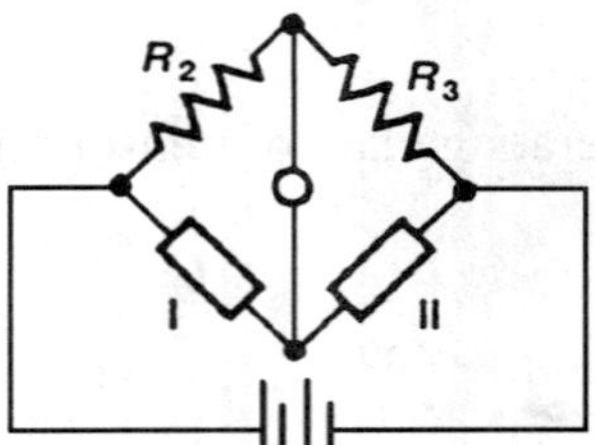

Exhibit 4

Solution

Both of the gauges are mounted on the top of the cantilever beam, thus the top (I) gauge is stretched directly proportional to the elongation of the upper fibers of the beam and is connected to the #1 position on the Wheatstone bridge. The second gauge (II) is perpendicular to the first gauge and thus reads a negative effect proportional to Poisson's ratio and is connected to the #4 position on the Wheatstone bridge. The output of the arrangement produces an effect equal to the output of one gauge plus the Poisson's ratio effect.

$$\text{output} = [1 + (a/b)\ \mu]$$

Note that the output is temperature compensated because the temperature will affect each gauge exactly the same and the values are subtracted from each other.

Example 10.4

Determine the output characteristics of the four-strain gauge cantilever beam shown in Exhibit 4. The numbers adjacent to the strain gauges indicate which resistance positions they are attached to.

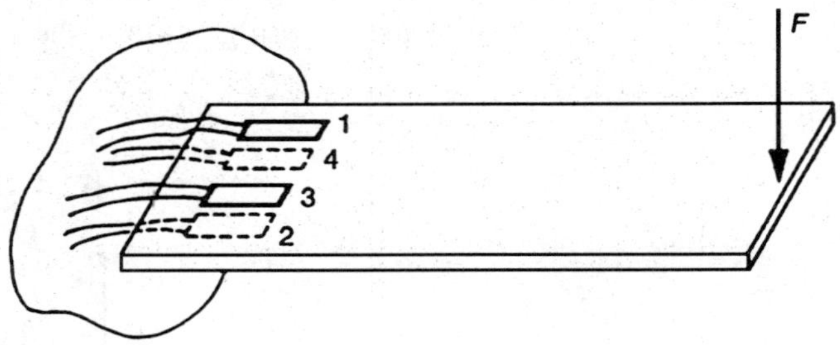

Exhibit 5

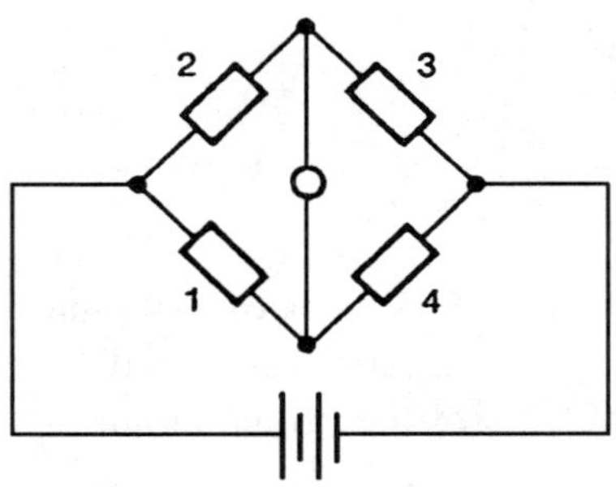

Exhibit 6

Solution

Note that gauges 1 and 3 are mounted directly over gauges 4 and 2, respectively. The characteristics are as follows:

a) Gauge 4 compensates for the temperature effect of 1, and gauge 2 compensates the temperature effect of 3.

b) For the load indicated, gauges 1 and 3 reflect a positive effect and read into positive arms, gauges 2 and 4 (the bottom gauges) reflect a negative effect and read into negative arms giving a net effect of 4 positive strain gauges.

c) Any axial load on the system would be compensated out giving a net effect of zero.

Example 10.5

Two resistance based strain gages are bonded upon two pieces of metal as shown. The first gage is mounted axially on a circular bar and attached to position #1 on the bridge circuit, and the second gauge is mounted on a different piece of metal and is wired into position #4 on the bridge circuit. A half bridge output will produce an output voltage in response to the following loads or environmental changes.

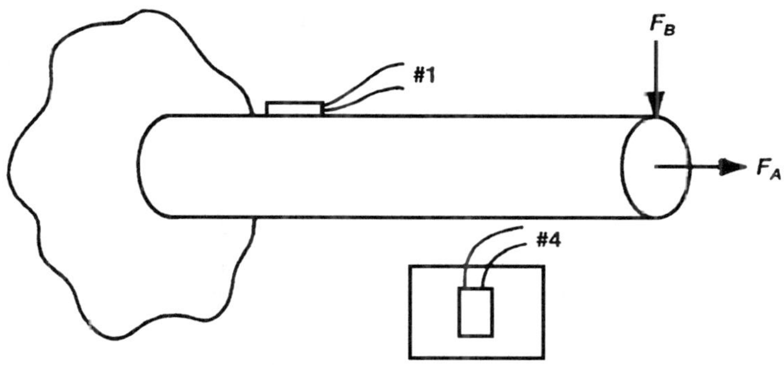

Exhibit 7

Solution

If it can be assumed that both of the gages are mounted in the same environment, there will be no temperature effect. The output then will be a summation of the effects caused by both the axial load F_A and the bending load F_B.

PROBLEMS

The following information may be used for problems 10.1 through 10.4.

Four resistance based strain gages are bonded to the metals as shown in Exhibit 10.1. Two are axially bonded to the circular bar and two to the additional piece of metal. All four of the strain gauges are subjected to the same environmental conditions. The gauges may be wired to a Wheatstone bridge in a one-quarter, one-half or a full bridge configuration.

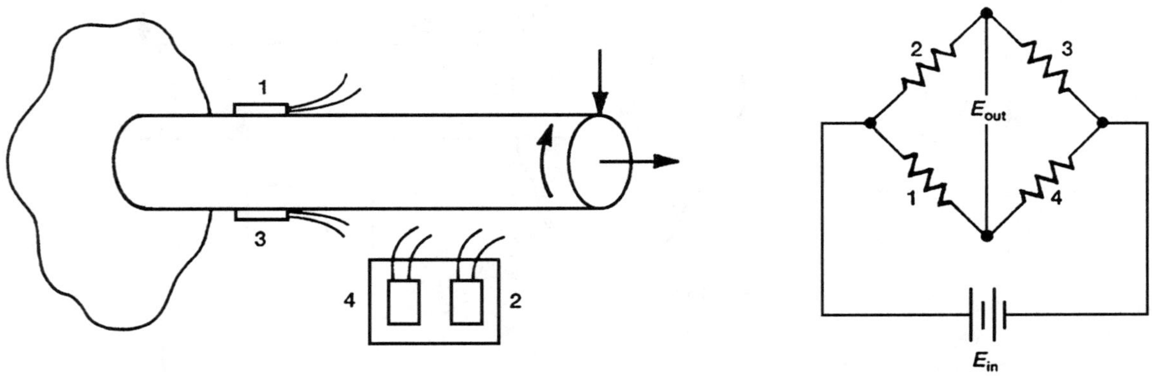

Exhibit 10.1

10.1 A quarter bridge utilizing only gauge 1 will produce an output voltage in response to the following conditions.
   a. Torsional load (τ) and temperature effects
   b. Axial load (F_A), bending load (F_B), and temperature effects
   c. Axial load (F_A) only
   d. Axial load (F_A), bending load (F_B), and temperature compensated

10.2 A one half bridge utilizing gauges 1 and 2 will produce an output voltage in response to the following conditions.
   a. Axial load (F_A), bending load (F_B), and temperature compensated
   b. Axial load (F_A), bending load (F_B), and temperature effects
   c. Torsional load (τ) only
   d. Torsional load (τ) and axial load (F_A) and temperature effects

10.3 A full bridge has gauges 1 and 3 attached to positions #1 and #3, and gauges 2 and 4 are attached to positions #2 and #4 on the Wheatstone bridge. This system will produce an output voltage that is equivalent to:
   a. Axial load (F_A) with temperature compensation
   b. Axial load (F_A) bending load (F_B), and temperature affects
   c. Absolutely nothing
   d. Bending load (F_B) and temperature effects

10.4 A one half bridge has gauges 1 and 3 attached to positions #1 and #4. If the voltage in to the circuit is equal to 10 volts, the strain gauges are 350 ohms, and the resistance change due to the loading is 1%, what is the output voltage?
   a. 1.0 volts
   b. 0.15 volts
   c. 0.05 volts
   d. 0.01 volts

SOLUTIONS

10.1 **b.** Because this is a quarter bridge, only one strain gage is connected to the bridge circuit. The strain gauge in question is mounted at the top of the bar and thus will be stretched both when the bar is pulled in tension by the axial load and when the bar is bent due to the bending load. Notice also that there is no compensating gauge, thus any temperature effects will confound the output voltage.

10.2 **a.** The one half bridge uses the top gage and a temperature compensating gauge.

10.3 **d.** The full bridge has both strain gauges mounted in the positive positions of the bridge circuit. Therefore, the axial load will be measured but the bending will be subtracted out as well as any of the temperature effects.

10.4 **c.** The one half bridge sends the top gauge to the positive arm of the bridge and the bottom gauge to the negative arm of the bridge. The results are temperature compensation and bending shall be measured. The axial load effect is zero. Using the equation for

$$\text{G.F.} = (\Delta R/R)/(\Delta L/L) = (\Delta R/R),$$

and

$$E_{out} = E_{in} \times (\varepsilon_1 - \varepsilon_4)/2$$

we get from the *FE Handbook*.

$$E_{out} = (10 \text{ volts})(3.5/350)/2 = .05 \text{ volts}$$

CHAPTER 11

Material Behavior/Processing

OUTLINE

Of all the different types of materials used by the mechanical engineer, steel is still probably the most important, especially in the tonnage used.

STEEL

Steel is a heat-treatable alloy, primarily an alloy of iron and carbon. The lower limit for a mild steel is about 0.05%, though the lower limit might be defined as that value below which iron carbide cannot be thrown out of solid solution, which is below 0.01%. The upper theoretical limit is 1.7%, which corresponds to the point in the iron-carbon equilibrium diagram beyond which iron carbide cannot be wholly held in solution at any temperature. The lowest carbon alloys are only slightly, and usually only with difficulty, affected by heat treatment, and are termed *irons* rather than *steels*.

Other elements are almost always included in ordinary steel, some by design, and some because it is too difficult or expensive to remove them. Excess amounts of sulphur and phosphorous, for example, are quit deleterious. Too much sulphur results in a condition termed hot-shortness, meaning that the steel becomes brittle at high temperatures. Too much phosphorous, on the other hand, causes cold-shortness. Steel is not a uniform substance like pure gold, but is made up of components—ferrite and cementite. Ferrite is BCC iron in the crystalline form that exists at room temperature in slowly cooled carbon steel. Cementite is a compound of carbon and iron, Fe_3C. When a carbon steel is cooled at a slow rate from a red heat, cementite and ferrite form a laminar composition, which resembles mother of pearl and is termed pearlite. Pearlite contains 0.85% C which is the eutectoid composition of steel. A hypoeutectoid steel is one containing less than 0.85% C and when cooled will contain free ferrite. A slowly cooled hypereutectoid steel will contain pearlite and excess cementite.

AUSTENITE

When carbon steel is heated above its transformation temperature the, pearlite changes to austenite. Austenite has a face centered cubic lattice and thus contains more atoms per cell, though the packing factor is greater for the FCC cell than for the BCC cell. Austenite is nonmagnetic and is more dense than ferrite. Austenite may exist at room temperature only when its transformation has been fully suppressed. Manganese, nickel, and chromium are used to suppress the transformation. Stainless steels, for example, which contain more than 6% nickel and more than 24% of nickel and chromium combined, are austenitic at room temperature, and are nonmagnetic.

MARTENSITE

When austenite is cooled rapidly to a temperature of 200 to 300 °C, it forms a very hard structure called martensite. This is a skewed tetragonal lattice that forms a needlelike structure. Martensite is less dense than pearlite, so there is a slight increase in volume when a steel is fully hardened. If the cooling is too rapid, the metal does not have time to absorb the increase in volume, and may rupture. Even if the cooling rate is lower than the rate that causes rupture, the increase in volume may still result in severe internal stresses, which must be relieved if the part is to be safely loaded. Such parts are heated to a temperature below the lower transformation temperature and "stress relieved."

Elements are added to steel to increase its toughness, hardness, corrosion resistance, or other property. As an example, plain carbon steel with a BCC structure becomes very brittle at low temperatures. At –0 °C, a wrought-iron pipe becomes as brittle as glass and will shatter if stuck by a hammer. But austenitic steels do not exhibit the low-temperature embrittlement phenomenon.

Example 11.1

A design for an army tank for use in Antarctica is designed with cleats of high-carbon steel. The ambient temperature is expected to drop as low as –0°C. What should be recommended by a design review?

Solution

Since the tank treads will be in contact with the ground, could reach as low a temperature as the forecast minus 80 °C, and will also be subject to continuous shock loading there is a high probability that the treads, being made of a BCC steel, would fracture. They should be replaced with an abrasion-resistant nonbrittle metal.

NONFERROUS METALS AND ALLOYS

This classification includes all metals in which iron is not present in large quantities. It includes such metals as copper, aluminum, magnesium, and zinc. Also included are more exotic metals such as silver, gold, tungsten, and others that, though used in relatively small amounts, are very important for specialized purposes. Nonferrous metals are generally used for parts requiring special fabrication and where the ease of fabrication outweighs the higher material costs. Other factors such as density, stiffness, corrosion resistance, electric properties, and color can also affect selection.

MODULUS OF ELASTICITY

The modulus of elasticity is a measure of the elastic deformation of a metal when it is stressed in tension or compression within its elastic limit. It is measured in Pa and equals pascals divided by m/m elongation.

Example 11.2

A 100 kg mass is supported by a 100 m long steel wire 2.5 mm diameter. How much would the wire stretch (strain) elastically?

Solution

The stress in the wire would equal

$$S = 100 \times 9.8066/(4.909 \times 10^{-6}) = 199.8 \text{ MPa}$$

$$\text{Strain, } \varepsilon = 199{,}800{,}000/(2.1 \times 10^{11}) = 95.14 \times 10^{-5} \text{ m/m}$$

For 100 m length the elongation would equal 95 mm

The stiffness of a part can sometimes be very important, and stiffness is a function of the modulus of elasticity, E. For example, the modulus of elasticity for steel, E_s, equals 2.1×10^{11} Pa, whereas E_a, the modulus of elasticity for aluminum is only 6.9×10^{10}, or 67% less.

Example 11.3

A steel cantilever beam holds 150 kg on the end. It is three meters long and ten cm high by five cm wide. To save mass it is proposed to replace the steel beam with an aluminum one of the same length and same proportions. To give the same deflection, (a) what would be the deflection of the end of the beam, (b) what would be the dimensions of the aluminum beam, and (c) what would be the percentage mass reduction? The density of steel is 7.9 g/cm^3 and that of aluminum is 2.7 g/cm^3.

Solution

(a) The deflection of the beam would equal $PL^3/(3EI)$

For the steel beam $I = b \cdot h^3/12 = 4.167 \times 10^{-6}$ m^4

$$\Delta h = 150 \times 9.8066 \times 3.00^3/(3 \times 2.1 \times 10^{11} \times 4.167 \times 10^{-6})$$

$$\Delta h = 0.01513 \text{ m or } 1.51 \text{ cm}$$

(b) For the aluminum beam to have the same stiffness as the steel beam, the product of $E \times I$ for the aluminum beam must be the same as for the steel beam. So $I_{a1} = 21/6.9 \times I_s$

$$I_{a1} = 12.682 \times 10^{-6} = b \times (2b)^3/12$$

Which gives $b = (19.023 \times 10^{-6})^{1/4} = 0.06604$

(c) The dimensions of the aluminum for the same stiffness would be 6.604 cm × 13.208 cm

The ratio of the masses would equal

$$6.604 \times 13.208 \times 2.70/(5.0 \times 10.0 \times 7.87) = 0.5985$$

There would be a 40 percent reduction in the mass of the beam.

CORROSION

The ability of a metal, or design, to resist corrosion is a very important factor in selecting a particular metal for a specific application. Corrosion causes millions of dollars of structural damage to metal components, and millions more due to surface damage or tarnishing to decorative items. The proper selection of materials or protective systems is very important to a mechanical design.

Corrosion may be defined as the destructive chemical or electrochemical reaction of a material and its environment. The most common type of corrosion is rust. Rust is a reddish-brown or orange coating on iron or steel due to oxidation of the surface when exposed to air or moisture. It is made up principally of hydrated ferric oxide. Rust is formed by the galvanic action of the iron and small particles of other metals contained in it in contact with acidic solutions formed by the dissolution of carbon dioxide in rainwater and dew. Moisture is always present in the atmosphere, so the potential for electrochemical corrosion always exists.

ELECTROCHEMICAL CORROSION

If two dissimilar metals in contact are joined by an electrolyte, they will form a galvanic corrosion cell and one will be dissolved or corroded. An electrolyte is a substance capable of carrying ions. It is typically an aqueous solution and can be formed from condensed moisture, dew, or rain and contaminated by small amounts of dirt, salts, acids, or alkalis from the atmosphere. The anode and cathode of a corrosion cell can even be parts of the same component with different potentials. The anodic material, the one with the highest potential in reference to hydrogen, as shown in the table of "Standard Oxidation Potentials for Corrosion Reactions" in the *FE Handbook*, will be corroded. For example if aluminum rivets should be used to join steel sheets, the rivets would be corroded and could fail if not protected. Conversely, if two aluminum plates should be bolted together with iron bolts, pitting due to corrosion would occur in the plates next to the bolts.

It is noted from the table that zinc has a higher oxidation potential than iron, and this explains why steel items are sometimes zinc coated—the zinc will corrode in preference to the steel and leave the steel uncorroded. The fact that the cathode is resistant to corrosion is made use of in protecting ship hulls by attaching sacrificial anodes of metals such as magnesium to the hull. Cathodic resistance to corrosion is also made use of in protecting buried pipelines by connecting them to the negative side of a DC current supply.

NONMETALLIC MATERIALS

There are many nonmetallic materials of importance to engineers, but most of them lie within the province of civil, chemical, and electrical engineers. However, mechanical engineers will find it of value to be aware of some of the properties of nonmetallic materials that can be of benefit in different mechanical designs. Many plastics can prove quite useful to the mechanical engineering designs. Plastics have the advantage of being more formable than any metallic material. In addition, plastics can have low density and can have special properties such as specially desired heat or electrical insulation, corrosion resistance, low friction coefficient, resistance to deterioration by moisture, good color range, or other property not obtainable in the required amount with a metal. A plastic is defined in the broadest sense as any nonmetallic material that can be molded to shape. Plastics can be molded, cast, or extruded. They can also be used for coatings and films. Plastics are generally lumped into two general classifications—thermoplastic and thermosetting. Thermoplastics are those which can begin to soften at a temperature as low as 60 °C and then can be molded without any change in chemical structure. Thermosetting materials undergo a chemical change when molded and cannot be re-softened by heating. Thermosetting plastics usually require considerably higher mold temperatures than thermoplastics, and the finished part can usually withstand much higher temperatures without deforming.

PROBLEMS

11.1 A design engineer should know what a metal's electromotive potential is because it tells

a. If the metal can be heat treated
b. What the atomic structure is
c. How easy the metal can be formed
d. Its resistance to corrosion

11.2 An extensometer is attached to a 15.0 cm length of a test piece of unknown metal in a laboratory. The test section is one cm square. It was then subjected to a tensile force of 20.0 kN and the elongation of the test section was measured. It was found to have increased in length by 0.008475 mm. What was the modulus of elasticity of the metal?

a. 21×10^{10} Pa
b. 6.9×10^{10} Pa
c. 35.4×10^{10} Pa
d. 14.8×10^{10} Pa

11.3 Plastically deforming steel at room temperature

a. Increases the ultimate strength.
b. Increases the corrosion resistance.
c. Makes it nonmagnetic.
d. Increases the yield strength.

SOLUTIONS

11.1 **d.** The electromotive potential of a metal indicates how susceptible it will be to corrosion in an application.

11.2 **c.** The modulus of elasticity is measured in pascals, or newtons per square meter. The test showed a strain of

$$\varepsilon = 0.008475/150 = 0.000565 \text{ mm/mm for a stress of}$$
$$S = 20{,}00/0.0001 = 200 \text{ MPa} \quad S = \varepsilon \cdot E \text{ so}$$
$$E = 20 \times 10^7/(5.65 \times 10^{-4}) = 3.54 \times 10^{11} \text{ Pa}$$

11.3 **d.** Cold working, or plastically deforming steel at room temperature increases the yield strength and reduces the ductility.

CHAPTER 12

Mechanical Design

OUTLINE

Mechanical design, or machine design as it is often called, combines all of the disciplines in the study of mechanical engineering. A typical design problem requires first the development of a mechanism to perform the desired function. Next it has to be demonstrated that the device will perform without failing and without imperiling the user or the public. Attention should also be paid to the cost of the device including material, manufacturing, operating, and maintenance costs.

Example 12.1

Design a square key for a gear wheel attached to a 2.5 cm diameter shaft to transmit 10.0 kw of power at a speed of 1750 rpm. The gear wheel is 1.5 cm thick. If the key is to be made of steel with an allowable shear stress of 200 MPa, what size key should be used if safety factor of three is required assuming shear stress controls?

Solution

The allowable stress on the key material would equal

$$200/3 = 66.67 \text{ MPa.}$$
$$10.0 \text{ kW} = 10.0 \text{ kJ/s and a joule is a N}\cdot\text{m.}$$
$$\text{Power} = 2\pi \times (\text{rev/s}) \times \text{Torque} = \text{watts}$$

The shaft rotates at a speed of 1750/60 = 29.17 rev/s and transmits energy at the rate of

$$10{,}000 \text{ J/s} = F \times \text{m/s}$$
$$F \times (\pi \times 0.025 \times 29.17) = 2.2910 \times F = 10{,}000 \text{ N}\cdot\text{m/s}$$

where
F = force on key in newtons
F = 10,000 watts/2.2910 = 4364.9 N

The required shear area of key = $4365/66.67 \times 10^{6} = 65.47 \times 10^{-6}$ m^2

The key would be 0.015 meters long (the thickness of the gear wheel) so the key would be 0.00006547/0.015 = 0.00436 meters, or the required key width would = 4.36 mm

INTERFERENCE FIT

Interference fits are some times desirable to provide a force fit. In such a case, the O.D. of the shaft will be larger than the I.D. of the mating part. For small interferences, this may be accomplished by forcing the shaft into the smaller diameter hole. For larger interferences, it will be necessary to expand the hole's diameter and/or reduce the shaft diameter by heating or cooling as, for example, shrinking rims onto railroad car wheels.

Example 12.2

If a car wheel is 60 cm in diameter and a rim is 0.9 mm smaller in I.D., to what temperature would it be necessary to heat the rim to provide a 0.100 mm clearance to assemble the rim on the wheel? Both members are made of steel. Dry ice at −79 °C is available to cool the wheel. The coefficient of thermal expansion for steel is 11.7×10^{-6} m/(m·°C)

Solution

The total ΔD required is 1.00 mm. For an ambient temperature of 20 °C, the wheel diameter would reduce

$$\Delta T = 90\,°\text{C} \quad \Delta D = 600 \times 99 \times 11.7 \times 10^{-6} = 0.6950 \text{ mm}$$

The rim diameter would then have to expand 0.3050 mm, which means that the increase in the temperature of the rim is

$$\Delta T = 0.3050/(600 \times 11.7 \times 10^{-6}) = 43.4\,°\text{C}$$

The rim would have to be heated to 63.4 °C.

Example 12.3

To examine Example 12.2 a little further, assume all deformation took place in the rim after assembly; that is, the wheel was strong enough not to deform after the rim was installed. What would be the stress in the rim after it had been shrunk onto the wheel?

Solution

$$\Delta D = 1 \text{ mm} \quad \text{Circumference} = \pi D \quad \text{so} \quad \Delta C = \pi \times 1.0 \text{ mm}$$
$$\varepsilon_c = \pi/(600\ \pi) = 1/600 \qquad \text{stress} = E \cdot \varepsilon_c = 2.1 \times 10^{11} \times 1/600 = 350 \text{ MPa}$$

Example 12.4

A 10-kW motor operates at speed of 1750 rpm. What diameter shaft should be used to transmit this power using a steel with a tensile strength of 572 MPa and a tensile yield strength of 496 MPa? Use a safety factor of 2.5, and assume the shear yield strength of the shaft steel is 0.60 times the yield strength in tension.

Solution

$$\text{Shear stress} = Tq \cdot c/J$$

where
$J = \pi D^4/32$, the polar moment of inertia and $c = D/2$

so

$$S_s = 16 \cdot Tq/(\pi \cdot D^3)$$
$$\text{Watt} = \text{joule/s} = \text{N} \cdot \text{m/s} \quad 1750 \text{ rpm} = 29.17 \text{ rev/s}$$
$$\text{Allowable shear stress} = 496 \times 0.60/2.5 = 119.04 \text{ MPa}$$
$$Tq = 10{,}000 \text{ watts}/(29.167 \text{ rev/s} \times 2 \cdot \pi) = 54.57 \text{ joules}$$
$$D = [16 \times Tq/(S_s \cdot \pi)]^{1/3} = 0.0133 \text{ m or } 1.33 \text{ cm}$$

Example 12.5

If it is desired to use a 2.00 cm O.D. hollow shaft for the conditions given in the above example, what would be the size of the bore to produce the same shear stress in the hollow shaft?

Solution

Since the stress is directly proportional to the polar moment of inertia, the J of the hollow shaft would have to be the same as for the solid shaft calculated previously.

$$J = \pi \cdot D^4/32 \quad \text{since } J_1 = J_2 \quad D^4 = D_{OD}^4 - D_{ID}^4$$
$$1.33^4 = 2^4 - D_{ID}^4 \quad D_{ID} = (16 - 3.129)^{1/4} = 1.894 \text{ cm},$$

which gives a wall thickness of 0.529 mm.

Example 12.6

In Example 12.5, a 1.500 cm solid shaft was selected. The power from the motor was delivered to the driven machine through a 60.00 cm diameter pulley. If the pulley were attached to the shaft at a distance of 15.00 cm from a bearing support, what would be the bending stress in the shaft (a), and what would be the total shear stress, τ, (b)?

Solution

The torque output of the motor is 54.57 J (N·m). The force acting on a belt at a distance of 60.00/2 cm would equal (1.00/0.300) × 54.57 = 181.9 N. The maximum bending moment would equal

$$0.15 \times 181.9 = 27.29 \text{ J}$$
$$S = M \cdot c/I,$$

where I is the transverse moment of inertia, and since $J = I_x + I_y$ it will equal one-half of the polar moment of inertia for a circular section.

$I_x = I_y$ so $I_x = \pi D^4/64 = 2.485 \times 10^{-9}\text{ m}^4$ $S = (27.29 \times 0.015/2)/(2.485 \times 10^{-9})$

S equals 82.36 MPa, bending stress (a) $J = 2 \times I_x = 4.970 \times 10^{-9}$

$$S_s = (54.57 \times 0.0150/2)/(4.970 \times 10^{-9}) = 82.35 \text{ MPa}$$
$$\text{Maximum shear stress } \tau = \left[s_s^2 + (S/2)^2\right]^{1/2} = 92.07 \text{ MPa}$$

The relationship for the maximum shear stress can be derived from the diagram of Mohr's Circle given in the *FE Handbook*, using only two tensile stresses (i.e., let $\sigma_2 = 0$). It would be a good idea for an examinee to study the section on Mohr's Circle before the exam if not already acquainted with it.

COMPOSITE BEAM

Example 12.7

How much stiffer would an aluminum beam 5.00 cm wide by 16.00 cm deep be if the bottom 1.00 cm were replace by steel? See Exhibit 1.

$$E_S = 2.1 \times 10^{11} \text{ Pa} \qquad E_{Al} = 6.9 \times 10^{10} \text{ Pa}$$

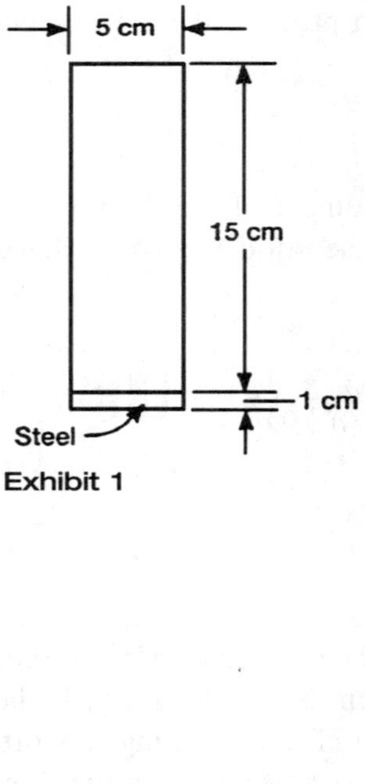

Exhibit 1

Solution

$$n = E_S/E_{Al} = 210/69 = 3.0435,$$

so the equivalent all aluminum beam would be as shown in the figure, a top section 5.00 cm by 15.00 cm with a 1.00 cm by n × 5 = 15.218 cm flange section at the bottom. The stiffness of a beam is proportional to its moment of inertia. $I = b \cdot h^3/12$ for a rectangular section. For the initial all aluminum beam

$$I = 5 \times 16^3/12 = 1706.67 \text{ cm}^4$$

If the bottom 1.00 cm was replaced by a steel strip, the location of the neutral axis would shift. Taking area moments about the top of the beam

$$5 \times 15 \times 7.5 + 1 \times 15.218 \times 15.5 = (75 + 15.218) \times y$$
$$y = 798.079/90.218 = 8.846 \text{ cm}$$

from the top of the beam is the location of the neutral axis of the composite beam.

The new I, I_c, can be determined by means of the parallel axis theorem, with the new neutral axis 8.846 cm from the top of the composite beam.

$$I_c = 5 \times 15^3/12 + [75 \times (8.846 - 7.5)^2] + (15.218 \times 1^3/12)$$
$$+ (15.218)(15.5 - 8.846)^2 = 2217.19 \text{ cm}^4$$
$$I_c/\text{I} = 2217.19/1,706.67 = 1.30$$

So the composite beam would be 30% stiffer.

Example 12.8

A swing is to be designed for a public park. The support beam will be 3.00 m high, and the swing will be designed to hold a 100-kg person. What should be the strength of the lines holding a swing seat? Since this is a public park, it has to be assumed that the swing will be subjected to uses (or misuses) far beyond what any one might imagine, so a safety factor of five has been decide upon.

Solution

The maximum load on the swing seat will occur at the bottom of the arc and will equal the gravitational force exerted by the swinger and the centrifugal force, F_c.

$$F_c = m \cdot v^2/R$$

Assume the swinger goes as high as the support beam, then

$$v = /(2gh) \quad \text{or} \quad v^2 = 2 \times 9.8066 \times 3.0 = 58.84 \text{ m}^2/\text{s}^2$$

assuming the swing seat is at the level of the ground when at rest.

$$F_c = 100 \times 58.84/3 = 1961 \text{ N}$$

The total force down would equal

$$1961 + 100 \times 9.8066 = 2942 \text{ N}$$

Assume that a swinger would hold onto only one side of the swing. The strength of each support line should equal 5 × 2942 = 14,710 N.

Example 12.9

A punch press is to be designed to punch 2.50 cm diameter holes in 1.0-cm thick mild steel plate. The shear strength of the steel is 250 MPa If 15% is added to the shearing force as a friction allowance, what force must the punch exert?

Solution

$$\text{Shear area} = 0.025 \times \pi \times 0.01 = 0.0007854 \text{ m}^2$$

$$\text{Force} = 250 \times 10^6 \times 785.4 \times 10^{-6} = 196.4 \text{ kN}$$

$$\text{Add allowance for friction} \quad F = 1.15 \times 196.4 = 225.9 \text{ kN}$$

Example 12.10

How much energy must be expended to punch one hole? The design calls for the punch press to be operated by a 400-watt motor running at a speed of 1750 rpm. The motor will drive a flywheel through a hydraulic clutch which is 85% efficient. Energy from the flywheel will be used to provide the required power to operate the press. The requirement is that the flywheel should not slow down by more than 20% to punch a hole. What should the moment of inertia of the flywheel be if rotates at a speed of 250 rpm before its energy is used to operate the punch press?

Solution

$$\text{Energy} = F \times \text{distance} = 225.9 \text{ kN} \times 0.01 \text{ m} = 2259 \text{ J}$$

KE of a rotating flywheel equals $\frac{1}{2} I\omega^2$

$$\omega_1 = (250/60) \times 2\pi = 26.180 \text{ rad/s}$$
$$\omega_2 = (200/60) \times 2\pi = 20.944 \text{ rad/s}$$
$$\Delta KE = \tfrac{1}{2} \cdot I \cdot (685.4 - 438.7) = 123.4 \times \text{I} = 2259 \text{ J}$$
$$I = 18.31 \text{ kg} \cdot \text{m}^2 \ (\text{N} \cdot \text{m} \cdot \text{s}^2)$$

Example 12.11

If a 65.0-cm diameter, steel flywheel of constant thickness is used, how thick should it be? The specific gravity of steel is 7.85.

Solution

$I = \frac{1}{2} MR^2 = 18.31 \quad R = 0.325 \text{ m} \quad M = 2 \times I/R^2 = 2 \times 18.31/0.1056 = 346.8 \text{ kg}$

The density of steel is 7850 kg/m^3, so the volume of the flywheel should be 346.8/7850 = 0.0442 m^3. The area = 0.3318 m^2 and required thickness = 13.33 cm.

MECHANICAL SPRINGS

The most common type of spring is the helical spring, made with round wire. Such springs are made to withstand either tensile or compressive loads. The spring rate is a function of the number of active coils, wire diameter, spring diameter, and the torsional modulus of the wire material. For steel wire the torsional or shear modules, G_s, equals 8.3×10^{10} Pa (see *Reference Handbook*). The apparent stress in the wire equals $S_t = 8 \cdot F \cdot D/(\pi d^3)$ equals torsional stress. However, this relationship does not include the effect of the curvature of the wire, which increases the stress in the wire by a factor K_s, which equals

$$(2C + 1)/(2C)$$

where
C = (mean coil diameter)/(wire diameter)

Example 12.12

A coil spring 4.0 cm in outside diameter made of 5.0-mm wire is to support a mass of 45 kg. What is the stress in the wire?

Solution

The load would equal 45.0 × 9.807 = 441.3 N. *D* is the mean spring diameter
From the *Handbook*,

$$\tau = K_s \times (8FD)/(\pi d^3)$$

where
d = wire diameter = 0.005 m
F = load 45 × g = 441.3 N
D = mean spring diameter = 0.040 – 0.005 = 0.035 m
K_s and C are as given previously, C = 0.035/0.005 = 7

The apparent stress,

$$S_t = (8FD)/(\pi d^3) = 8 \times 441.3 \times 0.035/(\pi \cdot 0.125 \times 10^{-6}) = 314.6 \text{ MPa}$$
$$K_s = (2 \times 7 + 1)/(2 \times 7) = 1.071$$
$$\text{The actual maximum stress} = 314.6 \times 1.071 = 336.9 \text{ MPa}$$

DEFLECTION

The deflection, ΔH, of a constant diameter circular spring is given by the relationship

$$\Delta H = 8FD^3N/Gd^4$$

where
ΔH = deflection
N = number of active coils

This comes from the equations given in the *Handbook*, such as $k = d^4 \times G/(8D^3N)$ and $F = kx$, where the deflection $x = F/k$, which gives, for the deflection, the relationship for the change in height given previously.

Example 12.13

For the spring in the previous example, a compression spring 4 cm in diameter made of 5-mm diameter wire, determine the number of active coils required for the spring to deflect 6.0 cm under the load of 45 kg.

Solution

The calculation to determine the deflection of the spring equals

$$\Delta H = 8FD^3\text{N}/Gd^4$$

where
$F = 45 \times g = 441.3$ N
$\Delta H = 0.060 = 8 \times 441.3 \times 0.035^3 \times N\,(8.3 \times 10^{10} \times 0.005^4)\ 3.113 = 0.151 \times N$
or number of active coils = 20.6, say 21.

For a compression spring the ends would probably have squared and ground ends giving one dead coil at each end, so the total number of coils would equal 23

SPRING LOAD

The load on a bolt supplied by a spring plus an additionally applied load will equal the load due to the force of the spring plus the additional load until the bolt has stretched (increased in length) enough to relieve the spring load. Another way of stating this is that the force acting on a bolt which is loaded by a spring plus an applied load equals the force applied by the spring before the additional load was applied plus the force of the applied load minus the reduction in the force applied by the spring due to the elongation of the bolt. When the assembly consists of an ordinary coiled spring the strain of the bolt due to the additionally applied load will be small and can usually be neglected. However, when the spring load is supplied by a composition gasket or a copper or aluminum washer or gasket, the reduction in the force applied by the spring can be appreciable.

Example 12.14

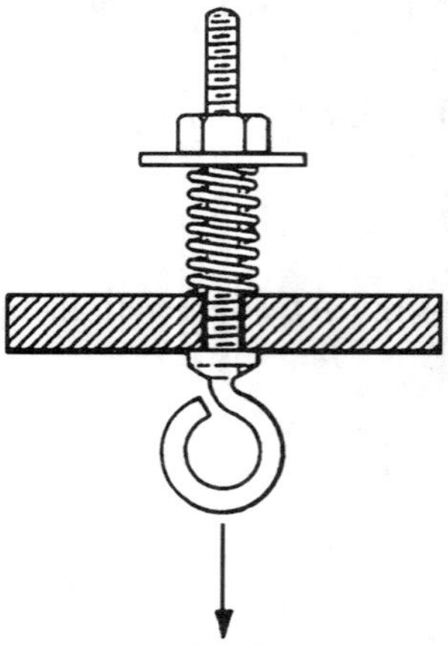

Exhibit 2

An eyebolt support is attached to a rigid steel beam as shown in Exhibit 2. A very stiff compression spring is used to preload the eyebolt. The spring is loaded by means of a nut screwed down on the end of the eyebolt until it exerts a force of 5000 N. What would be the force exerted on the shank of the bolt if a mass of 50 kg were attached to the eyebolt?

Solution

The force of gravity acting on the support by a mass of 500 kg would equal ($500 \times g$) or

$$F = 500 \times 9.8066 = 4908.3 \text{ N}$$

This is less than the upward force exerted by the spring, so the load held by the bolt would equal 5.0 kN.

Example 12.15

What would be the load held by the bolt if a mass of 800 kg were attached to the eye of the bolt?

Solution

$$F = 800 \times g = 7845 \text{ N}$$

This is greater than the initial load on the spring, so the spring would compress further, and the load on the bolt would be 7.85 kN.

When a bolt is tightened (preloaded) against a soft substanceb such as the condition shown in Exhibit 3, and a load P is applied, it will increase the load on the bolt, but not by the amount of the newly applied load. Stress and strain are proportional, the length of the bolt will increase, but the compression of the soft material will decrease by the same amount that the length of the bolt increases. The most common example of this is the case of a soft seal wherein the gasket material is compressed to seal the joint. The definition of a soft material is any material that has a lower coefficient of elasticity than that of the bolt, which is usually steel.

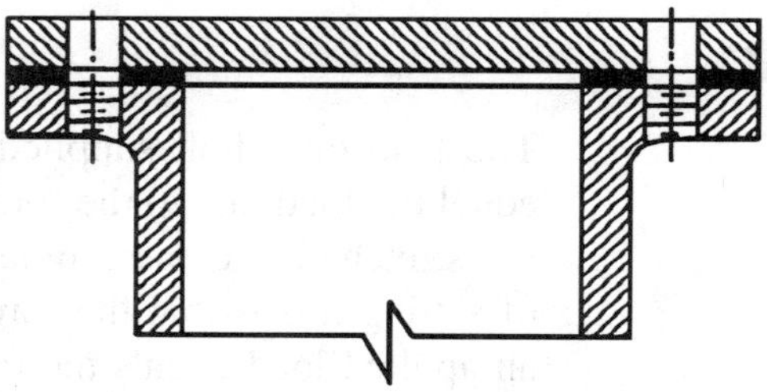

Exhibit 3

Example 12.16

Assume that a soft gasket which has a modulus of elasticity equal to 10.0×10^8 Pa is used to seal a pressure vessel as shown in the figure. The net area of the gasket (subtracting area of bolt holes) is 88 cm^2 and its thickness is 3.0 mm. The gasket is compressed by 12 steel bolts 15 cm long and 1.0 cm in diameter. Assume the joint has been carefully made up and that each bolt holds exactly 1/12 of the applied load. The preload on each bolt is 11.0 kN. If the force added to the joint due to pressurization of the vessel equals 54.0 kN, how much would the force acting on each bolt increase?

Solution

The bolt would strain an additional amount to withstand the added load, but the force exerted by the gasket would be reduced since the gasket would expand as the bolt lengthened. Both the bolt and the gasket can be treated as springs. The increase in the bolt load would equal the increase in bolt length, Δ, times the bolt-spring constant minus the gasket length decrease, Δ, times the gasket-spring constant. The magnitude of the added load would thus amount to

Added load $= \Delta \times (k_B + k_G)$ and the increase in the load held by the bold would equal $\Delta \times k_B$. The bolt-spring, k_B, constant equals $E_B A_B / L_B$, where E_B equals the modulus of elasticity of the bolt material, A_B, equals the cross-sectional area of the bolt, and L_B equals the free length of the bolt.

$$k_B = 2.1 \times 10^{11} \times 0.00007854/0.150 = 11.00 \times 10^7 \text{ N/m}$$

similar for the gasket, with $A_G = 0.0088/12 = 0.000733 \text{ m}^2$

$$k_G = 10.0 \times 10^8 \times 0.000733/0.0030 = 24.43 \times 10^7 \text{ N/M}$$
$$4.5 \text{ kN} = \Delta \times (1.10 + 2.44) \times 10^8$$
$$\Delta = 4{,}500 \text{ N}/(35.43 \times 10^7 \text{ N/m}) = 127.0 \times 10^{-7} \text{ m}$$

The increase in the load on each head bolt would equal $127.0 \times 10^{-7} \times 11.0 \times 10^7 = 1397$ N, which is considerably less than the load increase on the head of 4500 N or 54,000/12 = 4500 N/bolt.

It might be easier for some engineers who are completely unacquainted with the metric system to convert all the quantities to U.S. units, work the problem, and then covert the answer to metric units. There is an excellent table of conversion factors in the *Handbook* supplied to all examinees. Using the foregoing problem as an example

Bolt: Diameter = 0.3937 in Area = 0.1217 in^2 $L = 5.9055$ in
$E = 2.1 \times 10^{11}/6895 = 30.456 \times 10^6 \text{ lb/in}^2$
Gasket: Area = 1.1367 in^2 per bolt $L = 0.1181$ in.
$E = 10^9/6895 = 145{,}033$ psi

Then

$$k_B = 0.6276 \times 10^6 \text{ lb/in} \quad \text{and} \quad k_G = 1.395 \times 10^6 \text{ lb/in}$$

The added head load per bolt = 4500/4.448 = 1017 lb

$\Delta = 1017/(2.023 \times 10^6) = 503 \times 10^{-6}$ in or 0.0005 in the increase in the bolt load would equal $0.6267 \times 503 = 315.2$ lb, and converting to newtons, $315.2 \times 4.448 = 1402$ N, which is the same answer as obtained by the metric system calculation, considering slight errors in the conversion factors.

O-RING SEAL

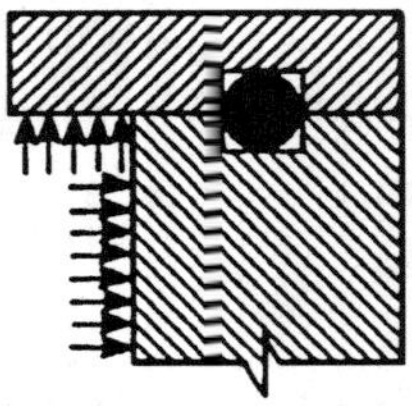

Figure 12.1

It is interesting to note that there would have been no increase in the load on the vessel-head bolts in Example 12.16 if an O-ring seal had been used; see Figure 12.1. This is a hard joint since the O-ring would not apply any load on the boltsb and thus the total load would equal only the original preload, at least until the pressure load exceeded the preload and the joint leaked.

FATIGUE AND ENDURANCE LIMITS

Metal parts that are subjected to alternating loads will often fail when the applied stress is well below the strength of the material as determined by a static tensile test. Such failure usually starts at a point of discontinuity such as a change in cross-section or a local surface blemish where the localized stress is greater than the endurance strength of the material. Large forgings such as, for example, a large high-pressure triplex pump, can fail because of a rough machining mark at a point that is subjected to a stress that is well below the static strength of the material. A fatigue failure is a progressive failure and occurs over a period of time. Allowance is made for such reduced operating stress through the application of a stress-concentration factor that is applied to the endurance limit of the metal (usually steel) of which the part is made. In an extreme case—a high strength, heat-treat steel operating in seawater—the endurance limit may drop to as low as 11% of the ultimate strength.

The stress-concentration factor depends upon a number of different factors and includes the condition of the surface (roughness, polishing the surface of parts subjected to repeated stress variations has resulted in a considerably longer life, even showing an endurance strength approaching the static strength of a steel for a mirror-polished part), a size factor (sudden changes in size, like a change in diameter or a hole in the part), load factor (whether the loading reverses from tensile to compressive or varies entirely in tension), temperature factor, the medium in which the part is to be used (air, water, sea water), and other factors which may affect a particular component.

The endurance limit of a given metal is usually obtained by a rotating test wherein a sample of the metal that has been machined to a specific size and shape (see Figure 12.2) has been rotated through many revolutions. Each revolution applies an equal tensile and compressive loading. When failure occurs, the stress and number of cycles to failure are recorded and plotted on a graph. The endurance limit is that stress at which the specimen will withstand an infinite number of such cycles. The stress and number of cycles to failure are plotted on an S-N diagram, usually a semi-log diagram. The curve of stress vs. failure usually flattens out for steels at about one to ten million cycles or more, but may not flatten out for

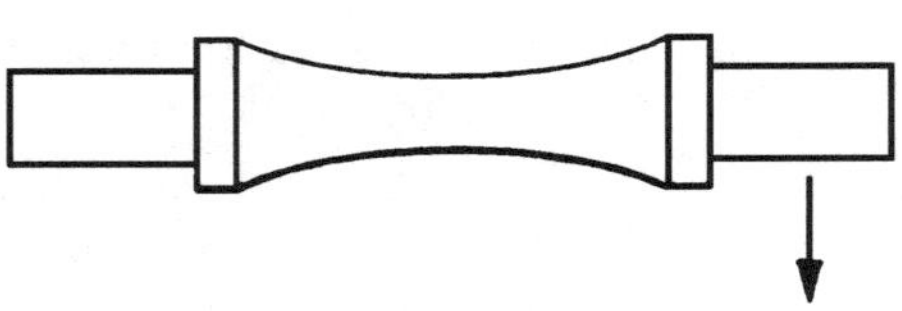

Figure 12.2

aluminum even at many million cycles. The endurance limit may be defined as that stress at which a standard test specimen may withstand an infinite number of cycles of reversed stress. For ferritic carbon steels the endurance limit will usually average out to about 60% of the ultimate tensile strength, for pearlitic steels to about 40% of the UTS, and for plain carbon martensitic steels about 25%. For alloy martensitic alloy steels the ratio will average out to about 35%.

For design purposes this stress must be further modified (reduced) to take into account the stress-concentration effects previously noted. One design method is that of the modified Goodman relation

$$S_a/S_e + S_m/S_{ult} = 1/\mathrm{N}$$

where

N = safety factor
S_a = the alternating tensile stress
S_e = endurance limit
S_m = mean tensile stress, and
S_{ult} = ult. tensile strength

Another design method that is more conservative is with the Soderberg relation in which the value for the yield strength is used in place of the ultimate strength, giving

$$S_a/S_e + S_m/S_y = 1$$

where

S_y = yield strength

For shear or torsional stresses the Goodman relationship becomes

$$S_{vs}/S_{ns} + S_{ms}/S_{ults} = 1/N$$

where

S_{vs} is the variable shear stress
S_{ns} the endurance limit in shear
S_{ms} the mean shear stress
S_{ults} the ultimate shear strength
N the safety factor

The Soderberg relationship would be the same except the ultimate shear strength would be replaced by the shear yield stress S_{ys}.

Example 12.17

A medium-carbon steel with S_u = 590 MPa and S_y = 380 MPa is to be used for the design of a shaft that is to be subjected to varying torques ranging from +340 J to −115 J. What diameter should be specified if a safety factor of 2.0 is desired? The shaft will have an ordinary smooth surface, that is, it will not be a polished surface, nor will it have any rough machine marks.

Solution

Data for shearing stress for a specific steel may not be available, so in that case it must be estimated. The yield in shear is about 0.60 times yield in tension, so for this steel

$$S_{ys} = 0.60 \times 380 = 228 \text{ MPa}.$$

The endurance limit in tension is approximately half the ultimate or about 295 MPa. The endurance limit in torsion for steel has been found to equal approximately 60% in both tests and by distortion energy failure theory for polished specimens, so

$$S'_{ns} = 0.60 \times 295 = 177 \text{ MPa.}$$

But this value must be further modified for surface effect and size effect. The surface effect for hot-rolled steel with a strength of 590 MPa is found from a graph to equal some 60%, and the size effect for a rod with a diameter between 13 and 50 mm equals some 85%. This gives a value for $S_{ns} = 0.60 \times 0.85 \times 177 =$ 90.3 MPa.

The mean torsional loading equals

$$T_m = [340 + (-115)]/2 = 112.5 \text{ J}$$

and the variable torque equals

$$T_v = [340 - (-115)]/2 = 227.5 \text{ J}$$

The corresponding stresses are then, from

$$S = Tc/J' \text{ where } J'/c = \pi D^3/16, \text{ so } S = 16T/(\pi D^3)$$
$$S_{ms} = 112.5 \times 16/\pi D^3 = 572.96/D^3 \text{ Pa}$$
$$S_{vs} = 227.5 \times 16/\pi D^3 = 1159.65/D^3 \text{ Pa}$$

Using the Soderberg relationship for a more conservative design gives

$$1/N = S_{ms}/S_{ys} + S_{vs}/S_{ns}$$
$$1/2 = 572.96/(228 \cdot D^3) + 1159.65/(90.3 \cdot D^3)$$
$$D^3 = (5.03 + 26.68)/1{,}000{,}000$$
$$D = 0.0317 \text{ m or } 31.7 \text{ mm}$$

Example 12.18

Estimate the endurance of a 4.0 cm diameter cold-drawn steel bar for 99% reliability, if it is made of steel with an ultimate tensile strength of 400 MPa.

Solution

The term *endurance limit* is generally considered to mean the bending endurance limit unless otherwise specified. From a table of recorded data we find that the size factor for this diameter equals 0.869. Also, for this size bar, the surface finish factor will equal 0.84. From another source it is found that the reliability factor for 99% reliability equals 0.814. Since we do not have any endurance test results for this particular steel we can estimate its endurance limit as 50% of the ultimate strength, or

$$S'_e = 0.5 \times 400 = 200 \text{ MPa}$$

combining these factors gives—

$$S_e = 0.869 \times 0.84 \times 0.814 \times 200 = 0.594 \times 200 = 119 \text{ MPa}$$

LOW CYCLE FATIGUE

There is another type of fatigue termed *low cycle fatigue* in which failure occurs when a metal is stressed beyond its yield point in tension and then compressed beyond its yield point when the loading is reversed, thus causing the item to first to elongate plastically and then to shorten plastically, or vice versa.

Example **12.19**

One place where such failures have occurred is in the exhaust system of a diesel engine. If the exhaust pipe of a diesel engine is rigidly held at a point one meter from its attachment to the exhaust manifold, what would happen to the stainless steel pipe when the diesel was operated at full load for a full shift?

Solution

The stainless steel exhaust pipe of the diesel near the exhaust manifold would probably become a bright red and its temperature would approach some 600 °C or more. The coefficient of thermal expansion for stainless steel is $16.92 \times 10^{-6}/°C$. The pipe would tend to expand in length

$$\varepsilon = 580 \times 16.92 \times 10^{-6} = 0.00981 \text{ m/m}$$

This would imply a compressive stress of

$$S = E\delta = 0.00981 \times 1.93 \times 10^{11} = 1894 \text{ MPa.}$$

However, the yield strength of the stainless steel at room temperature is only 275.8 MPa, and this would decrease to as low as 130 MPa at the higher temperature, so the fixed-end piece of pipe would compress. The time at temperature is long enough for the pipe to creep to an equilibrium condition at an applied stress of 130 MPa. At the higher temperature

$$\varepsilon = 130 \times 10^{6}/1.93 \times 10^{11} = 0.000674 \text{ m/m}$$

so the pipe would yield plastically— $0.00981 - 0.000674 = 0.00914$ m/m or 9.14 mm over the one meter fixed length. When the diesel engine was turned off, the pipe would cool and would attempt to shrink an amount equal to

$$\varepsilon = 580 \times 16.92 \times 10^{-6} = 0.00981 \text{ m/m,}$$

but this would indicate a stress of

$$S = E \times \varepsilon = 0.00981 \times 1.93 \times 10^{11} = 1894 \text{ Mpa.}$$

However, as previously noted, the yield strength of the stainless steel at room temperature is only 275.8 MPa, so the pipe would elongate plastically. This sequence of events would occur every time the diesel engine was operated, and after a number of such cycles (many less than the endurance limit) the pipe would rupture. It is interesting to note that such a design was actually constructed and such a failure did occur.

SCREW THREADS

Example **12.20**

A power screw to raise a load is designed with a single square thread and a major diameter of 35.0 mm. The pitch is 5.0 mm. The bearing surface of the nut is 50 mm in diameter where it fits against the end of the structure to provide the resisting force. It is to be used to raise a load of 815 kg. It is assumed that the coefficient of frication for all surfaces equals 0.10. What would be the efficiency of the power screw, and what would be the stress in the screw?

Solution

The pitch is 5 mm, so the width and depth of a square thread would equal 2.5 mm. The mean diameter would then equal OD minus the thread height or, $d_m = 35.0 - 2.5 = 32.25$ mm. When the screw is rotated, each turn will raise the mass 5.0 mm. The distance of the screw wedge would equal the circumference at the circumference at the pitch diameter or $\pi \times d_m$ and the thread lead angle

$$\lambda = \tan^{-1}/(10/2\pi d_m) = 2.825° \qquad \lambda = 2.825°.$$

Probably the easiest way to analyze this problem is with the aid of the friction angle, μ, where the friction angle adds to, or subtracts from the mechanical angle to give an equivalent frictionless system. The friction angle equals the $\tan^{-1}$ of the friction factor. A friction factor of 0.10 gives a friction angle equal to $\tan^{-1} 0.10 = 5.711°$. The total equivalent frictionless wedge angle would equal $(\mu - \lambda)$ or 8.536° for raising the mass and $(\mu - \lambda) = 2.886°$ for lowering the mass. It might be noted here that when the friction angle is greater than the lead angle, the screw will be statistically stable, but if the lead angle is greater than the friction angle the force of the load will turn the screw and the load would lower unless the screw were held to keep it from turning. The torque required to turn the screw, considering only thread pitch and thread friction would equal

$$Tq_1 = (d_m/2) \cdot \tan(\mu - \lambda) \cdot F$$
$$\text{from } 2\pi Tq = \pi d_m \cdot \tan(\mu - \lambda) \cdot F$$

Torque would also be necessary to overcome the frictional resistance to turning due to the friction force acting between the bearing surface of the nut and the support structure.

$$Tq_2 = F \cdot \mu \cdot (D_{od} + D_{id})/2 = F \times 0.1 \times (50 + 35)/2 = 4.25F$$

the units in this case equal mm · N so to state it in joules or Nm $Tq_2 = 0.00425 \cdot F$ joules

The force, F, exerted by a load of 815 kg would equal

$$F = 815 \times 9.807 = 7993 \text{ N so } Tq_2 = 33.970 \text{ Nm}$$
$$Tq_1 = (32.25/2) \text{ mm} \times \tan 8.536° \times 7993 \text{ N}$$
$$= 19{,}345 \text{ Nmm, or } 19.345 \text{ Nm} = 19.345 \text{ J}$$

The total torque required to raise the load would equal

$$\text{Torque} = Tq_1 + Tq_2 = 19.345 + 33.970 = 53.315 \text{ Nm} = 53.315 \text{ J}$$

Similarly, torque required to lower the load would equal

$$3.970 + 257.774 \times \tan 2.866° = 46.875 \text{ J}$$

The efficiency of the power screw would equal $F \cdot \text{pitch}/2\pi Tq$ so

$$e = 7993 \times 0.005/(2\pi \cdot 53.315) = 11.93\%$$

If a ball bearings thrust bearing were placed under the head of the bolt, assuming the same dimensions and a friction factor of 0.01, the torque i_2 would be reduced to 3.397 J and the total torque would equal 22.742 J. The efficiency would increase to

$$7993 \times 0.005/(2\pi \cdot 22.742) = 27.97\%$$

The shear area per thread would equal π times the root diameter (major diameter −2 times thread height) times the width of the thread

$$\text{Shear area} = \pi \cdot (0.035 - 0.005) \cdot 0.0025 = 2.356 \times 10^{-4}\text{m}^2$$

Stress for one thread of engagement

$$S_s = 7993/(2.356 \times 10^{-4}) = 33.93 \text{ MPA}$$

Which is a relatively low shear stress for steel, so, since there would be more than one thread of engagement, the screw thread is more than strong enough for the job.

BEAM DEFLECTION

Example 12.21

A walkway is made from two steel tubes (pipes) 10 cm O.D. by 1.0 cm wall, spaced 60 cm apart and covered with wooden strips. It is used to span a 10 m ditch. If the mass of the wooden strips is ignored, what would be the maximum deflection if a man plus a wheelbarrow with a combined mass of 250 kg were to go across the bridge?

Solution

The maximum deflection would occur when the man was at the center of the span. The deflection

$$\delta_{max} = P \times \ell^3/(48 \cdot E \cdot I)$$

This can be derived from the relationship for deflection for a simply supported beam with $b = \ell/2$ in the *FE Handbook*. I for one tube equals $(\pi/64) \times (OD^4 - ID^4)$

$$I = 0.04909 \times (1.000 \cdot 10^{-4} - 0.4096 \cdot 10^{-4})$$
$$I = 2.898 \times 10^{-6} \text{ per tube so } \quad I_{total} = 5.7966 \times 10^{-6}\text{ m}^4$$
$$P = 250 \times 9.807 = 2452 \text{ N}$$
$$\delta_{max} = 2{,}452 \times 1000/(48 \times 2.1 \times 10^{11} \times 5.7966 \times 10^{-6}) = 4.20 \text{ cm}$$
$$\delta_{max} = 4.20 \text{ cm}$$

Example 12.22

What is the maximum stress in the span in Example 12.21?

Solution

This is a simply supported beam so the load on the support at each end would equal $P/2$ and the maximum moment would occur at the center of the span.

$$M_{max} = \ell/2 \times P/2 = S = M \cdot c/I$$
$$= (10 \times .0500 \times 2{,}452/4)/(5.7966 \cdot 10^{-6}) = 52.88 \text{ MPA}$$

PROBLEMS

12.1 A factor of safety is
a. Yield stress divided by design stress.
b. Ultimate stress divided by design stress.
c. Ultimate stress divided by yield stress.
d. Maximum expected load divided by average load.

12.2 A swing is to be designed for a public park. The support beam will be 3.00 m high, and the swing will be designed to hold a 100-kg person. What should be the strength of the lines holding a swing seat? Since this is a public park, it has to be assumed that the swing will be subjected to uses (or misuses) far beyond what any one might imagine, so a safety factor of five has been decided upon.

a. 12.2 kN c. 13.9 kN
b. 14.7 kN d. 15.6 kN

12.3 A design requirement for a pogo stick is a spring that will compress 18 cm when a 100-kg person drops one meter. What is the required spring constant?

a. 65.7 kN/m c. 68.8 kN/m
b. 71.4 kN/m d. 75.2 kN/m

12.4 It is proposed to use a standard gasoline engine with a compression ratio of 6:1 for an industrial operation. A test is conducted to determine the efficiency of the engine. Gasoline with a density of 0.70 and a heating value of 44.432 MJ/kg is used for the test. A torque device measures 200 N at a distance of one meter at an engine speed of 1500 r/min. The engine uses 3 ¾ liters of gasoline in 15 min. while developing that torque. Which of the following most nearly equals the thermal efficiency of the engine?

a. 19.8% c. 24.2%
b. 28.3% d. 32.1%

12.5 A transformer core is to be built up of sheet-steel laminations, using two different strip widths, x and y, so that the resultant symmetrical cross section will fit within a circle of diameter D. As plant engineer you are asked to determine the values of x and y in terms of D so that the cross section of the core will have maximum value. See Exhibit 12.5. The correct values for the X and Y dimensions are as follows:

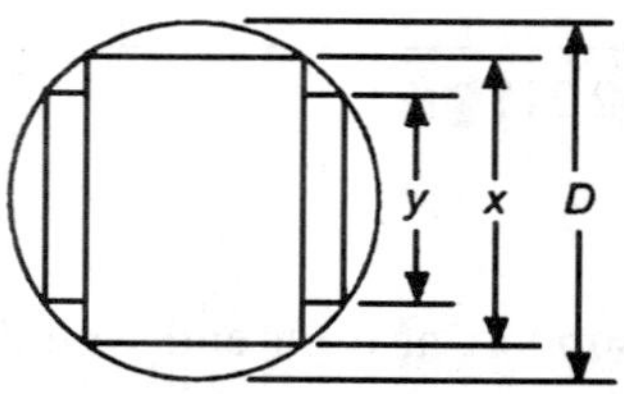

Exhibit 12.5

a. $y = 0.53D$ $x = 0.85D$
b. $y = 0.75D$ $x = 0.62D$
c. $y = 0.48D$ $x = 0.87D$
d. $y = 0.58D$ $x = 0.78D$

12.6 A jackscrew having threads spaced 3.2 mm apart, 312.5 threads/m, is to be operated manually by a handle. The point at which the handle is gripped to move is 60 cm from the axis of the screw. If friction is disregarded, what force must be applied at the grip on the handle to just raise a load of 1050 kg?

a. 7.6 N c. 8.4 N
b. 9.0 N d. 9.7 N

12.7 A proposed design contains three wires, each having a cross-sectional area of 1.30 cm^2 and the same unstressed length of 5.0 meters at 20 °C, which hang side by side in the same plane. The outer wires are copper. The middle wire, equidistant from each of the others, is steel. Given: (see *Handbook*)

$$E \text{ for steel} = 2.1 \times 10^{11} \text{ Pa} \quad \text{and} \quad E \text{ for copper} = 1.17 \times 10^{11} \text{ Pa}.$$

If a mass of 450 kg is gradually picked up by the three wires, the part of the load carried by the steel wire would most nearly equal which of the following?

a. 225 kg c. 150 kg
b. 213 kg d. 275 kg

12.8 The assembly described in the above problem will be located in a test bay in which the temperature can rise. Referring to the data in the above problem, to what temperature would the system have to rise for the steel wire to support the entire load?

The coefficient of linear thermal expansion for steel

$$\alpha_{\text{Steel}} = 11.7 \times 10^{-6}/°\text{C}$$

The coefficient of linear thermal expansion for copper $\alpha_{\text{Cu}} = 16.7 \times 10^{-6}/°\text{C}$
Other data are as given in problem 12.7.

a. 48 °C c. 53 °C
b. 56 °C d. 61 °C

12.9 A penstock 3.0 m in diameter carrying 7.0 cubic meters of water per second to a generating plant bends through a 45° angle just before it enters the plant. Which of the values below most nearly equals the force that the support post at the bend must be designed to withstand? Assume that the discharge is free (exhaust pressure equals zero gauge.) See Exhibit 12.9.

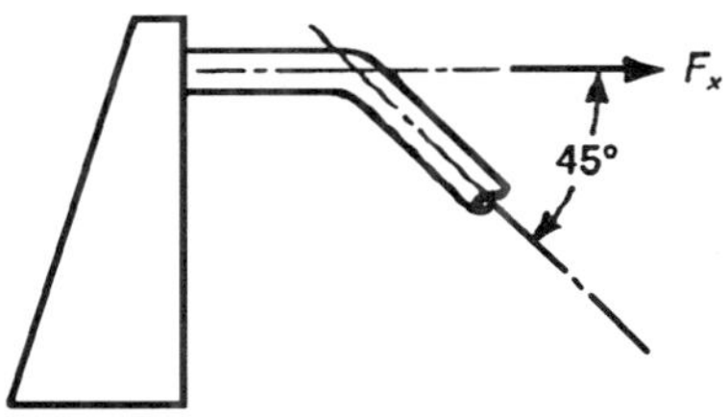

Exhibit 12.9

a. 5260 N c. 5950 N
b. 6650 N d. 6930 N

12.10 A mine elevator cage with a mass of 2.5 tons empty is descending at a rate of 32 km/hr when the hoist mechanism jams and stops suddenly. If the cable is steel with a cross-sectional area of 6.5 cm^2 and the cage is 1500 meters below the hoisting drum when the drum jams, which of the following most nearly equals the stress in the cable? (E for hoist cable equals 84,000 × 10^6 Pa, neglect the weight of the cable)

a. 165 MPa c. 173 MPa
b. 188 MPa d. 226 MPa

12.11 Shown in Exhibit 12.11 is a vertical tank, open to the atmosphere at the top. At the bottom of the tank is a valve that initially is closed, and the tank is filled to a level L with a liquid of zero viscosity. How long would it take the tank to empty if some one should mistakenly open the valve and fluid escaped through a round edged orifice ($C = 1.0$) which has an area 0.010 times the area of the tank ($A/a = 100$) and the height of the liquid in the tank is two meters?

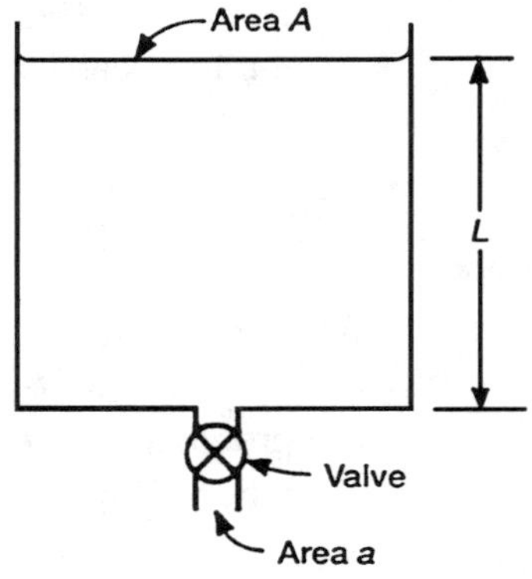

Exhibit 12.11

a. 97 s
b. 76 s
c. 82 s
d. 64 s

12.12 A coupling connecting a motor to a pump transmits 3.36 kW at 1750 rpm. Two 6.35-mm diameter pins hold the coupling together. What stress must the pins withstand when the motor transmitting the design power at the rated speed? See Exhibit 12.12.

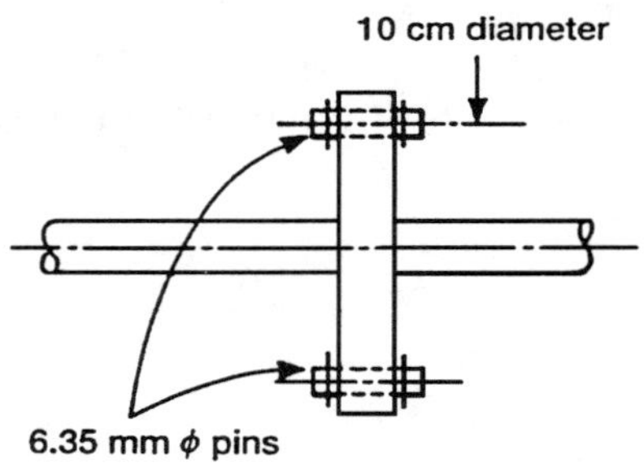

Exhibit 12.12

a. 6.89 Mpa
b. 5.79 Mpa
c. 6.21 Mpa
d. 4.89 MPa

12.13 A water jet is used on a mixing tank as shown in Exhibit 12.13. When the gate valve is full open the pressure gage reads 276 kPa. At what velocity is the water discharging through the jet when the water surface is 61 cm above the centerline of the jet? Assume no head loss in the line or jet.

a. 28.5 m/s
b. 23.8 m/s
c. 25.2 m/s
d. 22.1 m/s

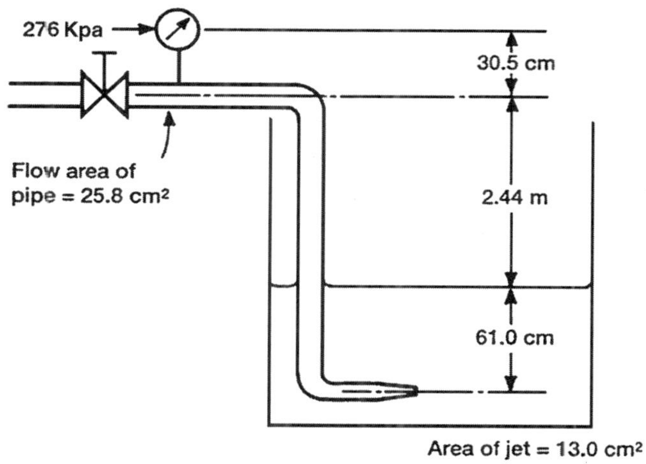

Exhibit 12.13

12.14 Determine which of the following is most nearly equal to the amount of heat which would be lost through a window which measured 60 cm by 75 cm if the inside temperature were 25 °C greater than the temperature outside. The window glass is 3.0 mm thick, and the surface coefficients, h, and glass conductivity, k are as follows

$$h_i = 8.52 \text{ W/m}^2 \cdot \text{K}$$
$$h_o = 34.07 \text{ W/m}^2 \cdot \text{K}$$
$$k = 1.04 \text{ W/m} \cdot \text{K}$$

a. Q = 58 W
b. Q = 71 W
c. Q = 64 W
d. Q = 75 W

12.15 A rigid bar is to be held by three wires of length L, with a modulus of elasticity, E, and a cross-sectional area, A. They are to be spaced a distance of a apart as shown in Exhibit 12.15a. All three wires are the same length. Which of the following is the maximum held by a wire if a force, P, is applied at a distance of $a/2$ from one wire as shown in the figure? Neglect the mass of the bar.

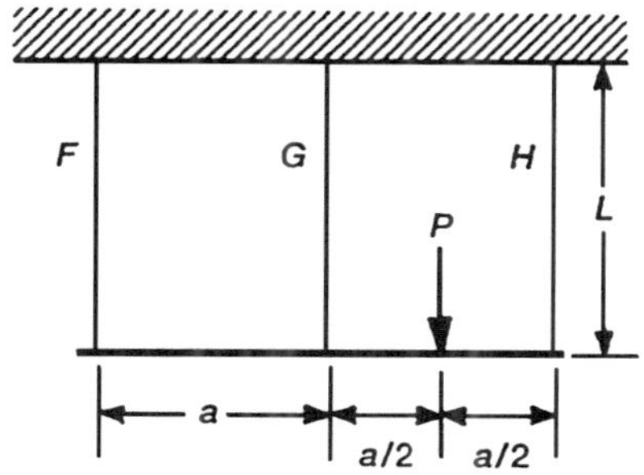

Exhibit 12.15a

a. F = P/2
b. F = 5P/7
c. F = 7P/12
d. F = 3P/10

12.16 The plant engineer in a steam power plant that burns a hydrocarbon fuel makes an Orsat analysis of the stack gas. The analysis shows the following:

$$11.3\%\ CO_2,\ 4.5\%\ O_2,\quad \text{and} \quad 84.2\%\ N_2$$

The discharge temperature of the stack gas is 315 °C and the pressure is 29.65 kPa gauge. The specific heats at constant pressure for the products of the combustion are: CO_2, 0.846 kJ(kg · K) O_2, 0.918 kJ/(kg · K), N_2, 1.04 kJ/(kg · K) and H_2O (steam), 1.989 kJ/(kg · K).

The fuel is a member of the methane series ($C_nH_{2n}+_2$)
Which of the following is the chemical formula of the fuel?

a. C_6H_{14} c. C_5H_{12}
b. C_4H_{10} d. C_2H_6

12.17 Using the data from Problem 12.16, how much excess air in percent, was used in the combustion process?

a. 10% c. 17%
b. 25% d. 32%

12.18 Determine what size load can be lifted with a hydraulic jack that has a 3.8 cm diameter ram if the diameter of the piston has a diameter of 1.6 cm and the force on the jack handle is limited to 135 N. The jack handle mechanism has a mechanical advantage of 15:1. Neglect any friction that may act.

a. 1162 kg c. 1239 kg
b. 1294 kg d. 1331 kg

12.19 An engineer is required to lay out a linear distance of 320 meters. The tape to be used is a 100 meter tape which is actually 99.992 meters long at 20 °C and is 99.999 meters long at 26 °C. Which of the following is the distance to be indicated by the tape to give a true distance of 320 meters if the ambient temperature on the day the measuring takes place is 33 °C?

a. 320.000 meters c. 319.974 meters
b. 320.023 meters d. 319.981 meters

12.20 A threaded bolt has a root diameter of 1.6 cm. The limiting stress for the bolt material is 344.8 MPa. Which of the following is the maximum load that can be safely supported by the bolt?

a. 7980 kg c. 7750 kg
b. 7450 kg d. 7070 kg

12.21 If a body has a mass of 100 kg in air, but tips a scale at 25 kg when suspended in fresh water, its specific gravity is closest to which of the following?

a. 1.52 c. 1.33
b. 1.27 d. 0.98

12.22 A torsion member is made of steel 15 cm in diameter. To save mass it is proposed to replace it with a hollow bar of the same type of steel with a 10 cm diameter hole through its center. What should the outside diameter of the hollow bar be to provide the same torsional stiffness?

a. 15.7 cm c. 16.2 cm
b. 16.7 cm d. 17.3 cm

12.23 Air is compressed rapidly in an insulated cylinder to 1/4 its initial volume. What will a gage show as the final pressure if the initial pressure is 13.79 kPa gage?

a. 575 kPa
b. 660 kPa
c. 620 kPa
d. 700 kPa

12.24 A solid steel flywheel 150 cm in diameter rotates at a speed of 600 rpm about an axis through its center. If it has a mass of 1460 kg, how much kinetic energy does it possess?

a. 692 kJ
b. 810 kJ
c. 775 kJ
d. 853 kJ

12.25 The surface temperatures of the faces of a 15 cm thick slab of material are 77°C and 20°C. What is the rate of heat loss, Q, through the slab if the thermal conductivity, k, of the slab equals 1.039 W/m · K?

a. 365 W/m^2
b. 432 W/m^2
c. 395 W/m^2
d. 451 W/m^2

SOLUTIONS

12.1 **b.** The factor of safety is defined as the ultimate stress divided by the design stress.

12.2 **b.** The maximum load on the swing seat will occur at the bottom of the arc and will equal the gravitational force exerted by the swinger and the centrifugal force, F_c.

$$F_c = m \cdot v^2/R$$

Assume the swinger goes as high as the support beam, then

$$v = /(2gh) \quad \text{or} \quad v^2 = 2 \times 9.8066 \times 3.0 = 58.84 \text{ m}^2/\text{s}^2$$

assuming the swing seat is at the level of the ground when at rest.

$$F_c = 100 \times 58.84/3 = 1961 \text{ N}$$

The total force down would equal

$$1961 + 100 \times 9.8066 = 2942 \text{ N}$$

Assume that a swinger would hold onto only one side of the swing. The strength of each support line should equal 5 × 2942 = 14,710 N.

12.3 **b.** The energy to be absorbed equals

$$100 \times 9.8066 \times (1 + 0.18) = 1{,}57.18 \text{ J}$$
$$1{,}157.18 = \tfrac{1}{2} \times k \times 0.18^2 = 71.431 \text{ kN/m}$$

12.4 **c.** The ordinary gasoline engine operates on the Otto cycle, see *Handbook*.

The power output of the engine equals $(1500/60) \times 2\pi \times 200 =$ 31,400 Nm/s = 31.4 kW or 31.4 kJ/s.

The heat supplied equals $3.75 \times 0.70 \times 44.432 \times 10^6/(15 \times 60) =$ 129,600 J/s.

$$\text{Efficiency} = 31.4/129.6 \times 100 = 24.2 \text{ percent.}$$

12.5 **a.** Refer to Exhibit 12.5. The system is symmetrical, so

$$x^2 + y^2 = D^2 \qquad A = xy + y(x - y)$$
$$x = /(D^2 - y^2)$$

Set $dA/dy = 0$ then

$$A = 2xy - y^2 = 2y/(D^2 - y^2) - y^2 = 2/(y^2D^2 - y^4) - y^2 \; dA/dy$$
$$= (2/2)(2yD^2 - 4y^3)/(y^2D^2 - y^4) - 2y$$

which reduces to $yD^2 - 2y^3 - 2y/(y^2D^2 - 4y^4) = 0$

$$yD^2 - 2y^3 = 2y/(y^2D^2 - y^4)$$

divide through by y and square both sides

$$D^4 - 4D^2y^2 + 4y^4 = y^2D^2 - y^4$$

which reduces to

$$D^4 - 5D^2y^2 + 5y^4 = 0$$

solve for y^2 using the general solution for a quadratic equation as given in the *Handbook*

$$y^2 = 0.5D^2 \pm 0.224D^2$$

which gives $y = 0.851D$ or $y = 0.525D$

Substitution in the equation for x gives that

$$x = 0.525D \quad \text{for } y = 0.851D \quad \text{and} \quad x = 0.851D \quad \text{for} \quad y = 0.525D$$

12.6 **b.** Work per revolution equals $F \times 2\pi \times 0.60 = 3.770$ N·m or 3.770 J
Work also equals $1050 \times g \times 0.0032 = 33.95$ J so $F = 33.95/3.77 = 9.01$ N

12.7 **b.** The load held by a wire will equal

$$\text{Stress} \times \text{area} = E \cdot \epsilon \times A$$

The elongations of the three wires will be the same, so the strains will be equal. So

$$450 \times g = (2 \times E_{cu} + E_{\text{Steel}}) \times \epsilon \times 0.00013$$
$$4413 = (2.34 + 2.1) \times 10^{11} \times \epsilon \times 0.00013 = \epsilon \times 577.2 \times 10^6$$

which gives $\epsilon = 7.646 \times 10^{-6}$ m/m

Each copper wire will thus hold

$$1.17 \times 7.646 \times 10^{-6} \times 0.00013 = 1163 \text{ N or } 118.6 \text{ kg}$$

The steel wire will hold

$$2.1 \times 7.646 \times 10^{-6} \times 0.00013 = 2087 \text{ N} \quad \text{or} \quad 212.8 \text{ kg}$$

12.8 **c.** In order for all the load to be held by the single steel wire, the strain due to the load alone would have to equal

$$450 \times g/(E \times A) = 4413/(2.1 \times 10^{11} \times 0.00013)$$
$$= 4413/(27.3 \times 10^6) = 0.000162 \text{ m/m}$$

The total strain in the steel wire will equal $0.000162 + \alpha_{\text{Steel}}\,\Delta T$

Strain in copper wire due to thermal expansion alone will equal $\alpha_{Cu}\Delta T$

The two strains will be equal $0.000162 + 11.7 \times 10^{-6} \times \Delta T = 16.7 \times 10^{-6} \times \Delta T$

Which gives $\Delta T = 0.000162/(5.0 \times 10^{-6}) = 32.4$ C

The temperature would have to rise $20 + 32.4 = 53.4$ C

12.9 **d.** Flow area equals

$$\pi/4 \times 3^2 = 7.069 \text{ m}^2 \quad Q = 7.00 \text{ m}^3/\text{s} \quad V = 7.00/7.069 = 0.990 \text{ m/s}$$

Take the axis of the pipe into the bend as the X-axis. The force the water would exert on the bend would equal

$$\text{mass} \times v = (7.00 \times 1{,}000) \text{ kg/s} \times 0.990 \text{ m/s} = 6930 \text{ kg} \times \text{m/s}^2$$
$$= 6930 \text{ N into the bend}$$

The reaction forces on the discharge side of the bend would equal

$$6930\cos 45° = 6930 \times 0.707 = 4900 \text{ N}$$

The vertical reaction force on the bend would equal

$$6930 \sin 45° = 4900 \text{ N upward, or in a positive direction}$$

The force that the elbow would exert on the support post, R, would equal

$$(4900^2 + 4900^2) = 6930 \text{ newtons; see Exhibit 12.9}$$

12.10 **c.** Treat the cable as a spring, k = N/meter. Energy absorbed by spring = $\frac{1}{2}k(\Delta L)^2$ where ΔL is the elongation of the spring due to the applied load. One metric ton equals 1000 kg.

Energy that must be absorbed by spring = $\frac{1}{2}mv^2 + (mg \times \Delta L)$

$$KE + \text{change in } PE$$

Next determine the spring constant of the cable

$$\text{Stress} = P/A \qquad A = 6.5 \text{ cm}^2 = 0.00065 \text{ m}^2$$

For one kg load,

$$\text{stress} = 1.00 \times g/0.00065 = 15.09 \text{ kPa}$$
$$\text{Stress} = E \times \varepsilon \quad \text{or} \quad \varepsilon = S/E$$

Change in length of the spring

$$1500 \times \varepsilon \text{ for a load of 1.0 kg or 9.807 N} \quad \varepsilon = 15{,}090/E$$
$$\Delta L = 1500\ \varepsilon$$

which equals $1500 \times 15{,}090/(84{,}000 \times 10^6) = 0.0002695\ m$

For mass of 1.0 kg, $k = 9.807 \times 1/0.0002695 = 36{,}390$ N/m, which equals the spring constant of the cable spring.

The KE of the hoist cage = $\frac{1}{2}mv^2 = \frac{1}{2} \times 2500 \times (32{,}000/3600)^2 =$ 98.76 kJ or 98,760 Nm

$$98{,}760 + (2500 \times g \times \Delta L) = \tfrac{1}{2} \times 36{,}390 \times (\Delta L)^2 = 18{,}195\ (\Delta L)^2,$$

which reduces to

$$(\Delta L)^2 - 1.347\ \Delta L - 5.428 = 0$$

Using the general solution for a quadratic equation, see *Handbook*,

$$\Delta L = [1.347 \pm /(1.814 + 21{,}712)]/2 = (1.347 + 4.850)/2 = 3.099,$$

which gives that the elongation of the supporting cable would be 3.099 m. The strain, ε would equal 3.099/1500 = 0.002066 m/m.

The stress would be $E \times \varepsilon = 84{,}000 \times 10^6 \times 0.002066 = 173.5 \times 10^6$ Pa or 173.5 MPa.

12.11 **d.** Call the height of the liquid above the discharge h. The velocity of efflux will then be $v = C/(2gh)$.

The rate at which the liquid drains from the tank would thus equal $a \times v$ or dQ/dt m^3/s.

The rate at which the liquid drains from the tank would also equal the velocity of the surface of the liquid times the cross-sectional area of the tank or $dQ/dt = A \times dh/dt$ m^3/s.

The height, h, is decreasing so dh/dt is negative. The two equivalent rates of flow can be set equal to one another

$$a \times v = -A \times dh/dt \quad \text{or} \quad a \times C/(2gh) = -A \times (dh/dt)$$

Rearranging terms gives

$$dt = -A/[a \times C \times /(2g)] \times h^{-1/2}dh$$

Integration gives

$$t = -2 \times h^{1/2} \times A/[a \times C \times /(2g)] + \text{constant of integration}$$

At $t = 0$, $H = L$, which givesconstant $= 2AL^{1/2}/[a \times C\,/(2g)]$. The length of time to empty is the time required for h to reach zero.

For $A/a = 100$ and $L = 2.0$ meters,

$$C = 1.00 \quad t = 0 + \text{constant} \quad \text{or} \quad t = 2 \times 100 \times /2/(2g)$$
$$\text{or} \quad t = 200/g = 63.9 \text{ seconds}$$

12.12 **b.** A power of 3360 watts equals 3360 J/s or 3360 N·m/s. 1750 rpm = 29.17 rev/s.

Power $= 2\pi NT$ so these values correspond to a torque of $3360/(29.17 \times 2\pi) = 18.33$ N·m

The pins are at a distance of 10/2 = 5.0 cm from the axis of rotation (see Exhibit 12.12 or 0.050 meter).

The force acting on the pins equals 18.33/0.050 = 366.6 N, or 183.3 N/pin. The area of a pin equals $0.00635^2 \times \pi/4 = 31.67 \times 10^{-6}$ m^2

$$\text{stress in pin} = 183.3/(31.67 \times 10^{-6}) = 5.79 \text{ MPa}$$

12.13 **a.** Using the data in Exhibit 12.13, Bernoulli's equation can be applied between the point at which the gauge is attached to the supply pipe and the discharge to solve for the discharge velocity.

$$P_1/\rho g + v_1^2/2g + z_1 + h_A = P_2/\rho g + v_2^2/2g + z_2 + h_L$$

At point (1) the pressure in the pipe will equal 276 kPa plus 0.305 m of water.

$$P_1/\rho g = 276{,}000/(1000 \times 9.807) + 0.305 = 28.45 \text{ meters of water}$$
$$P_2/\rho g = 0.61 \text{ meter of water}$$
$$P_1/\rho g - P_2/\rho g = 27.84 \text{ meters} \quad h_A \text{ and } h_L \text{ are zero} \quad z_1 \text{ minus } z_2$$
$$\text{equals } 2.44 + 0.61 = 3.05 \text{ m}$$
$$v_1 = Q/0.00258 \text{ and } v_2 = Q/0.0013 \text{ so } v_1 = 0.5039\, v_2 \text{ and } v_1^2 = 0.254\, v_2^2$$

Combining values and substituting in Bernoulli's equation gives

$$27.84 \text{ m} + 3.05 \text{ m} = 0.746\, v_2^2/2g$$
$$27.84 \text{ m} + 3.05 \text{ m} = 0.746\, v_2^2/2g \quad V_1^2 = 41.41 \times 2 \times 9.807 = 812.22$$
$$V_1 = 28.49 \text{ m/s}$$

12.14 **d.** The overall heat transfer coefficient can be calculated from the relationship

$$1/U = 1/h_i + L/k + 1/h_o$$

So

$$1/U = 1/8.52 + 0.003/1.04 + 1/34.07 = 0.117 + 0.00288 + 0.0294 = 0.1493$$
$$\text{so } U = 6.698 \text{ W/m}^2 \cdot \text{K}$$

The area of the window equals

$$0.60 \times 0.75 = 0.45 \text{ m}^2$$

The heat loss, Q, would equal $6.698 \times 0.45 \times 25 = 75.35$ watts.

12.15 **c.** Label the three members F, G, and H as shown in the problem figure. It is assumed that all three wires are in tension. They will all stretch somewhat, and since they are all the same length, same cross-sectional area, and same material, i.e., are identical, the load held by each wire will be proportional to the amount it is stretched. Knowing that the bar is rigid, a deflection diagram can be drawn, see Exhibit 12.15b.

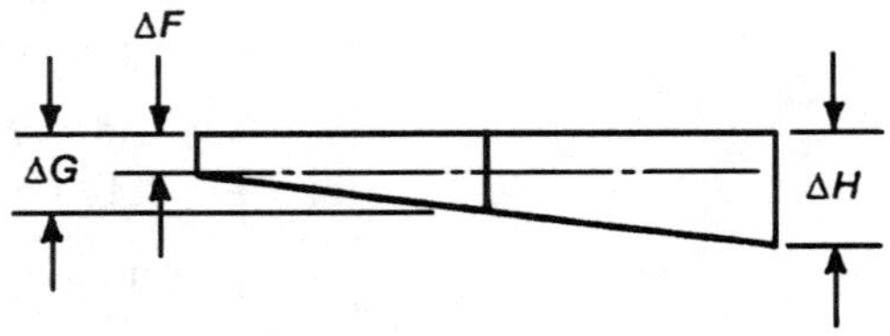

Exhibit 12.15b

Looking at the deflection diagram it is seen that—

$$\Delta G = \Delta F + D \qquad \text{(Equation No. 1)}$$

and

$$\Delta H = \Delta F + 2D \qquad \text{(Equation No. 2)}$$

In addition it is known that the force exerted by F,

$$F_F = \Delta F/L \times AE$$

also, similarly

$$F_G = \Delta G/L \times AE \quad \text{and} \quad F_H = \Delta H/L \times AE$$

summing the vertical forces and setting then equal to zero gives

$$P = AE/L(\Delta F + \Delta G + \Delta H) = AE/L(3\Delta F + 3D) \qquad \text{(Equation No. 3)}$$

Taking moments about the point where wire F is connected to the bar gives

$$F_G \times a - 3/2a \times P \times 2a \times F_H = 0 \qquad F_G - 3/2P + 2 \times F_H = 0$$

Then

$$\Delta G/L \times AE + 2 \times \Delta H/L \times AE = 3/2 \times P \quad \text{and} \quad \Delta G + 2 \times \Delta H = 3/2 \times PL/AE \qquad \text{(Equation No. 4)}$$

Taking moments about the point where G connects to the bar gives

$$F_H \times a - F_F \times a - P \times a/2 = 0 \quad \text{giving } \Delta H - \Delta F = PL/2AE \qquad \text{(Equation No. 5)}$$

Solving Equations 1, 2, 3, 4, and 5 gives

$$3\Delta F = D$$

Substituting for D in Equation No. 3 gives

$$P = AE/L \times (12 \times \Delta F) \text{ since } F_F = (\Delta F/L) \times AE$$

Then

$$F_F = P/12 \;\; F_G = P/3 \quad \text{and} \quad F_H = 7P/12$$

12.16 **a.** The combustion equation would be of the following form

$$C_xH_y + O_2 + 84.2\ N_2 \rightarrow 11.3\ CO_2 + 4.50\ O_2 + 84.2\ N_2$$

The nitrogen does not burn, so the amount of pure nitrogen would be the same after combustion as before. Air is approximately 79% nitrogen and 21% oxygen by volume. The volumetric ratio of nitrogen to oxygen is thus 0.79/0.21 = 3.76 so that ratio would have existed in the air supplied. Thus there would have been 84.2/3.76 = 22.39 moles of oxygen supplied for each 100 moles of air. (For the purpose of calculation 100 moles of stack gas is considered.)

Since only 4.5 moles of uncombined oxygen is in the stack gas and 11.3 moles is accounted for in the carbon dioxide 22.39 – 4.5 – 11.3 = 6.59 moles of oxygen is unaccounted for and would have existed in the form of water. Water is H_2O so the molecular weight of water is 2 + 16 = 18 (see the *Handbook*) and since there would have been 2 × 6.59 = 13.18 moles of atomic oxygen supplied for the combustion, there would be 13.18 moles of water vapor in the stack gas. There would then have been 26.36 moles of hydrogen in the fuel that was burned, assuming complete combustion.

This means that 11.3 moles of carbon would have combined with 26.36 moles of hydrogen, or the ratio of hydrogen in the fuel to carbon, H/C = 26.36/11.3 = 2.33. So $(2n + 2)/n = 2.33$ (This is a form of a Diophantine equation; that is, an equation whose solution is in integers). The possibilities are few (only four) so it can be easily solved. For n = 2 we get 6/2 = 3.33. For n = 4 we get 10/4 = 2.5. For n = 5, 12/5 = 2.4 For n = 6 14/6 = 2.33, so the answer is C_6H_{14}.

12.17 **b.** There were 84.2/3.76 = 22.39 moles of oxygen in the input air. The combustion equation can be written as $C_{11.3}H_{26.36} + O_2 \rightarrow 11.3\ CO_2 + 13.18\ H_2O$ so 11.3 + 13.18/2 = 17.89 moles of oxygen were required for combustion.

$$22.39/17.89 = 1.25 \quad 25\%\ \text{excess air}$$

12.18 **a.** The area of piston equals $1.6^2 \times \pi/4 = 2.01\ \text{cm}^2 = 2.01 \times 10^{-4}\ \text{m}^2$. The force acting on the piston would equal 15 × 135 = 2025 N, so the pressure under the ram would equal

$$2025/(2.01 \times 10^{-4}) = 10.07\ \text{MPa}$$

The force exerted by the ram would thus equal

$$10{,}070{,}000 \times (3.8^2 \times \pi/4 \times 10^{-4}\ \text{m}^2) = 11.4\ \text{kN} = 1162\ \text{kg}$$

12.19 **b.** The tape increases 99.999 – 99.992 = 0.007 meters in length with an increase of 6°C in temperature. It would, then, increase (7/6) × 0.007 = 0.0082 meters for an additional increase in temperature of 7°C. At the measuring conditions the tape would actually measure a distance of only 100 meters times 100/(99.999 + 0.0082) = 100/100.0072 = 0.999928 or 99.9928 meters when 100.000 meters was indicated. As a result, 0.0072 meters would have to be added to every tape length to obtain an accurate measurement. Thus 3.2 × 0.0072 = 0.023 meters would have to be added to the indicated length of the tape to obtain the desired dimension, meaning that the tape would have to indicate a distance of 320 + 0.023 = 320.023 meters.

12.20 **d.** The root area equals

$$1.6^2 \times \pi/4 = 2.011\ \text{cm}^2 = 2.011 \times 10^{-4}\ \text{m}^2$$

The permissible load thus equals

$$2.011 \times 10^{-4}\ \text{m}^2 \times 344.8 \times 10^6\ \text{Pa} = 69{,}339\ \text{kN}$$

which corresponds to 69,339/g = 7069 kg.

12.21 **c.** The body shows an apparent mass of 25 kg when weighed in fresh water, so it displaces 75 kg of fresh water. It thus has a volume of 75/1000 = 0.075 m^3 and has a density of

$$100\ \text{kg}/0.075 = 1333\ \text{kg/m}^3$$

so its specific gravity equals 1333/1000 = 1.33.

12.22 **a.** From the *Handbook* we see that $\phi = TL/GJ$ or torsional stiffness $= T/\phi = GJ/L$, where J equals the polar moment of inertia and G is the shear modulus. Since G and L will be the same, the torsional stiffness in this case will be proportional to the polar moment of inertia. From the table in the *Handbook*, taking the outer radius as a and the radius of the hole as b, we find that for the solid bar

$$J = \pi \times 7.5^4/2 = \pi \times 3164/2 = 4970\ \text{cm}^4$$

The J of the hollow bar must be the same for the torsional stiffness to be the same.

$$J = \pi(a^4 - b^4)/2 = 4970 \qquad 4970 \times 2/\pi = a^4 - 5^4$$
$$a = [3164 + 625]^{1/4} = 7.846$$

The outside diameter of the hollow torsion bar would thus equal 15.692 cm.

12.23 **d.** The air is compressed rapidly in an insulated cylinder so no heat is transferred during the process, thus the process is adiabatic and PV^k = constant

$$P_2 = P_1 \times (V_1/V_2)^{1.4} \text{ since } k \text{ for air equals 1.4}$$
$$V_1/V_2 = 4 \qquad P_1 = 13.79 + 101.3 = 115.09\ \text{kPa absolute}$$
$$P_2 = 115.09 \times 4^{1.4} = 115.09 \times 6.964$$
$$P_2 = 801.5\ \text{kPa absolute or 700.2 kPa gauge}$$

12.24 **b.** The kinetic energy for a rotating mass equals $\frac{1}{2} \times I \times \omega^2$. From the *Handbook* the polar moment of inertia about an axis through its center for a rotating cylinder equals $MR^2/2$

$$I \text{ for the flywheel thus equals } 1460 \times 0.75^2/2 = 410.6$$
$$\omega = (600/60) \times 2\pi = 62.83\ \text{radians/s}$$
$$I = \tfrac{1}{2} \times I \times \omega^2 = 0.5 \times 410.6 \times 62.83^2 = 810{,}444 \text{ or } 810\ \text{kJ}$$

12.25 **c.** The rate of heat transfer through the slab will equal

$$k \times \Delta T/L = \text{watts/m}^2$$

k is given as 1.039 W/m · K

$$\Delta T = 77 - 20 = 57\,^\circ\text{C} \quad L = 0.15\ \text{m} \quad Q = 1.039 \times 57/0.15 = 394.8\ \text{W/m}^2$$

APPENDIX

Afternoon Sample Examination

INSTRUCTIONS FOR AFTERNOON SESSION

1. You have four hours to work on the afternoon session. You may use the *Fundamentals of Engineering Supplied-Reference Handbook* as your *only* reference. Do not write in this handbook.

2. Answer every question. There is no penalty for guessing.

3. Work rapidly and use your time effectively. If you do not know the correct answer, skip it and return to it later.

4. Some problems are presented in both metric and English units. Solve either problem.

5. Mark your answer sheet carefully. Fill in the answer space completely. No marks on the workbook will be evaluated. Multiple answers receive no credit. If you make a mistake, erase completely.

   Work 60 afternoon problems in four hours.

FUNDAMENTALS OF ENGINEERING EXAM

AFTERNOON SESSION

ⒶⒷⒸ Fill in the circle that matches your exam booklet

1 ⒶⒷⒸⒹ	16 ⒶⒷⒸⒹ	31 ⒶⒷⒸⒹ	46 ⒶⒷⒸⒹ
2 ⒶⒷⒸⒹ	17 ⒶⒷⒸⒹ	32 ⒶⒷⒸⒹ	47 ⒶⒷⒸⒹ
3 ⒶⒷⒸⒹ	18 ⒶⒷⒸⒹ	33 ⒶⒷⒸⒹ	48 ⒶⒷⒸⒹ
4 ⒶⒷⒸⒹ	19 ⒶⒷⒸⒹ	34 ⒶⒷⒸⒹ	49 ⒶⒷⒸⒹ
5 ⒶⒷⒸⒹ	20 ⒶⒷⒸⒹ	35 ⒶⒷⒸⒹ	50 ⒶⒷⒸⒹ
6 ⒶⒷⒸⒹ	21 ⒶⒷⒸⒹ	36 ⒶⒷⒸⒹ	51 ⒶⒷⒸⒹ
7 ⒶⒷⒸⒹ	22 ⒶⒷⒸⒹ	37 ⒶⒷⒸⒹ	52 ⒶⒷⒸⒹ
8 ⒶⒷⒸⒹ	23 ⒶⒷⒸⒹ	38 ⒶⒷⒸⒹ	53 ⒶⒷⒸⒹ
9 ⒶⒷⒸⒹ	24 ⒶⒷⒸⒹ	39 ⒶⒷⒸⒹ	54 ⒶⒷⒸⒹ
10 ⒶⒷⒸⒹ	25 ⒶⒷⒸⒹ	40 ⒶⒷⒸⒹ	55 ⒶⒷⒸⒹ
11 ⒶⒷⒸⒹ	26 ⒶⒷⒸⒹ	41 ⒶⒷⒸⒹ	56 ⒶⒷⒸⒹ
12 ⒶⒷⒸⒹ	27 ⒶⒷⒸⒹ	42 ⒶⒷⒸⒹ	57 ⒶⒷⒸⒹ
13 ⒶⒷⒸⒹ	28 ⒶⒷⒸⒹ	43 ⒶⒷⒸⒹ	58 ⒶⒷⒸⒹ
14 ⒶⒷⒸⒹ	29 ⒶⒷⒸⒹ	44 ⒶⒷⒸⒹ	59 ⒶⒷⒸⒹ
15 ⒶⒷⒸⒹ	30 ⒶⒷⒸⒹ	45 ⒶⒷⒸⒹ	60 ⒶⒷⒸⒹ

DO NOT WRITE IN BLANK AREAS

PROBLEMS

1. A coupling consisting of two circular flat plates welded to the ends of two shafts is to be designed using 2.54 cm diameter bolts to connect one plate to the other. The bolts are to be located at a distance of 0.1524 m from the axes of the two shafts. If the coupling is to transmit 4026.8 kW at a shaft speed of 1200 rpm, how many bolts should be used in the coupling if the allowable shear stress for the bolts is 103.43 Mpa?

   a. two c. four
   b. three d. five

2. A hollow steel shaft 2.54 m long is to be designed to transmit a torque of 33.900 kJ. The shear modulus of elasticity $G = 82.740$ GPa for steel. The total angle of twist in the length of the shaft must not exceed 3° and the stress in the shaft must not exceed 110.32 MPa. What would be the ratio of outside diameter (d) to inside diameter (d)?

   a. 1.92 c. 1.56
   b. 1.74 d. 1.65

3. Four bolts are used to hold a bracket to a base as shown in Exhibit 3. The maximum shear load that can be applied to a bolt is 8896 N. Based on bolt strength, what is the maximum load that can be applied to the bracket at the point F?

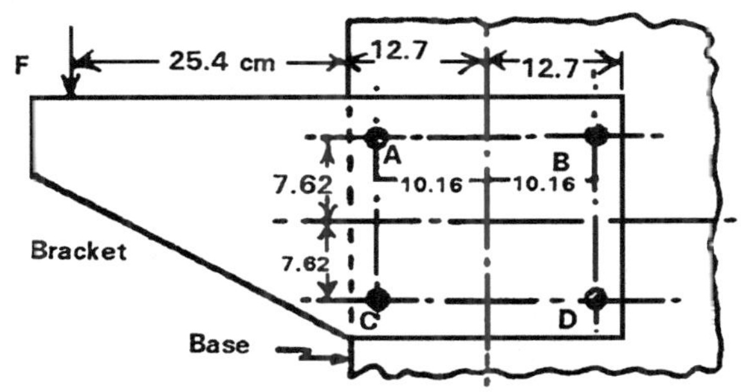

Exhibit 3

   a. 8896 N c. 9602 N
   b. 9249 N d. 9955 N

4. A 19.05 mm diameter rod is used to hold up a wooden beam. The pitch of the rod thread is 2.54 mm. The rod passes through the beam and support is affected by a nut on the lower end. The root diameter of the thread is 15.75 mm. The tensile strength of the steel is 137.9 MPa. The compressive strength of the wood is 2.379 MPa. What diameter of washer should be placed between the nut and the wooden beam so that the wood is not over-stressed when the rod is carrying its maximum allowable load?

   a. 97.3 mm c. 123.7 mm
   b. 110.5 mm d. 136.9 mm

5. A brake shoe is pressed against the surface of a 25.4 cm diameter drum rotating at 1250 rpm with a force of 111.20 N. If the coefficient of friction between drum and cylinder is 0.21, what power is dissipated in the form of heat by the drag of the brake shoe?

a. 388.2 W
b. 394.9 W
c. 401.6 W
d. 408.8 W

6. A freight car with a total mass of 68,027 kg breaks loose at the top of a 304.8 m long 5° grade. The rolling resistance friction factor equals 0.050. It hits a bumper formed of a combination of heavy-duty springs at the bottom of the grade. What must be the spring constant of the bumper assembly to bring the freight car to a stop in 1.0668 m? (Assume bumper track is horizontal.)

a. 13.34 MN/m
b. 13.12 MN/m
c. 12.90 MN/m
d. 12.69 Mn/m

7. A flat circular disk 0.500 m in diameter is made of steel (sp. gr. 7.87) and is 20 mm thick. It is attached solidly at its center to a 10.0 mm diameter steel rod 1.0 m long and is oriented horizontally. The upper end of the attached rod is rigidly held by a massive support. What is the period of vibration of this torsional pendulum?

a. 0.52 s
b. 0.68 s
c. 0.84 s
d. 1.00 s

8. What is the tension in the rope for the system shown in Exhibit 8? (Assume no friction and weightless pulleys.)

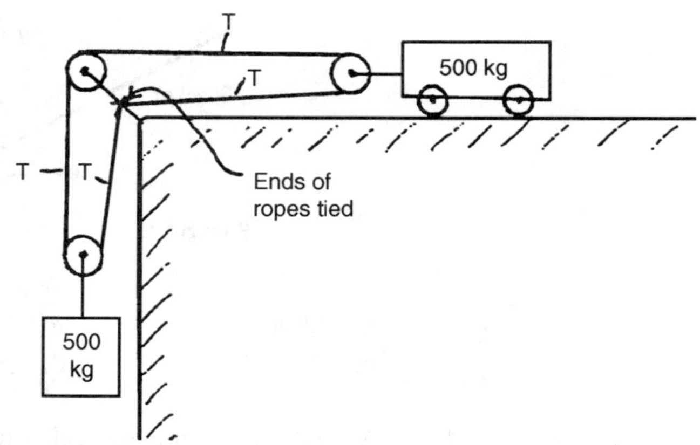

Exhibit 8

a. 1.03 kN
b. 1.23 kN
c. 1.43 kN
d. 1.63 kN

9. Air at 103,425 Pa absolute with a volume of 0.2831 m^3 is isentropically compressed by transfer of 63,300 J until the temperature is 190.6 C. Assume $k = 1.4$, $R_{air} = 286.82$ J/kg-K. What is the change in internal energy?

a. 69,848 J
b. 63,200 J
c. 57,196 J
d. 51,805 J

10. A vessel with a volume of 1.416 m^3 is being filled with air, which is to be considered a perfect gas. At time zero the temperature in the tank is 115.56°C and the pressure is 1.379 MPa abs. The pressure is increasing at a rate of 137.9 kPa/s and the temperature is increasing at the rate 27.78°C/s. What is the instantaneous rate of airflow into the tank *at this instant* of time?
    a. 0.36 kg/s
    b. 0.41 kg/s
    c. 0.45 kg/s
    d. 0.50 kg/s

11. A fluid at 689.5 kPa and a specific volume of 0.250 m^3/kg enters an apparatus with a velocity of 152.4 m/s. Heat the radiation losses in the apparatus equal 23,263 J/kg of fluid supplied. The fluid leaves the apparatus at a pressure of 137.90 kPa with a specific volume of 0.9365 m^3kg and a velocity of 304.79 m/s. In the apparatus the shaft work done by the fluid is equal to 582,938 J/kg. What is the change in the internal energy of the fluid?
    a. –598 kJ/kg
    b. –561 kJ/kg
    c. –523 kJ/kg
    d. –486 kJ/kg

12. Methane (CH_4) enters an ideal nozzle at a static pressure of 689.5 kPa abs and a static temperature of 26.67°C. The entering velocity is 121.9 m/s. The nozzle discharges 0.7256 kg/s into a vessel in which the static pressure is 551.6 kPa abs. For methane $k = 1.30$ and $c_v = 1.687$ kJ/kg·K. What is the exit velocity from the nozzle?
    a. 207 m/s
    b. 230 m/s
    c. 256 m/s
    d. 284 m/s

13. A heavy fuel oil with a specific gravity of 0.94 and a viscosity of 2.01 poise is to be pumped at the rate of 2000 liters/min through a 30 cm I.D. pipe. What would be the friction factor?
    a. 0.033
    b. 0.055
    c. 0.077
    d. 0.099

14. An ore carrier 120 m long by 1 m wide displaces 8500 m^3 of fresh water. It is moved into a lock in fresh water that is 140 m long by 15 m wide and then is loaded with 3500 metric tons of ore. What will be the increase in the depth of the water in the lock after the ship is loaded?
    a. 0.83 m
    b. 1.67 m
    c. 2.33 m
    d. 5.71m

15. A careless hunter shoots a 7.00 mm diameter hole in the vertical side of a water tank one meter above the ground. The level of the water in the tank is 10.0 m above the ground. How far from the base of the tank will the water strike the ground?
    a. 2.67 m
    b. 4.75 m
    c. 6.00 m
    d. 7.88 m

16. The water level of a reservoir is 150 m above a power plant. A 30.0-cm I.D. pipe 200 m long connects the reservoir to the plant. If the friction factor for flow through the pipe is 0.02, how much water will flow to the plant?
    a. 0.80 m^3/s
    b. 1.0 m^3/s
    c. 1.2 m^3/s
    d. 1.4 m^3/s

17. A 45,341.5 kg mass is held up by a 101.6-mm diameter steel cylinder surrounded by a copper tube of equal length with an outside diameter of 203.2 mm, which fits snugly around the steel cylinder. The weight is distributed uniformly over the surfaces of the supporting members. What is the stress in the copper tube?

E (steel) = 206.85 Gpa E (copper) = 110.32 GPa

a. 10.563 MPa
b. 10.908 MPa
c. 11.253 MPa
d. 11.597 MPa

18. What is the magnitude of the force acting on member CD as shown in Exhibit 18?

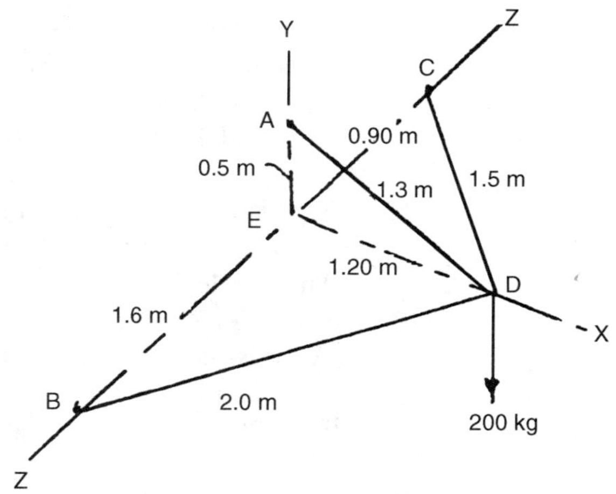

Exhibit 18

a. 3700 N
b. 3766 N
c. 4132 N
d. 4198 N

19. A 25.0-mm thick steel plate 0.25 m wide is loaded as shown. What is the maximum stress in the plate? See Exhibit 19.

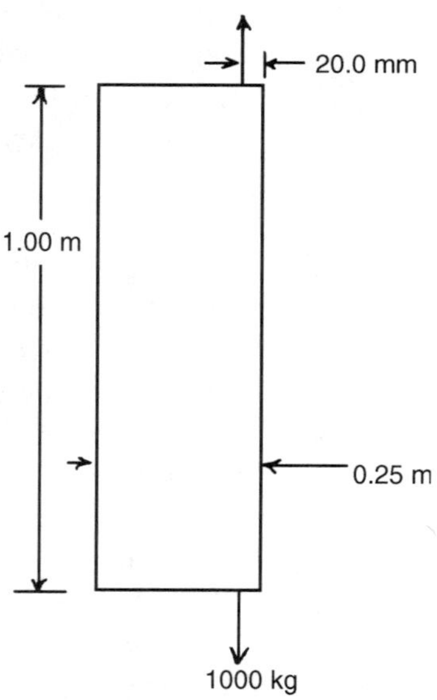

Exhibit 19

a. 1.60 MPa
b. 3.95 MPa
c. 4.22 MPa
d. 5.52 MPa

20. A closed-end tube 152.4-mm O.D. with a 2.54-mm wall thickness contains a fluid at a pressure of 3.448 MPa. It is subjected to a torsional load of 678 Nm. What is the maximum principal stress in the tube?
   a. 24.5 MPa c. 99.9 MPa
   b. 74.5 MPa d. 101 MPa

21. From an instrument that is considered to be a secondary standard with an error less than +/–0.5% and from another instrument with an error less than +/– is calibrated from the secondary standard, statistically what is the probably error of the second instrument?
   a. 0.5% c. 1.50%
   b. 0.75% d. 1.12%

22. Two different transducers are to be used at different times to drive a strip-chart recorder and the interfacing between the transducers will be an op-amp (see Exhibit 22). Assume the op-amp has a 1-megohm resistor in feedback across the op-amp, also assume the full scale expected output of the first transducer will be less than 500 mv and the second one less than 200 mv. If the recorder has a full-scale display of 10 volts, what should R_1 and R_2 be (for a full-scale display on the recorder)?

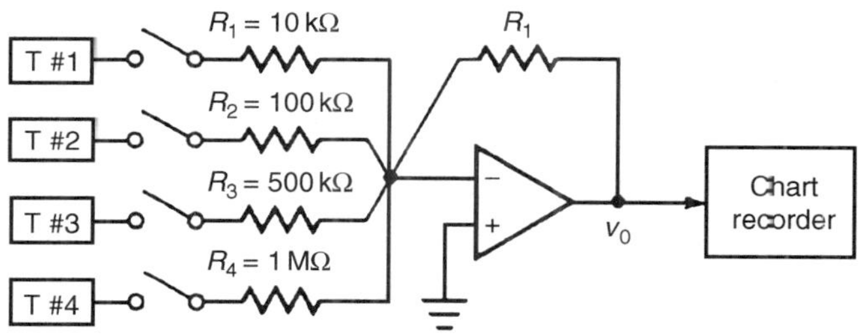

Exhibit 22

   a. $R_1 = 10\text{k}\Omega$, $R_2 = 20\text{k}\Omega$
   b. $R_1 = 30\text{k}\Omega$, $R_2 = 10\text{k}\Omega$
   c. $R_1 = 50\text{k}\Omega$, $R_2 = 20\text{k}\Omega$
   d. $R_1 = 80\text{k}\Omega$, $R_2 = 40\text{k}\Omega$

23. It is necessary to measure a temperature range from –40° to + 120°C and it is important to know the temperature within a quarter of a degree. A certain linear transducer is capable of producing a variation of voltage over this range. Assume an A/D converter will produce a binary output that will represent the temperature readings. Determine the number of bits (n rating of the A/D converter) needed.
   a. 9 c. 8
   b. 10 d. 16

24. Why is it not practicable to join aluminum wire to copper wire for an electrical connection by twisting the two wires together?
   a. It would not be mechanically strong.
   b. It would corrode in time and provide poor conduction.
   c. Electrical connections should be soldered.
   d. Aluminum wire is not as ductile as copper wire.

25. Aluminum bolts 2 cm in diameter and 500 cm long are used to hold the head on a pressure vessel. One bolt has a sensitive extensometer attached to the body of the bolt. It measures an elongation of 0.00400 cm in a 10-cm length of the bolt when it is tightened. How much force is exerted by the bolt?
   a. 8670 N
   b. 8770 N
   c. 8870 N
   d. 8970 N

26. Two masses of 100 kg are suspended by wires that are 5 mm in diameter. One wire is of aluminum and the other is of steel. The wires are 10 meters long. How much lower will the mass held by the aluminum wire be?
   a. 4.37 mm
   b. 4.87 mm
   c. 5.37 mm
   d. 5.87 mm

27. An aluminum pipe is buried in the earth between two buildings. What would you suggest to prevent excessive corrosion of the pipe?
   a. Bury an excess scrap copper wire alongside the aluminum pipe.
   b. Run the aluminum pipe alongside an existing iron pipe water line.
   c. Place pieces of junk magnesium in the ditch beside the aluminum pipe.
   d. Replace the earth around the pipe with sand.

28. What would be the expected increase in efficiency of an Otto-cycle gasoline engine if the compression ratio were increased from 6.5 to 8.0?
   a. 3.8%
   b. 5.9%
   c. 7.2%
   d. 8.5%

29. A homeowner plans to use solar heat to warm his house. He plans to heat water, store it, and then circulate the warm water through coils to heat his living area. The design calls for 5.0 kW of roof heating to supply the reserve heat for the water reservoir. If his solar panels will provide a 20% conversion of incident energy to heat water running through piping attached to the panels, and it is estimated that in his area the radiant solar energy will average 1.35 kW/m^2, how large a roof area must he cover with the solar panels?
   a. 18.5 m^2
   b. 20.7 m^2
   c. 22.9 m^2
   d. 25.0 m^2

30. A production line moves in steps of two meters. Between each two steps, there is a dwell time of one-and-a-half minutes. During that time it is necessary to install thirty cap screws at one location. Time study tests have shown that an operator can easily locate, thread in, and tighten ten cap screws in one minute. How many operators should be assigned to this station?
   a. One
   b. Two
   c. Two and a half (one operator to divide time between two adjacent stations)
   d. Three

31. Five-liter containers are to be filled with oil with a specific gravity of 0.85. The contents must be kept within 0.10% of the specified amount. The filling system is controlled by a scale that automatically cuts off the flow of oil when the prescribed amount has been put into the container. What should be the accuracy of the scale?

a. 0.0353 N c. 0.0472 N
b. 0.0417 N d. 0.0516 N

32. Containers to hold 20.0 liters of a liquid are filled on a production line that moves in pulses controlled by a Geneva motion device. At each stop a nozzle lowers into a container, opening a valve in the nozzle. Then a cylinder filled with the exact amount of the liquid is emptied into the container by a piston inside the cylinder. If it requires one second to insert the nozzle and one second to remove it, at what rate could the containers be filled providing the liquid velocity is limited to 6.00 m/s and the nozzle has an I.D. of 5.00 cm?

a. 10.2/min c. 14.2/min
b. 12.2/min d. 16.2/min

33. What should be the width of a composition belt with thickness of 9.0 mm to deliver 100 kW of power to a driven pulley in a sawmill? The following design criteria control:

1. Belt material ultimate strength, 27.5 MPa
2. Use a safety factor of ten
3. Pulley diameters of 60.0 cm (both pulleys)
4. Belt velocity of 1,200 meters per minute
5. The belt is initially tightened with 4.5 kN tension on each side of the driving pulley

a. 23 cm c. 33 cm
b. 28 cm d. 38 cm

34. What would be the stress in a 2.0-cm diameter shaft that transmitted two kW of power at 1750 RPM?

a. 5.45 MPa c. 6.45 MPa
b. 5.95 MPa d. 6.95 MPa

35. What size engine is required to propel a 3-ton vehicle up a 6% grade at 90 km/hr?

a. 34 kW c. 44 kW
b. 39 kW d. 49 kW

36. How much force would a 75 kg pilot exert on his plane seat at the top of a vertical loop that has a radius of 300 meters if the speed of the plane were 300 km/hr?

a. 1136 N c. 1536 N
b. 1336 N d. 1736 N

37. Two 1.5-kg masses are connected by a massless string hanging over a smooth frictionless peg. A third mass of 1.5 kg is added to one of the other masses. Once the system is released, what is the force on the peg?

a. 44 N c. 42 N
b. 39 N d. 37 N

38. What theoretical power is required for the isothermal compression of 25 cubic meters of air per minute from one atmosphere to 830 kPa absolute pressure.

a. 94 kW c. 83 kW
b. 89 kW d. 79 kW

39. A rigid container holds 1.5 kg of air at 1.75 kPa and 40 C. What is the volume of the container?

a. 0.57 m^3 c. 0.77 m^3
b. 0.67 m^3 d. 0.87 m^3

40. Heat is lost through 1 m^2 of furnace wall at the rate of 1.639 kW. The temperature on the inside of the furnace wall is 1040°C, and it is 0.500 m thick. The average thermal conductivity of the furnace wall is 1.056 W/m · K. What is the temperature of the outside surface of the wall?

a. 264°C c. 274°C
b. 269°C d. 279°C

41. A counterflow heat exchanger operates with hot liquid entering at 200°C and leaving at 160°C. The liquid to be heated enters at 40°C and leaves at 140°C. What is the log mean temperature difference between the hot and cold liquids?

a. 87°C c. 97°C
b. 92°C d. 102°C

42. A beaker of liquid is set on a bench to cool. If the temperature of the liquid falls from 70°C to 60°C in 12 minutes, how much longer will it take for the liquid to cool to 50°C? The ambient temperature is 20°C. Neglect extraneous losses.

a. 13.5 min c. 14.9 min
b. 14.2 min d. 15.5 min

43. The wall of a house is made up of a layer of plaster over brick. The coefficients of conductance are as follows:

Film coefficient for indoor surface	9.37 W/m^2K
Plaster layer	26.35 W/m^2K
Brick outer layer	3.58 W/m^2K
Film coefficient for outer surface	34.07 W/m^2K

What is the overall heat transfer coefficient for the wall?

a. 2.21 W/m^2K c. 1.46 W/m^2K
b. 1.84 W/m^2K d. 1.32 W/m^2K

44. For the conditions in the above problem, what would be the rate of heat flow through 14 square meters of wall area if the outside temperature were –2.0 °C and the inside temperature were 21 °C?
   a. 650 W c. 770 W
   b. 710 W d. 810 W

45. Air in the amount of 280 m^3 per minute is to be heated from 20 °C to 75 °C in a tubular heater by means of saturated steam at 107 °C condensing on the outside of the tubes. How much steam would be required if the heat of vaporization of steam at 107 °C is 2.24 MJ/kg?
   a. 423 kg/hr c. 473 kg/hr
   b. 448 kg/hr d. 498 kg/hr

46. A 15 cm diameter cylinder has a mass of 1.10 kg. It is placed in a cylindrical barrel that is 30 cm inside diameter and is partly full of water. The inserted cylinder floats. How much will the surface of the water in the barrel rise?
   a. zero c. 1.56 cm
   b. 0.82 cm d. 2.10 cm

47. A jet of water 2.5 cm in diameter flows at the rate of 0.500 m^3/min. It impinges on a flat surface perpendicular to the axis of the stream. What force is exerted on the surface?
   a. 142 N c. 152 N
   b. 147 N d. 157 N

48. A steel band is shrunk onto a 30-cm diameter aluminum cylinder. The band is 2.00 mm thick and the diametral interference fit is 0.0108 mm. What is the stress in the steel band?
   a. 6.56 MPa c. 8.56 Mpa
   b. 7.56 MPa d. 9.56 MPa

49. A cable extends upward from a post at an angle of 30° with the vertical. It passes over a pulley at a point two meters away horizontally (3.464 m higher) and supports a mass of 900 kg. The post is two meters high (from the ground to the point of attachment of the cable) and is set in concrete. The section modulus of the post is 78.64 cm^3, and its cross-sectional area equals 18.5 cm^2. What is the maximum stress in the pole?
   a. 108 MPa c. 116 Mpa
   b. 112 MPa d. 120 MPa

50. A distance of 1579.560 meters was measured with a steel tape on a cold day at –5.5 °C. The tape was calibrated at 20 °C. The coefficient of thermal expansion for steel is 11.7×10^{-6}/°C. What was the true distance?
   a. 1579.560 m c. 1581.321 m
   b. 1579.089 m d. 1580.031 m

51. A pressure gage on a tank reads 689.64 kPa. The gage was calibrated on a dead-weight gage tester and found to read 0.025% low over the range from 600 to 700 kPa. What was the true pressure in the tank?
   a. 688.47 kPa c. 689.64 kPa
   b. 689.47 kPa d. 689.81 kPa

52. A U-tube mercury manometer is to be used to measure an estimated differential pressure of 88 cm of water in a water system. Using a safety factor of 1.5, how long should the measurable section of the U-tube be?
a. 8.33 cm
b. 9.04 cm
c. 9.71 cm
d. 10.47 cm

53. A cantilever beam 5.0 cm wide by 10.0 cm high and 3.0 m long holds a 700 kg load on its end. What would be the percentage increase in the deflection if the beam were made of aluminum rather than steel?
a. 194%
b. 204%
c. 254%
d. 304%

54. A diesel-engine-driven generator produces 10.0 kW of power. The generator has an efficiency of 96%. If the diesel engine operates with an efficiency of 34%, how many liters of fuel would it consume in an eight-hour shift? Diesel fuel has a specific gravity of 0.81 and a heating value of 42.57 MJ/kg.
a. 26 ℓ
b. 30 ℓ
c. 41 ℓ
d. 35 ℓ

55. Outside air at 5 C is heated to 25 C before it is introduced into an occupied room. If the relative humidity of the outside air is 70%, what would be the relative humidity, RH, of the heated air?
a. 65%
b. 42%
c. 31%
d. 18%

56. How much moisture is contained in one m^3 of dry air if the dry bulb temperature is 29°C and the wet bulb temperature measures 20°C?
a. 9.6 g
b. 11.5 g
c. 13.5 g
d. 15.6 g

57. A refrigerant leaves the condenser as a liquid at 26°C and expands into the evaporator, which is at 1.5°C. Data from a table show a refrigerating effect of 130.85 J/kg. What would be the mass flow rate for a refrigeration requirement of 52.75 kW?
a. 20.72 kg/min
b. 22.09 kg/min
c. 23.17 kg/min
d. 24.19 kg/min

58. What size motor would be required to operate a fan that is to supply 30.0 m^3/min of air against a static pressure of 12.0 cm of water pressure if the fan efficiency is 78%? Assume standard conditions of one atmosphere and 20 C.
a. 654 W
b. 694 W
c. 754 W
d. 794 W

59. What size motor would be required to pump 350 ℓ/min of fluid with a density of 1.20 against a pressure of 360 kPa if the efficiency of the pump is 85%?
a. 2.25 kW
b. 2.50 kW
c. 2.75 kW
d. 3.00 kW

60. A centrifugal pump delivers 1,200 ℓ/min of water at a discharge pressure of 165 kPa and a suction pressure of 5.0 cm of mercury. The motor draws 6.30 kW of power. What is the overall efficiency of the motor-pump combination?

a. 55%	c. 65%
b. 60%	d. 70%

End of Exam. Check your work.

SOLUTIONS

After most solutions, a chapter or chapter/section reference is indicated where you can turn to review the concepts relevant to the problem. If your solution to a problem is incorrect, you may want to review the referenced material.

1. **c.** Watt = Joule/s = Nm/s so 4026.8 kW = 4.0268 MNm/s

1200 rpm = 20 rps, and a point 0.1524 m out from the axis of the shaft would travel at a velocity

$$v = 20 \times 2\pi \times 0.1524 = 19.15 \text{ m/s}$$

The cross-sectional are of a bolt would equal

$$0.0254^2 \times \pi/4 = 506.7 \times 10^{-6} \text{ m}^2$$

Allowable force on one bolt is

$$506.7 \times 10^{-6} \times 103.4 \text{ MN/m}^2 = 52.393 \text{ kN/bolt}$$
$$(4.0268 \text{ MNm/s})(19.15 \text{ m/s}) = 0.2103 \text{ MN at bolt circle}$$
$$210{,}300/52{,}393 = 4.01 \text{ bolts required, so use four bolts.}$$

(Chapter 7; "Torsion" and Chapter 12; "Low Cycle Fatigue")

2. **d.** Stress in the outer fibers of a shaft subject to torsion = TC/J'

where $J' = \pi \times \text{Diam}^4/32$

$$110.32 \text{ MPa} = (33.900 \text{ kJ} \times D/2)/J'$$
$$D/J' = 6{,}508.55$$

$$3° \text{ angle of twist} = 3 \times \pi/180 = 0.05236 \text{ radian}$$

$$\theta = T\ell/(GJ') = 33.900 \text{ kJ} \times 2.54 \text{ m}/(82.740 \text{ GPa} \times J')$$
$$0.05236 = 1.0407 \times 10^{-6}/J'$$
$$J' = 1.9876 \times 10^{-5}$$
$$D = 6{,}508.55 \times 1.9876 \times 10^{-5} = 0.1294 \text{ m}$$

The diameter of the bore can be determined with the aid of the relationship for the stress

$$Tc/J'$$

where

$$J' = \pi(D^4 - d^4)/32$$
$$J' = 1.9876 \times 10^{-5} = (\pi/32) \times (D^4 - d^4)$$
$$20.2455 \times 10^{-5} = 24.037 \times 10^{-5} - d^4$$
$$d = .078\text{m}$$
$$D/d = 1.65$$

(Chapter 7; "Torsion")

3. **b.** Each bolt will withstand one-quarter of the vertical load F plus the force applied due to the moment resulting from the applied load. The applied moment will equal 0.381 Nm where 0.381 meter is the distance from the point of application of the force perpendicular to the horizontal distance

to the centroid of the bolt pattern. The moment will be resisted equally by the four bolts, and the force exerted on a bolt will be perpendicular to the line from the bolt to the centroid. For bolts *A* and *B* the applied force will have a component in the same direction as the applied force, so the total loads on bolts *A* and *B* will be greater than the total loads on bolts *B* and *D*. The moment force applied to each bolt will equal

$$0.381F/(4 \times 0.127) = 0.750F$$

where 0.127 m is the distance of the moment arm to a bolt from the centroid of the bolt pattern

The moment-producing force acting on bolt *A* will act downward at an angle of 36.87° from the vertical. The two components of this force will equal

$$F_y = 0.80 \times 0.75\ F = 0.60\ F$$
$$F_x = 0.60 \times 0.75\ F = 0.45\ F$$

The total vertical force acting on bolt *A* would equal

$$0.25F + 0.60\ F = 0.85\ F \text{ and horizontal force} = 0.45\ F$$

The total force acting on bolt A

$$= F \times \sqrt{(0.85^2 + 0.45^2)} = 0.9618\ F$$

So

$$F = 8896/0.9618 = 9249\ \text{N}$$

(Chapter 12; "Low Cycle Fatigue")

4. **d.** The maximum load that can be held by the rod equals

$$15.75^2 \times 10^{-6} \times \pi/4 \times 137.9 \times 10^6 = 26{,}867\ \text{N}$$

The area of the washer contacting the wood must equal

$$26{,}867/(2.379 \times 10^6) = 0.011298\ \text{m}^2$$

The washer will have a 19.05-mm diameter hole in its center so the outer diameter of the washer would equal

$$\sqrt{(0.011298 + 19.05^2 \times \pi/4 \times 10^{-6}) - /0.7854} = 0.1370\ \text{m} = 137\ \text{mm}$$

(Chapter 12; "Low Cycle Fatigue")

5. **a.** The velocity of the surface of the drum—

$$\text{vel} = 1250/60 \times \pi \times 0.254 = 16.624\ \text{m/s}$$

Frictional force exerted by brake shoe on surface of drum

$$F = 111.20 \times 0.21 = 23.352\ \text{N}$$

Power dissipated = 23.352 × 16.624 = 388.20 J/s or 388.20 watts.

(Chapter 12; "Interference Fit")

6. **a.** Force perpendicular to track exerted by car = kgf × cos 5°

$$68{,}027 \times 0.996195 = 67.768 \text{ kgf} = 664.57 \text{ kN}$$

Friction resisting force = 0.05 × 664.57 = 33.229 kN

Gravitational force acting on car parallel to track

$$F = 68{,}027 \text{ kgf} \times \sin 5° = 68{,}027 \times 0.087156$$
$$= 5{,}929 \text{ kgf or } 58.143 \text{ kN}$$

Net force acting on car parallel to track = 58.143 kN – 33.229 = 24.914 kN

This force acts over a distance of 304.8 m producing a total amount of energy at impact of 7.5938 MJ

Energy absorbed = $\int$FDS where force exerted by spring = ks

Energy absorbed = $\frac{1}{2}ks^2 = \frac{1}{2}k \times 1.0668^2 = 0.56903k$

so $k = 7.5938 \times 10^6$ Nm/0.56903 = 13.345 MN/m.

(Chapter 12; "Interference Fit," "Spring Load")

7. **b.** Mass of disk = $(\pi/4) \times 50.0^2 \times 2.0 \times 7.87 = 30.91$ kg

The moment of inertia of circular disk about an axis through its center and perpendicular to the disk = $\frac{1}{2}MR^2$.

$$I = \tfrac{1}{2}MR^2 = \tfrac{1}{2} \times 30.91 \times 0.25^2 = 0.9659 \text{ kg/m}^2$$
$$\omega = \sqrt{(GJ/IL)}$$

where

G = shear modulus of the steel rod = 8.3×10^{10} Pa

J = Polar moment of inertia of the rod = $\frac{1}{2}\pi r^4$

$J = \frac{1}{2}\pi \times 0.0050^4 = 9.818 \times 10^{-10}$

I = Moment of inertia of the disk = 0.9659

$\omega = \sqrt{([8.3\times10^{10} \times 9.818\times10^{-10})/(0.9659\times1.00)]} = \sqrt{83.37}$

ω = 9.186/s angular frequency

$f = \omega/(2\pi) = 1.462 \quad \tau = 1/f = 0.684$ s

(Chapter 8; "Simple Harmonic Motion")

8. **b.** Net force acting on the cart to cause motion

$2T$ to the left resisted by the inertial or d'Alembert force of $m_c a$

$$\text{Net Force} = 2T - 500a = 0$$
$$\text{giving } 2T = 500a$$

The force applied by the suspended mass

$$\text{Force down} = m_s g \text{ resisted by } 2T + m_s a$$

or

$$2T = 500 \times 9.8066 - 500a \quad \text{but} \quad 2T = 500a$$

so

$$1000a = 4.9033 \text{ kN} \quad \text{and} \quad a = 4.9033 \text{ m/s}^2$$
$$T = (500/2) \times 4.9033 = 1225.8 \text{ N or } 1.23 \text{ kN}$$

(Chapter 8; "Linear Motion")

9. **b.**

$$\text{Work} = (P_2V_2 - P_1V_1)/(1 - k) = 63{,}300 \text{ J}$$
$$63{,}300 \times (-0.40) = P_2V_2 - 103{,}425 \text{ Pa} \times 0.2831 \text{ m}^3$$
$$P_2V_2 = 29{,}280 + 25{,}320 = 54{,}600 \text{ J}$$
$$P_2V_2 = m_2RT_2 = m_g \times 286.8 \times (190.6 + 273.2) = m_g \times 133{,}018$$

where
m_g = mass of gas

$$T_1 = P_1V_1/m_gR = 103{,}425 \times 0.2831/(0.410 \times 286.32) = 249.0 \text{ K}$$
$$T_2 = 190.6 + 273.2 = 463.8 \text{ K}$$
$$\Delta T = 463.8 - 249.0 = 214.8 \text{ K}$$
$$\Delta U = m_gc_v\Delta T = 0.410 \times 718 \times 214.8 = 63{,}233 \text{ J}$$

10. **d.**

$$PV = m_gRT$$
$$V = \text{constant} = 1.416 \text{ m}^3 \text{ and } R \text{ constant}$$

where
m_g = mass of gas

Differentiate and divide by dt

$$VdP/dt = Rm_gdT/dt + RT\,dm_g/dt \quad \text{mass of gas at time zero}$$
$$m_g = PV/RT = 1.379 \times 10^6 \times 1.416/(286.7 \times 388.7) = 17.52 \text{ kg}$$
$$VdP/dt = 1.416 \times 137.9 \text{ kPa/s}$$
$$= 286.7 \text{ J/kg K} \times 17.52 \times 27.78 + 286.7 \times 388.7 \times dm_g/dt$$
$$dm_g/dt = 0.500 \text{ kg/s}$$

(Chapter 1; general entropy equation, Example 1.3)

11. **a.** Use the general energy equation
For steady state (turbines etc.) for one kg

$$h_1 + Vel_1^2/2 + q = h_2 + Vel_2^2/2 + \text{Work} \quad \text{but } h = u + Pv$$
$$u_1 + P_1v_1 + Vel_1^2/2 + q = u_2 + P_2v_2 + Vel_2^2/2 + \text{Work}$$
$$Vel_1^2/2 = 152.4^2/2 = 11{,}613 \text{ J/kg}$$

Energy in

$$11{,}613 + 172{,}375 - 23{,}263 + u_1 = 160{,}725 + u_1 \text{ J/kg}$$
$$Vel_2^2/2 = 304.79^2/2 = 46{,}448 \text{ J/kg}$$
$$P_2v_2 = 137{,}900 \times 0.9365 = 129{,}143 \text{ J/kg}$$

Energy out

$$46{,}448 + 129{,}143 + 582{,}938 + u_2 = 758{,}529 + u_2 \text{ J/kg}$$
$$u_1 - u_2 = -597{,}804 \text{ J/kg}$$

(Chapter 1; "General Energy Equation")

12. **d.** For steady state

$$h_1 + Vel_1^2/2 = h_2 + Vel_2^2/2$$
$$\Delta h = c_p \Delta T$$
$$c_p = c_v \times k = 1.687 \times 1.30 = 2.1931 \text{ kJ/kg·K}$$
$$T_2 = T_1 \times (P_2/P_1)^{(k-1)/k} = (26.67 + 273) \times (551.6/589.5)^{0.2308}$$
$$T_2 = 299.67 \times 0.800^{0.2308} = 284.63 \text{ K}$$
$$\Delta T = 299.67 - 284.63 = 15.04$$
$$c_p \times \Delta T + 2.1931 \times 15.04 = 32.98 \text{ kJ/kg·K}$$
$$\Delta h + 2.1931 \times 15.04 = 32.980 \text{ kJ/kg K}$$
$$Vel_2^2/2 - Vel_1^2/2 = \Delta h = 32{,}980$$
$$Vel_2^2 = 121.9^2 + (2 \times 32{,}980) = 80{,}820$$
$$Vel_2 = 284 \text{ m/s}$$

(Chapter 1; "General Energy Equation")

13. **d.** The friction factor will be a function of the Reynolds number.

$$Re = \rho D v/\mu$$
$$\rho = 0.94 \times 1000 = 940 \text{ kg/m}^3$$
$$D = 0.300 \text{ m} \quad A = 0.707 \text{ m}^2$$
$$v = (2{,}000/1{,}000)/(0.707 \times 60) = 0.0471 \text{ m/s}$$
$$\mu = 201 \times 0.001 = 0.201 \text{ Pa·s}$$
$$Re = 940 \times 0.300 \times 0.0471/(0.201/g) = 648$$

This indicates that the flow will be well into the laminar-flow range so

$$f = 64/Re = 64/648 = 0.0988$$

(Chapter 5; "Flow Through Pipes")

14. **b.** The ship displaces 8500 m^3 of fresh water before loading. This would cause a rise of 8.500/(140 × 15) = 4.0476 m in the lock. The loaded ship would displace 8500 + 3500 = 12,000 m^3 of water and would cause a total rise of 12,000/2100 = 5.714 m in the lock, so the depth would increase by 1.667 m.

(Chapter 5; "Buoyancy")

15. **c.** The velocity of the water out of the tank will equal

$$v = \sqrt{(2gh)} = \sqrt{(29.80669.00)} = 13.286 \text{ m/s}$$

The jet is 1.00 m above the ground.

$$h = v_o t + \tfrac{1}{2} a t^2$$

For this case the vertical velocity at time zero is zero. The time for the water to fall 1.00 m

$$t = \sqrt{2(h/g)} = 0.452 \text{ s}$$

The distance from the base of the tank the water will strike

$$s = 0.452 \times 13.286 = 6.00 \text{ m}$$

(Chapter 5; "Hydrodynamics")

16. **b.** Apply Bernoulli's equation between the surface of the reservoir and the discharge from the pipe. There is no energy added and the velocity of the surface of the water is zero. The pressures at the surface and at the pipe discharge are both atmospheric, so they cancel. The only term remaining on the left side of the equation is the difference in elevation between the surface of the reservoir and the outlet of the pipe, $Z_1 - Z_2 =$ 150 m. On the right hand side of the equation would be the velocity term $v^2/2g$ and the work done in overcoming pipe friction

$$h_L = fL/D \times v^2/2g \qquad fL/D = 0.02 \times 200/0.30 \times v^2/2g$$

so $$150 = (1 + 13.33) \times v^2/2g \quad \text{and} \quad v = 14.33 \text{ m/s}$$

$$Q = 14.33 \times 0.0707 = 1.013 \text{ m}^3\text{/s}$$

(Chapter 5; "Flow through Pipes")

17. **c.** The deflection for the steel cylinder and the copper tube will be equal and since the lengths are the same, the unit deflections, δ, will also be equal.

$$\text{Stress} = E\delta$$

$$\text{Steel area} = 101.6^2 \times \pi/4 = 8107 \text{ mm}^2 = 0.008107 \text{ m}^2$$

$$\text{Force exerted by steel cylinder} = 0.00810734 \times E_s \times \delta$$

$$\text{Copper area} = (203.2^2 - 101.6^2) \times \pi/4 = 24{,}322 \text{ mm}^2 = 0.024322 \text{ m}^2$$

$$\text{Force exerted by mass } F = 45{,}341.5 \times 9.8066 = 444.717 \text{ kN}$$

$$\text{Force exerted by copper tube} = 0.024322 \times E_c \times \delta$$

$$444{,}717 = 0.008107 \times 206.85 \times 10^9 \times \delta$$

$$+ 0.024322 \times 110.32 \times 10^9 \times \delta = 4.3601 \times 10^9 \times \delta$$

$$\delta = 444{,}717/(4.3601 \times 10^9) = 101{,}997 \times 10^{-9} \text{ m/m}$$

$$\text{Stress in copper tube} = 101{,}997 \times 10^{-9} \times 110.32 \times 10^9 = 11.253 \text{ Mpa}$$

(Chapter 7 introduction; Example 7.1; Chapter 12, "Beam Deflection")

18. **b.** The force exerted by a 200 kg mass = 200 × 9.8066 = 1961 N
Force in member AD = 13/5 × 1961 = 5099 N
Force in member ED = X component of force in AD
= 1.2/1.3 × 5099 = 4,707 N

The Z components of forces in members CD and BD in the X-Z plane are equal

$$(0.9/1.5) \times F_{CD} = (1.6/2.0) \times F_{BD}$$

so force in BD = 0.75 × Force in CD

The sum of the X components of forces CD and BD equal the X component of the force in AD

$$(1.2/1.5) \times F_{CD} + (1.2/2.0) \times F_{BD} = (1.2/1.3) \times F_{AD}$$

$$0.800\ CD + 0.600 \times 0.750 \times CD = 4707 \text{ N}$$

$$\text{Force in } CD = 4707/1.250 = 3765.6 \text{ N}$$

(Chapter 7 introduction; Example 7.1)

19. **d.** The maximum stress will occur at the edge of the plate and will be made up of the direct tensile stress due to the applied load plus the bending stress due to the eccentricity of the load.

$$F/A = 1000 \times 9.8066/(0.250 \times 0.025) = 1.569 \text{ MPa}$$

The stress due to the eccentric load $S = Mc/I$

$$M = (0.125 - 0.020) \times 9806.6 = 1029.7 \text{ Nm or } 1029.7 \text{ J}$$
$$c = 0.250/2 = 0.125$$
$$I = bh^3/12 = 0.025 \times 0.250^3/12 = 32.55\ \mu\text{m}^4$$
$$\text{Bending stress} = 1029.7 \times 0.125/(32.55 \times 10^{-6}) = 3954 \text{ MPa}$$
$$\text{The maximum tensile stress} = 1.569 + 3.954 = 5.523 \text{ MPa}$$

(Chapter 7 introduction; Example 7.1)

20. **d.** The maximum principal stress will be calculated using the method of Mohr's Circle as given in the *FE Handbook*
Assume a thin-walled tube then $\sigma = PD/2t$

The tube I.D. = 152.4 – 5.08 = 147.32 mm or 0.14732 m

Hoop stress

$$\sigma_y = (3.448 \times 10^6 \times 0.14732)/0.00508 = 99.992 \text{ MPa}$$

Longitudinal stress, $\sigma_x = P \times A_{ID}/\text{Area of tube wall}$

$$A_{ID} = \pi/4 \times 0.14732^2 = 0.017046 \text{ m}^2$$
$$\text{Area of tube wall} = \pi/4 \times (O.D.^2 - I.D.^2) = 0.001195 \text{ m}^2$$
$$\sigma_x = (3.448 \times 10^6 \times 0.017046)/0.001195 = 49.184 \text{ MPa}$$

Shear stress $s_s = TcJ$

where

$$J = \pi/32 \times (O.D.^4 - I.D.^4).$$
$$J = 6.7152 \times 10^{-6} \text{ m}^4$$
$$s_s = 678 \times 0.0762/6.7152 \times 10^{-6} = 7.6935 \text{ MPa}$$

From the figure of Mohr's Circle in the *FE Handbook*

$$\tau_{\max} = \sqrt{\left[s_s^2 + \{(\sigma_x - \sigma_y)/2\}^2\right]} = \sqrt{(7.6935^2 + 25.404^2)}$$
$$= 26.45 \text{ MPa}$$

$$\sigma_1 = (\sigma_x + \sigma_y)/2 + \tau_{\max} = 74.588 + 26.45 = 101 \text{ MPa}$$

(Chapter 7; "Hoop Stress," "Mohr's Circle")

21. **d.**

$$e_c = \sqrt{e_1^2 + e_2^2 + e_3^2 \cdots e_n^2} \qquad e_c = [(0.5)^2 + (1)^2]^{1/2} = 1.12,$$

therefore, statistically, the error is less than ±1.12%.

(Equation from "Instrumentation" section, *FE Handbook*)

22. **c.** The gain needed for the first transducer is 20 and the second is 50; therefore, (ignoring the minus sign of the inverting circuit) the equations are

$$\text{Gain}_1 = R_f/R_1 = 20,\ R_1 = 1 \times 10^6/20 = 50 \text{ k ohms},$$
$$\text{Gain}_2 = R_f/R_2 = 50,\ R_2 = 1 \times 10^6/50 = 20 \text{ k ohms}.$$

(Chapter 10; "Strain Gages")

23. **b.** The number of increments or resolutions must be at least $160 \times 4 = 640$. Therefore, $2^n > 640$. $n = 9$ gives 512, whereas $n = 10$ gives 1024. Therefore, $n = 10$ is required.

24. **b.** High mechanical strength is not needed in an electrical connection. Soldering is not necessary for low-current applications when two copper wires are connected and joined with a twist-on connector, and aluminum cannot be soldered to copper. The ductility of the conductors is not important in an electrical connection. From the table of oxidation potentials in the *FE Handbook* it is seen that aluminum has a much higher potential than copper. Thus it would form the anode of a galvanic cell and corrode. Due to the large potential difference, the corrosion would proceed rapidly when an electrolyte formed from the moisture in the air. It might also be noted that the corrosion would result in a poor connection, and would result in a high resistance in the connection that could result in a fire.

(Chapter 11; "Electrochemical Corrosion")

25. **a.** Measured strain is

$$\varepsilon = 0.00400/10 = 0.000400 \text{ cm/cm}$$
$$\text{Stress} = \varepsilon \cdot E = 0.000400 \times 6.9 \times 10^{10} = 27.60 \text{ MPa}$$
$$\text{Area of bolt} = 0.0003142 \text{ m}^2$$
$$F = S \times A = 0.3142 \times 27.6 \text{ kPa} = 8672 \text{ N}$$

(Chapter 7; "Strain")

26. **b.** The stresses in the two wires will be equal.

$$S = 980.66(1.9635 \times 10^{-5}) = 50.00 \text{ MPa}$$
$$\varepsilon_s = 50.00 \times 10^6/(2.1 \times 10^{11}) = 238.1 \times 10^{-6} \text{ m/m}$$
$$\Delta L \text{ for steel wire} = 2.381 \text{ mm}$$
$$\varepsilon_{Al} = 50.00 \times 10^6/(6.9 \times 10^{10}) = 724.5 \times 10^{-6} \text{ m/m}$$
$$\Delta L \text{ for aluminum wire} = 7.246 \text{ mm}$$

The mass held by the aluminum wire will be 4.87 mm below the mass held by the steel wire.

(Chapter 7; Introduction, Example 7.1)

27. **c.** A study of the table of oxidation potentials in the *FE Handbook* shows that aluminum has a higher oxidation potential than copper or iron, so (b), and especially (a), would only accelerate the corrosion of the aluminum pipe. Magnesium has a higher oxidation potential so it would act as a sacrificial anode and protect the aluminum pipe. Selection (d) would do nothing to protect the aluminum pipe.

(Chapter 11; "Corrosion")

28. **c.** From the *FE Handbook* the efficiency of an Otto-cycle engine is given as $\eta = 1 - r^{1-k}$, where r equals the compression ratio.

For $r = 6.5$

$$\eta = 1 - 6.5^{-0.4} = 1 - 0.473 = 0.527 \text{ or } 52.7\%$$

For $r = 8.0$

$$\eta = 1 - 8.0^{-0.4} = 1 - 0.435 = 0.565 \text{ or } 56.5\%$$

There would be an increase of 3.80 percentage points or an increase of 7.21% in the engine efficiency.

(Chapter 2; "Otto Cycle")

29. **a.** The energy available for heating the water equals

$$0.20 \times 1.35 = 0.270 \text{ kW/m}^2 \text{ of solar paneling}$$
$$5.00/0.270 = 18.5 \text{ m}^2$$

(Chapter 1; "General Energy Equation")

30. **b.** If one operator can easily handle ten cap screws in one minute, then one operator can handle 15 in a minute-and-a-half. Two operators could then handle the requirement of 30 in a minute-and-a-half.

31. **b.** The nominal mass of oil would equal

$5 \times 1.00 \times 0.85 = 4.250$ kg, which would produce a force of 41.68 N

For a 0.10% accuracy the weighing device would have to be accurate to within 0.0417 N.

32. **d.** The rate of flow through the nozzle equals

$$6.00 \times 0.001964 = 0.0118 \text{ m}^3\text{/s or } 11.8 \text{ L/s}$$
$$\text{Filling time equals } 20/11.8 = 1.695 \text{ s}$$

To this would have to be added 2.0 s insertion and removal time giving a total of 3.695 s, giving a theoretical rate of 60/3.695 = 16.24 per minute.

(Chapter 5; "Flow Through Pipes")

33. **b.** The maximum safe design stress equals 2.75 MPa. Assume the increase in belt tension on one side of the pulley equals the decrease in tension on the other side.

$$\text{Power} = Tq \times \omega = (F_1 - F_2)r \times \omega \quad \omega = v/r$$
$$P = (F_1 - F_2) \times v$$
$$[(4.5 + \Delta) - (4.5 - \Delta)]\text{kN} \times 20 \text{ m/s} = 100 \text{ kW} \quad \Delta = 2500 \text{ N}$$

Maximum required tension in belt = 7.00 kN

Required belt area = 7.00/2,750 = 0.00255 m^2

Thickness = 0.0090 m, so width = 28.3 cm

34. **d.** Two kW equals 2000 N · m/s

1750 RPM/60 = 29.167 rev/s

Torque = $2000/(2\pi \times 29.167) = 10.91$ N · m or 10.91 J

Stress = Tc/(polar moment of inertia) or

$S = 2T/(\pi r^3) = 6.946$ MPa

(Chapter 7: "Torsion")

35. **c.** The car moves at the rate of 25 m/s. A 6% grade gives an angle of 3.434°, so when the car travels 100 meters on the road it rises 6.00 m. The engine must produce energy in the amount of

$$25 \times 6/100 \times 3000 \text{ kg} \times 9.8066 \text{ N/kg} = 44{,}050 \text{ W}$$

The power requirement is 44.05 kW

(Chapter 8; "D'Alembert Principle," "Friction," "Kinetic Energy," "Potential Energy")

36. **d.** Centripetal force equals mv^2/r

$$\text{Force} = 75 \times 83.33^2/300 = 1736 \text{ N}$$

(Chapter 8; "Flight of a Projectile," "Rotary Motion," "Force")

37. **b.** Initially the two masses are at rest and the tension in the cord equals 1.5 × 9.8066 = 14.71 N so the force acting on the peg equals 29.42 N. When the extra 1.5 kg is added to one of the masses and the system is released the larger mass will accelerate downward and the smaller mass will accelerate upward. The tension in the string will equal

$$T = 3.00 \cdot g - 3.00 \cdot a = 1.5 \cdot g + 1.5 \cdot a$$
$$a = 3.2689 \text{ m/s}^2$$
$$T = 19.613 \text{ N} \quad \text{force on peg} = 2T = 39.23 \text{ N}$$

(Chapter 8; "Linear Motion")

38. **b.** The gas law states $PV = mRT$ if the temperature is constant, then $PV =$ constant.

Work = $\int P \cdot dV$, which equals, for an isothermal process

$W = PV \ln(V_2/V_1)$ work done by the gas

For constant T, $V_2/V_1 = P_1/P_2$

For work done on the gas

$$W = 101.3 \text{ kPa} \times 25 \times \ln(830/101.3) = 5327 \text{ kJ/min}$$
$$\text{Power required} = 5{,}327{,}000/60 = 88{,}783 \text{ watts or } 88.8 \text{ kW}$$

(Chapter 1; "General Energy Equation")

39. **c.** Assume air is a perfect gas, then

$$PV = mRT \; R_{air} = 8314/29{,}000 = (0.2867/\text{K})\text{joules}$$
$$V = mRT/P = 1.5 \times 0.2867 \times 313/175 = 0.7692 \text{ m}^3$$

(Chapter 1; "General Energy Equation")

40. **a.** The rate of heat flow equals

$$Q = kA\Delta T/\Delta L \quad 1639 = 1.056 \times 1.0 \text{ m}^2 \times \Delta T/0.500$$
$$\Delta T = 776°\text{C Outside wall temperature equals}$$
$$T = 1040 - 776 = 264°\text{C}$$

This could also be calculated as

$$5{,}900{,}000 \text{ MJ/hr} = 3{,}800 \text{ (J/hr·m·K)} \times 1.0 \text{ m}^2 \times \Delta T/0.500$$
$$\text{giving } \Delta T = 776°\text{C as before.}$$

(Chapter 4; "Conduction")

41. **a.** Log mean temperature difference for a counterflow heat exchanger equals

$$(\max \Delta T - \min \Delta T)/\ln(\max \Delta T/\min \Delta T)$$

max ΔT equals hot leaving temperature minus cold entering temperature
For this case

$$\max \Delta T = 160 - 40 = 120°\text{C}$$

min ΔT equals hot entering temperature minus heated liquid leaving temperature.

For this case

$$\min \Delta T = 200 - 140 = 60°\text{C}$$

log mean temperature difference = (120 − 60)/ln(120/60) = 86.6°C

(Chapter 4; "Heat Exchangers")

42. **d.** Newton's law of cooling is expressed as

$$dT/dt = -k \times (T - T_{amb})$$

where
T = temperature
t = time

which gives

$$\int dT/(T - T_{amb}) = -k \int dt \text{ from } T_o \text{ to } T$$
$$\ln(T - T_{amb})/(T_o - T_{amb}) = -kt \text{ so}$$
$$e^{-k12} = (60 - 20)/(70 - 20) = 0.800$$

which gives $k = 0.0186$ per minute

$$e^{-0.0186t} = (50 - 20)/60 - 20) = 0.750$$

which gives $t = 15.5$ min

(Chapter 4; "Conduction")

43. **a.**

$$1/U = 1/9.37 + 1/26.35 + 1/3.58 + 1/34.07 = 0.4534$$
$$U = 2.206 \text{ W/m}^2\text{K}$$

(Chapter 4; "Series Composite Wall")

44. **b.** The rate of heat flow

$$Q = U \times A \times \Delta T = 2.206 \times 14 \times 23 = 710.33 \text{ W}$$

(Chapter 4; "Conduction," "Series Composite Wall")

45. **d.** The mass of air would equal

$$m = PV/RT = 101.3 \times 280/[(8.314/29) \times 293] = 337.7 \text{ kg/min}$$
$$Q = 337 \times 554 \times 1.00 = 18{,}574 \text{ kJ/min}$$
$$18{,}574{,}000 \times 60/2{,}240{,}000 = 498 \text{ kg/hr}$$

(Chapter 4; Circular Pipes")

46. **c.** The floating cylinder will displace 1.10 kg of water, or 1100 cm^3, since the density of water is one gram per cm^3. The inside area of the barrel is 706.9 cm^2. So the surface of the water would rise

$$1.100/706.9 = 1.56 \text{ cm.}$$

(Chapter 5; "Buoyancy")

47. **a.** The velocity of the jet equals

$$v = (500{,}000/60)/(2.5^2 \times \pi/4) = 1698 \text{ cm/s}$$
$$\text{mass flow rate} = 8.333 \text{ kg/s}$$
$$F = ma = 8.333 \times 16.98 \text{ m/s} = 141.5 \text{ N}$$

(Chapter 5; "Momentum Force")

48. **b.** The circumference equals pi times the diameter so the circumferential strain in the steel band equals $0.00108 \times \pi/(30 \times \pi) = 3{,}600 \times 10^{-5}$ m/m ore = $0.00108/30 = 3.6C0 \times 10^{-5}$.

$$\text{Stress} = ee = 3.6 \times 10^{-5} \times 2.1 \times 10^{11} = 7.56 \text{ MPa}$$

(Chapter 7; "Shrink Fit")

49. **c.** The tension in the cable will equal

$$T = 900 \times 9.8066 = 8826 \text{ N}$$

This will apply a lateral force of 4413 N on the post and a vertical force of 7644 N.

The lateral force will produce a moment of 8826 Nm at the base of the post. The section modulus, Z, equals I/c.

Bending stress

$$S = M/Z = 4413 \times 2/(78.64 \times 10^{-6}) = 112.23 \text{ MPa}$$

to this must be added the tensile stress due to the vertical component of the applied force

$$S = 7644/(18.4 \times 10^{-4}) = 4132 \text{ kPa}$$

Maximum tensile stress = 112.23 + 4.13 = 116.36 kPa

(Chapter 7; "Stress Analysis")

50. **d.** The tape shrunk so the distance indicated by the tape was longer than the actual distance. The shrinkage equaled

$$\Delta L = 25.5 \times 11.7 \times 10^{-6} \times 1579.56 = 0.4713 \text{ m}$$

The true distance thus equaled 1579.560 + 0.4713 = 1580.031 m

(Chapter 7; "Thermal Stresses")

51. **d.** Since the gage reads low, the actual pressure will be higher than the indicated pressure. The error equals 0.00025 × 689.64 kPa = 0.172 kPa. The true pressure equals 689.64 + 0.172 = 689.81 kPa.

52. **d.** The density of mercury from the *FE Handbook* is 13,560 kg/m^3, or 13.6 times that of water. Since water will be in one leg of the U-tube, the differential pressure will equal 12.6 times the difference in the height of the mercury above that of the water. The difference in the heights of the two columns should equal

$$\Delta H = 88/12.6 = 6.98 \text{ cm}$$

For a safety factor of 1.5 the differential ΔH = 10.47 cm

(Chapter 5; "Manometry")

53. **b.** The maximum deflection equals

$$\Delta H = PL^3/3EI$$

The only change in the two beams would be the material. The deflection of a steel beam would equal

$$\text{Constant}/(21 \times 10^{10})$$

An aluminum beam would deflect 3.043 times as much or 2.043 times more.

The deflection would increase 204 percent.

(Chapter 7; "Shrink Fit")

54. **a.** The net efficiency of the generating system is 0.96 × 0.34 = 0.3264
The heat consumed would equal (10,000 J/s)/0.3264 = 30.637 kJ/s
The diesel fuel contains

$$0.81 \times 42.57 = 34.482 \text{ MJ/L of energy}$$
$$\text{Vol} = 30{,}637 \times 3600 \times 8/34{,}482{,}000 = 25.59\text{L}$$

(Chapter 2; "Diesel/Dual Cycle")

55. **d.** On the psychrometric chart in the *FE Handbook* locate the intersection of the 5°C temperature line and the 70% RH line. Follow the horizontal line showing moisture content to where it crosses the 25°C temperature line and read the RH at 25°C. The RH at 25°C equals 18%.

(Chapter 3; "Psychrometrics")

56. **c.** From the psychrometric diagram in the *FE Handbook* determine the intersection of the 29°C dry bulb temperature and the 20°C wet bulb temperature. Follow the horizontal line of saturation temperature to the right and read 11 grams of moisture per kg of dry air for the dew point temperature of 15°C. Dry air at 15°C and one atmosphere

$$m = PV/RT = 101{,}300 \times 1.00/[(8314/29) \times 288] = 1.227 \text{ kg/m}^3$$
$$\text{mass of moisture} = 11 \times 1.227 = 13.5 \text{ grams}$$

(Chapter 3; "Psychrometrics")

57. **d.** The refrigeration requirement is 52.75 kW or 3165 J/min. The mass rate of flow required would equal 3165/130.85 = 24.19 kg/min.

(Chapter 3; "Absorption Refrigeration," "Efficiency of Refrigeration Systems")

58. **c.** The density of air equals

$$\rho_{air} = PV/RT = 101{,}300 \times 1/[(8314/29) \times 293)]$$
$$= 1.206 \text{ kg/m}^3 \text{ and the flow rate 36.18 kg/min}$$
$$\Delta P = (1{,}000/1.206) \times 0.12 = 99.50 \text{ m}$$
$$\text{Power} = (36.18/60) \times 9.8066 \times 99.50 = 588 \text{ W}$$
$$\text{Motor requirement} = 588/0.78 = 754 \text{ W}$$

(Chapter 6; "Fans")

59. **b.** The mass rate of flow would be

$$350 \times 1.20 \times 1.0/60 = 7.00 \text{ kg/s}$$

The head against which the pump would act

$$\Delta H = 360{,}000[(1{,}200 \text{ kg/m}^3) \times 9.8066] = 30.59 \text{ m}$$
$$\text{output power} = (7.00 \times 9.8066)\text{N/s} \times 30.59 \text{ m} = 2100 \text{ W}$$
$$\text{Required motor power} = 2.100/0.85 = 2.47 \text{ kW}$$

(Chapter 6; "Pumps," "Pump Affinity Laws")

60. **a.** The negative pressure at the pump intake equals 5/76 × 101,300 = 6664 Pa (76 cm Hg equals one atmosphere). The pressure differential across the pump equals 165 + 6.66 = 171.66 kPa, which equals

$$171{,}660/(1{,}000 \times 9.8066) = 17.505 \text{ m}$$

$$\text{mass flow rate} = (1200/60) \times 1.00 = 20.0 \text{ kg/s}$$

$$\text{pump output power} = 20.0 \times 9.8066 \times 17.505 = 3433 \text{ W}$$

$$\text{Efficiency} = 3433/6.3 = 54.5\%$$

(Chapter 6; "Pumps," "Pump Affinity Laws")

INDEX